AF478706

Chemistry and Spectroscopy
of Interstellar Molecules

Chemistry
and Spectroscopy
of Interstellar Molecules

Edited by
Diethard K. Bohme
Eric Herbst
Norio Kaifu
Shuji Saito

UNIVERSITY OF TOKYO PRESS

© UNIVERSITY OF TOKYO PRESS, 1992
ISBN 4-13-068205-9
ISBN 0-86008-465-5
Printed in Japan

CONTENTS

Preface

I. Observational Astoronomy

II. Laboratory Spectroscopy

III. Chemical Kinetics and Theory

IV. Modelling

PREFACE

During the last fifteen to twenty years, more than eighty different gaseous molecules have been discovered in interstellar space. Located in between stars in giant clouds of gas and dust, these molecules range in complexity from diatomics to the thirteen-atom unsaturated nitrile, $HC_{11}N$. The list of known interstellar molecules includes many simple organic species such as formaldehyde, methanol, ethanol, acetaldehyde, dimethyl ether, and methyl formate. At the low temperatures (10 K - 50 K) of interstellar clouds, gaseous molecules emit in the radio and microwave regions of the electromagnetic spectrum at characteristic frequencies determined by their rotational energy level patterns. Radioastronomers measure these spectral frequencies in the same way that spectroscopists in the laboratory determine molecular spectra. In addition to determining what species are present in interstellar clouds, radioastronomers can deduce from the spectra the concentration of each molecular species and the physical conditions (temperature, overall gas density) in the clouds. Given the vast size of interstellar clouds, the amounts of some of the organic molecules seen in these regions far surpass the amounts of these species on earth.

The gaseous molecules discovered to date are a varied lot. In addition to well known organic molecules, a host of unusual ring species, neutral radicals, and molecular ions have been discovered. The abundance of unusual molecules has given impetus to a new field of molecular spectroscopy in which laboratory spectroscopists synthesize unusual molecules and determine their spectra in an attempt to determine whether these species are to be found in interstellar space. In addition to this impetus to molecular spectroscopy, the discovery of interstellar molecules has stimulated chemists to determine how they are synthesized in the low density cold interstellar clouds in which they are found. Both gas-phase mechanisms and surface mechanisms have been investigated. The major gas-phase reactions that appear to be important in the syntheses are ion-molecule reactions, since these reactions normally occur without activation energy. This discovery has given a boost to the field of ion-molecule reactions, and laboratory investigators all over the world have studied reactive systems of importance in the interstellar medium. Some experimentalists have managed to study ion-molecule reactions at temperatures as low as 2 K and below in an attempt to simulate the conditions in interstellar clouds. In addition, a variety of unusual types of reactions have been shown to be of importance in interstellar syntheses of organic molecules. One such type of reaction is radiative association, in which two smaller species literally stick together to form a larger species by emitting energy in the form of radiation. Theoretical treatments of this process and its importance in the interstellar medium have stimulated experimental investigators to attempt to measure such reactions and several systems have now actually been studied. Using rate constants both measured and determined theoretically, modellers have created large models of the chemistry of interstellar clouds and predicted concentrations of individual species with a fair degree of success.

Gas-phase reactions are not the only mechanism for the synthesis of interstellar molecules, however. Reactions on the surfaces of dust particles are also important. Dust particles are much harder to characterize than are gaseous molecules and our knowledge of their existence is not in doubt, nor is the existence of molecules adsorbed on their surface. These adsorbed molecules cannot be investigated by radio or microwave spectroscopy since they do not rotate, but they can be investigated by low resolution infra-red spectroscopy.

Spectral investigation in this wavelength region requires either that the molecules be warmer than normal or that a suitable infra-red continuum be available as a background for absorption studies. Both of these conditions can be met in regions of active star formation where the interstellar gas and dust are collapsing and where temperatures are higher than in the interstellar medium as a whole. Most recently, high resolution infra-red absorption studies of cool gas-phase molecules in front of warm background sources have also become practical. In any event, the synthesis of molecules on the surface of dust grains has been investigated by a variety of researchers although our lack of knowledge of surface chemistry at low temperatures has hindered progress.

In addition to the gas and dust phases of interstellars clouds, there is a possible third phase called PAH or polycyclic aromatic hydrocarbons. Looked at either as large gaseous molecules or as very small interstellar particles, the species appear to exist in a variety of sources based principally on functional group analysis of infrared spectral features. The probable existence of interstellar PAH's has provided an impetus for laboratory studies of the spectra and chemistry of these species.

Since the interstellar clouds in which molecules are located are in a state of slow collapse to form stars, they are also implicated in the formation of planetary systems such as our solar system. Indeed there is strong evidence of a phase in meteorites that consists of pristine interstellar material, undisturbed by the collapse of a precusor interstellar cloud. Thus it is quite probable that there is a strong connection between the interstellar chemistry of the cloud that collapsed to form our solar system and the current-day chemical composition of the solar system.

Interstellar chemistry is thus a vibrant field of interdisciplinary science in which observational astronomers, laboratory spectroscopists, laboratory chemists, surface chemists, theoreticians and modellers all contribute. The international scope of the field is also noteworthy, with important contributions having been made by European, American, Japanese, and Canadian investigators. It was therefore particularly appropriate to hold a symposium on "The Chemistry and Spectroscopy of Interstellar Molecules" at the International Chemical Congress of Pacific Basin Societies (PACIFICHEN) late in December of 1989. The symposium provided an opportunity to bring together astronomers, mathematicians, physicists and chemists for the first time under the banner of chemistry. This volume is a collection of the contributions to this symposium.

The symposium, which was initiated at the suggestion of Dr. T. Amano of the National Research Council of Canada, provided a golden opportunity to make known to the chemical community at large the richness, depth and breath of the field of interstellar chemistry and spectroscopy. Also it was intended to promote future interdisciplinary and international cooperation among astronomers, mathematicians, physicists and chemists.

The symposium was dedicated to Dr. Hiroko Suzuki, an interdisciplinary expert of interstellar chemistry and spectroscopy of the highest order, whose brilliant career came to an abrupt end as a result of a tragic accident in 1987. Dr. Suzuki excelled as an astronomer, as a chemist, and as a modeller, and so it was most appropriate to dedicate this symposium in her memory.

December, 1991

Diethard BOHME
Eric HERBST
Norio KAIFU
Shuji SAITO

I
OBSERVATIONAL ASTRONOMY

Hiroko Suzuki and Chemistry in Dark Clouds

Norio KAIFU

National Astronomical Observatory, Osawa 1-21-1, Mitaka, Tokyo 181, Japan

To combine the chemical evolution of dark cloud with its physical evolution was a constant and strong motif for late Hiroko Suzuki's efforts on astrochemistry. Her works, both theoretical and observational, are reviewed from this point of view, including interim results of an extensive molecular line survey for TMC1 which had been conducted by Hiroko Suzuki. In this survey work, about 60 transitions from 19 molecules and 9 isotope species were identified in the 35-50 GHz frequency area observed so far. Several new molecules, mainly carbon chains such as C6H, CCS and CCCS, were detected or confirmed through the survey works, and some 20 lines are remained unidentified. Discussions are extended to the progress lead from above theoretical and observational works, which are to reveal the relation between chemical and physical history in dark clouds, as predicted by Hiroko Suzuki.

I. PERSPECTIVE

Works by late Dr. Hiroko Suzuki cover wide area related to the interstellar molecules (1); formation theory of interstellar molecules including the large format computer synthesis, extensive observations with mm-wave telescopes which resulted the detection of a number of new exotic molecules, construction of a line data analysis software system (LINEPRC) which has been used in Nobeyama Radio Observatory for many years, a molecular line data-base system (ISLINE) and others. Through her such activities, we see a strong scientific motive; to understand the chemical evolution of dark cloud in connection with its physical evolution.

The base of Suzuki's such scientific perspective was formed in the atmosphere of the astro-nucleophysics group at the University of Kyoto under the leadership of Professor Chushiro Hayashi, where Suzuki had started her career as an astrophysicist. In 1973, a short after Suzuki started the theoretical study of interstellar molecule which was an exciting new field, she wrote in her review report (2);

"Knowing the physical condition in dark cloud we should be able to estimate the quantity of chemical products in it. The point we have to consider is whether the molecular formation and destruction processes are balanced with each other in the cloud or not, in other words, whether the chemistry in dark cloud is already in steady state.

If this is not the case, the abundance of interstellar molecules should depends on the cloud history, i.e. the process of cloud formation, the age since its creation, and physical conditions through which the cloud has passed, etc. This also means that we have a good possibility to understand the past history of the cloud by observing the present molecular abundance in it. To revealing the formation process of interstellar molecule is, therefore, our important target."

Suzuki developed this view through her works, firstly by means of the extensive computer simulation of molecular synthesis in dark clouds at the University of Kyoto. After she moved to the Nobeyama Radio Observatory in 1983, her efforts were concentrated on the observational works, and related laboratory experiments. By using Nobeyama 45-m mm-wave telescope, Suzuki, Kaifu, Ohishi and collaborators started the all-frequency scan survey of molecular lines in

TMC1, to obtain the basic and systematic understanding of the dark cloud chemistry.   This survey and related works provided new aspects to connect the observations and theories, as well as detections of many new molecules.   Suzuki's idea that the carbon-chain molecules might be the key species to understand the cloud history (3) thus became to be a practical observational target.

Just when we were starting to test the scenario of the evolution of the dark clouds, she suddenly passed away, in the very successful phase of her scientific activities.

Some papers presented in this symposium by Suzuki's collaborators (see, for example, (4) and (5)) demonstrate how we actually started to understand some of the relations between the chemical abundance and the physical history of dark clouds.

II. SYNTHESIS OF INTERSTELLAR MOLECULES

In 1976 Suzuki and collaborators published the first time-dependent calculation of molecular formation in dense dark cloud to test the time scale and principal process of chemical evolution (6).   They simulated the molecular formation in the contracting cloud with H2 density ranged from 10 to 107 cm-3, taking 35 species and 86 reactions into account.   Their result showed that (a) The reaction does not reach to the equilibrium within the cloud age, and (b) There are three principal stages in the chemical evolution of dark clouds, those are; optically thin stage   (CII phase, nH2 < 800 cm-3), transient stage (CI phase, 800 < nH2 < 5x103 cm-3), and optically thick stage (CO/SiI phase, nH2 > 5x103cm-3).

This result had encouraged Suzuki to develop a large format synthesis calculation.   The time dependent calculations published in 1979 (7) took a cloud model with fixed molecular hydrogen density nH2 =105 cm-3 and T = 30 K, and 2884 reactions among 234 species containing one or two of C, N, O, S, and Si atoms were included.   The size of her calculation had been conspicuously large among   similar works done before 1988.   A sample of published results is shown in Figure 1.   Here Suzuki concluded that C-containing molecules other than CO reach their abundance maximum at around several 107 years, and then move to the steady state because C atoms tend to be taken by CO molecules.   Therefore the C/CO ratio should be a good indication of cloud chemical evolution.

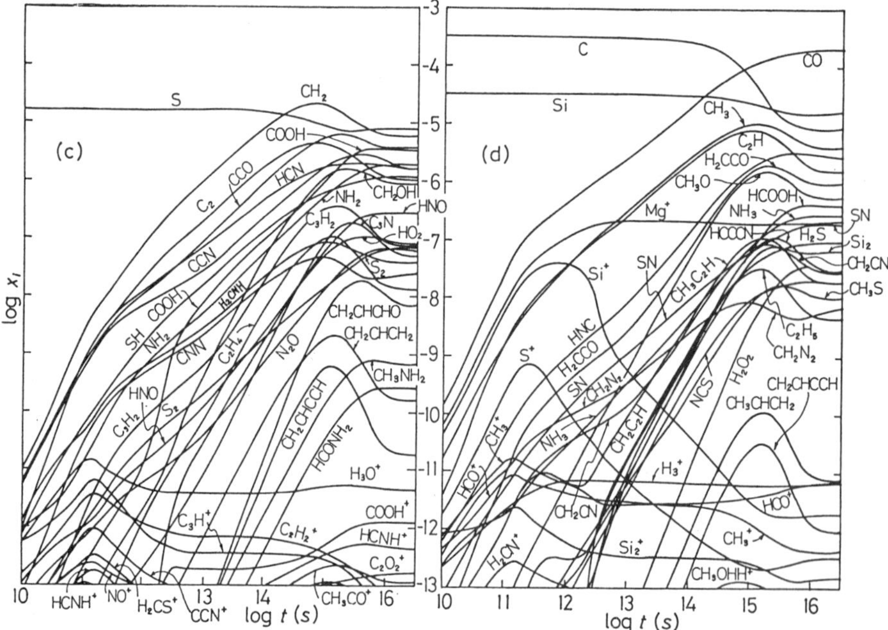

Figure 1.   Relative abundance $x_i = n_i / n(H)$ of produced molecules as a function of log t, where t is the time.   Taken from a part of Fig. 1 in Suzuki, 1979 (7).

Though the model did not include the cloud physical evolution, neutral-neutral reactions, photo-chemical reactions etclution, above outline of results of this work is still important,. Suzuki's program included the most of such extension for the following works. Recently Umebayashi constructed a new large-format program on the bases of Suzuki's program, adding new information of reaction rates and fixing up above and other extensions (Umebayashi, private communication).

"Synthesis of Chain Molecules in Regions with Partially Ionized Carbon" (2) was a follow-up and development of a paper published one year before (8). She proposed a growing process of carbon chain molecules by adding C atoms one by one, due to the photochemical reactions in the C+ region in early stage of the cloud contraction or in the cloud surface region. The chain growth factor was calculated to take its maximum value near the ionization front (Figure2), and Suzuki claimed that the rapid cloud contraction produces high abundance carbon chains, and thus the carbon chain molecules represent the early stage of cloud history. This view was strengthened by the detection of a number of exotic carbon chain molecules in the following molecular line survey.

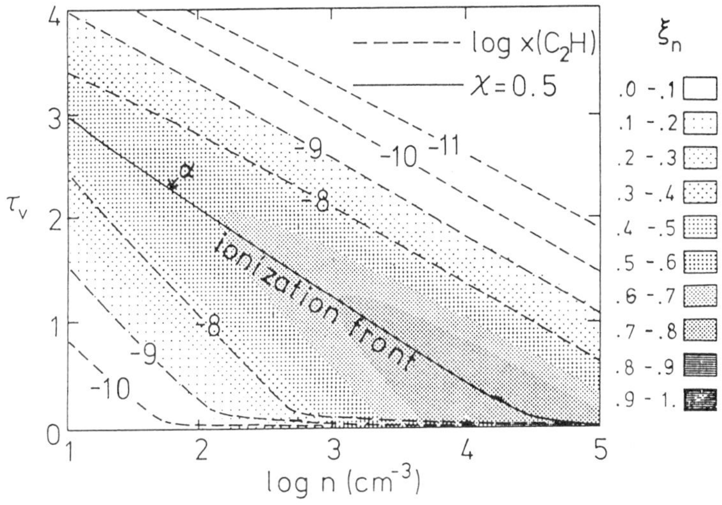

Figure 2.   Contours of x(C$_2$H) and the chain growing factor ξ_n , taken from Suzuki 1983 (8), Fig. 4. Contours of x(C$_2$H) are indicated by dashed lines, and coutours of ξ_n by shading.

III. NOBEYAMA MOLECULAR LINE SURVEY

All-frequency scan molecular line survey for a typical dark cloud, TMC1, using the Nobeyama 45-m telescope has started in 1984. The importance of such survey is obvious; we expect to obtain systematic information of the chemistry in dark cloud, the basic interstellar material from which stars and possibly the planetary systems are formed. Such survey method would yield the detection of unknown interstellar molecules, which are difficult to predict because we understand the interstellar chemistry only to limited extent. Consequently the new detection, when it is actually made, will add important new information to our knowledge of chemistry in dark clouds.

On the other hand there are technical difficulties in such survey. Very narrow line width in dark clouds (dvr = 0.2 to 0.5 km/s-1) and wide frequency range to cover inevitably require a very large number of spectral channel. Relatively weak line intensity in the cold dark clouds requires sensitive receivers and long observing time. Also the confusion becomes to be a considerable factor, and all these facts make the observations to be a time consuming work. At Nobeyama Radio Observatory the project started as a observatory project, using the 45-m telescope, large AOS (acousto-optical spectrometers) and wide-bandwidth sensitive receivers in the frequency region of 8 to 115 GHz.   The 45-m telescope provides the largest collecting area in the 35 to 115 GHz range, and thus fits to the dark cloud line survey from which the rich lines in the middle and long mm-wave region. are expected.   We used two sets of AOS simultaneously, one was a wide-band system with 250 KHz frequency resolution, 1600 output channels and 2 GHz total coverage.   Another was a high-resolution AOS system with 40 KHz resolution, 16000 channel and 320 MHz total coverage which can be divided into 8 independent spectrometers.   In the lower frequency observation a 80 MHz wide, digital FX type spectrometers with 4000 output channels was also used.   Wide band (2 GHz instantaneous bandwidth) HEMT amplifiers in the 8-35 GHz, SIS receivers in the 35-50 GHz and 70-115 GHz, and a cooled shottkey mixer receiver with 2 GHz bandwidth in the 100GHz regions were used for two different frequency bands simultaneously.   Though this system is a very powerful combination for the dark cloud line survey, we could cover only 70 % of the frequency range of the 8-50 GHz (Figure 3) and a fraction of the 70-115 GHz region so far.   The project is still under way, and we report here some of the obtained results.

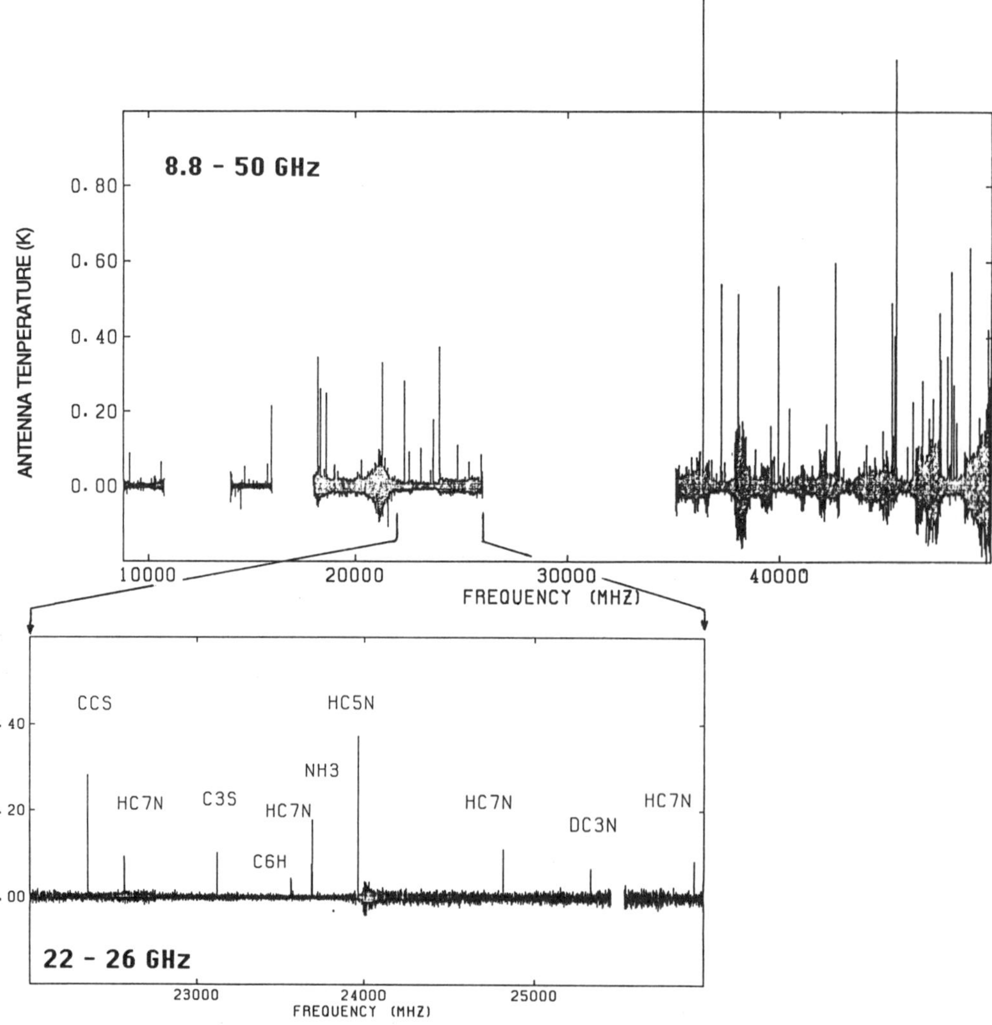

Figure 3.   Sample data from Nobeyama Molecular Line Survey in TMC1.   Condenced data for 8.8 to 50 GHz and 22-26 GHz part showing the identifications of observed lines.   Both are taken with the wideband(250 KHz resolution) AOS.   Note the richness of carbon chain molecules.

A number of lines which could not be identified to any known molecules were found as expected.   Among such unidentified lines those of C6H were the first to be identified to new molecule successfully (9).   The detection of C6H confirmed, together with the detection of C5H (10), the existence of a carbon chain series CnH with continuous integers n from 2 to 6.

The most exciting detection was that of CCS and CCCS, a new series of carbon chains containing sulphur.   U45379 was a strong unidentified line found while this survey observations in TMC1. As this was a single line the efforts to find the candidate molecule was difficult, and Suzuki reported only the detection of strong unidentified line (11).   Nearly 4 years after the detection the line was revealed by chance to be a transition of new molecule CCS, from a laboratory experiments by Saito and Yamamoto.   Identification of so far detected 7 strong unidentified lined in TMC1 to the transitions of CCS and CCCS immediately followed (Figure4) (12,13,14).   This detection provided Suzuki an idea of sequential formation of CnS molecules in the C+ and S+ region near the dark cloud surface (15).   Also, as the CCS transitions are relatively strong and have moderate optical depth, and as CCS was found to be abundant in the quiet dark clouds whereas it is deficient in the star-forming regions, CCS might be a good probe to trace the dark cloud chemical evolution.   Based on this idea Suzuki and collaborators started an extensive dark cloud survey with CCS, NH3 and some other molecules aiming a statistical study, especially of relative abundance of CCS and NH3 among many dark clouds(15).   Also, as the structure of dark cloud may keep evidences of its chemical history, the detailed observations of typical dark clouds with several key molecules are important.   From this view we started the detailed mapping observations of TMC1.   Recent results of both observations are reported in this symposium by Yamamoto and by Hirahara, respectively.

To summarize the data in 35-50 GHz area, which is relatively well-observed region, we have observed 60 transitions from 19 molecules and 9 isotope species, and nearly 20 unidentified lines. We expect that among them exsit several unknown molecules, probably relatively large organic molecules.

CH2CN, HCCCOH, H2CCC, H2CCCC and CCO were also found or confirmed in TMC1 in the course of this survey (16,17,18,19).

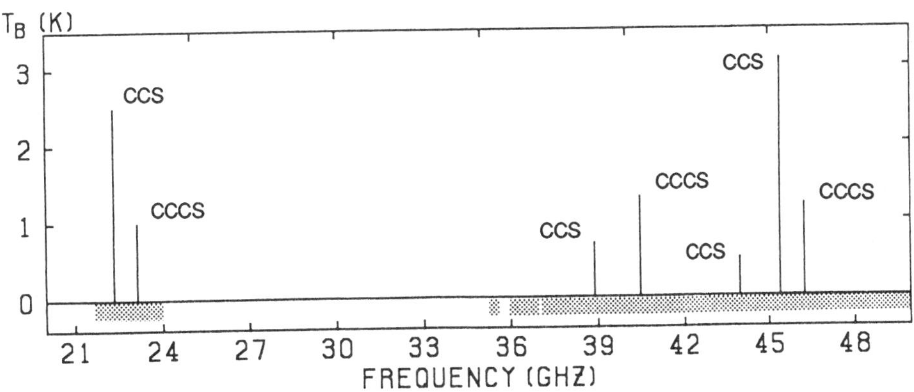

Figure 4.   A schematic spectrum showing the detected "unidentified lines" in TMC1 and their identification to new carbon chain molecules CCS and CCCS.   Taken from Kaifu et al., 1987 (12), Fig. 2, and (13,14).

Interstellar molecules provided enormous new aspects and tools to the astrophysics.   Star formation process hided deep in dark clouds, dynamics and kinetics of interstellar medium, structure and history of our and external galaxies, etc.   One of the most exciting field is, naturally, the interstellar molecules themselves and their reactions in the dark clouds.

For this new field the collaboration among astronomers, physicists, laboratory and theoretical chemists is essencially important.   Suzuki had connected people in these variety of fields, organized annual summer meetings at Nobeyama to discuss on the interstellar molecules, and created many collaborative works.   Nobeyama molecular line survey, for example, is a collaborative work from all above fields; H.Suzuki, N.Kaifu, M.Ohishi, S.Ishikawa, K.Miyazawa,

T.Miyaji, S.Saito, K.Kawaguchi, S.Yamamoto, H.Takano, Y.Hirahara, A.Murakami and W.Irvine.   This wide circle collaboration has yielded fruitful results, as described above briefly.

The perspective that we might be able to understand the chemical and physical evolution of molecular clouds was deeply discussed in the symposium "Chemistry in Dark Clouds" which was held in November 1988 at Kiyosato and dedicated to late Hiroko Suzuki, and an extensive and systematic collaboration study of interstellar medium and its evolution including wide field of astronomy, physics and chemistry in Japan was organized(20).   This collaborative work will start from 1991 under the financial support of the science research funds by the ministry of education and science , Japan.   In this work, we plan to finish the Nobeyama molecular line survey within 4 years by building another 16000 channel AOS.   A variety of mm-wave and IR observations, theoretical and laboratory projects are also planned.

I t is sad that we can not discuss on these exciting progresses with Hiroko Suzuki, but I believe she would appreciate our efforts to solve the dark cloud chemistry.

REFERENCES

1) Kaifu,N. (ed), *"Toward Interstellar Chemistry ; Works by Hiroko Suzuki"*
        University of Tokyo Press, (1989)
2) Suzuki,H., *proceedings of summer school of astronomy,* (1973) (in Japanese)
3) Susuki,H., *Astrophys.J.,* 272,579 (1983)
4) Yamamoto,S., *this proceedings.*
5) Hirahara,Y., *this proceedings.*
6) Suzuki,H.,Miki,S.,Sato,K.,Kiguchi,M.,and Nakagawa,Y., *Prog.Theor.Phys.,* 56,1111 (1976)
7) Suzuki,H., *Prog.Theor.Phys.,* 62,936 (1979)
8) Freed,K.F.,Oka,T.,and Suzuki,H., *Astrophys.J.,* 253,718 (1982)
9) Suzuki,H.,Ohishi,M.,Kaifu,N.,Ishikawa,S.,and Kasuga,T., *Publ.Astron.Soc.Japan,* 38,911 (1986)
10)   Cernicharo,J.,Guelin,M.,and Walmsley,C.M.,*Astron.Astrophys.(Letters),*172,L5.(1987)
11)   Susuki,H.,Kaifu,N.,Miyaji,T.,Morimoto,M.,and Ohishi,M., *Astrophys.J.,* 282,197 (1984)
12)   Kaifu,N.,Suzuki,H.,Ohishi,M.,Miyaji,T.,Ishikawa,S.,Kasuga,T.,Morimoto,M., and Saito,S.,
        Astrophys.J.(Letters),   317, L111 (1987)
13) Saito,S.,Kawaguchi,K.,Yamamoto,S.,Ohishi,M.,Suzuki,H.,and Kaifu,N.,
        Astrophys.J.,(Letters),   317, L115 (1987)
14) Yamamoto.S.,Saitp,S.,Kawaguchi,K.,Kaifu,N.,Suzuki,H.,and Ohishi,M.,
        Astrophys.J.(Letters), 317,L119 (1987)
15) Suzuki,H., in *"Proceedings of 4-th IAU Asian-Pacific Regional Meeting",* Beijing, 1987,
16) Saito,S.,Yamamoto,S.,Irvine,W.M.,Ziurys,L.M.,Suzuki,H.,Ohishi,M.,and Kaifu,N.,
        Astrophys.J.(Letters), 334,L113 (1988)
17) Irvine,W.M.,Friberg,P.,Hjalmarson,A.,Ishikawa,S.,Kaifu,N.,Kawaguchi,K., Madden,S.C.,
        Matthews, H.E.,Ohishi,M.,Saito,S.,Suzuki,H.,Thaddeus,P., Turner,B.E.,Yamamoto,S.,
        and Ziurys,L.M., *Astrophys.J.(Letters),* 334,L107 (1988)
18) Kawaguchi,K.,Kaifu,N.,Ohishi,M.,Ishikawa,S.,YamamotoS.,Saito,S.,Takano,S.,Murakami,A.,
        Vrtilek,J.M.,Gottlieb,C.A., Thaddeus,P.,and Irvine,W.M.,
        submitted to Publ.Astron.Soc.Japan, (1991)
19) Ohishi,M.,Suzuki,H.,Ishikawa,S.,Yamada,C.,Kanamori,H.,Irvine,W.M.,and Kaifu,N.,
        submitted to Astrophys.J.(Letters), (1991)
20) Kaifu,N. (ed), *"Chemistry and Physics of Dark Clouds",* proceedings of a
        symposium held in November 1988, Kiyosato. (in Japanese)

A Survey of Carbon-Chain Molecules and Ammonia Toward Dark Clouds

Satoshi YAMAMOTO

Department of Astrophysics, Nagoya University, Chikusa-ku, Nagoya 464, Japan

Survey observations of CCS, HC_3N, HC_5N, and NH_3 have been carried out toward 49 dark cloud cores. The CCS radical is found to be abundant in quiescent dark cloud cores, whereas it is deficient in star forming regions. The column density of CCS shows a good positive correlation with those of HC_3N and HC_5N, indicating that the sulfur chemistry is closely related to the carbon-chain chemistry in dark clouds. On the other hand, the column density of CCS shows no correlation with that of NH_3. A possible chemical model is proposed for the CCS production, and chemical changes in a dark cloud core due to formation of a low-mass star is discussed.
*The present study was initially proposed and carried out by Hiroko Suzuki, who regrettably passed away in late 1987.

INTRODUCTION

One of the most peculiar features in interstellar chemistry is the presence of various carbon-chain molecules, most of which are chemically unstable under terrestrial conditions. Since the recognition of HC_5N (1), four series of carbon chain molecules had been detected in molecular clouds; those were C_nH (n=1-6), C_nN (n=1,3), C_nO (n=1,3), and $HC_{2n+1}N$ (n=0-5). In 1987, the CCS radical was discovered in interstellar space on the basis of the laboratory detection of its rotational spectral lines (2). A strong unidentified line, U45379 (3,4), was assigned to the $J_N = 4_3 - 3_2$ transition of CCS. The discovery of CCS along with the subsequent detection of C_3S in the laboratory and in space (5) clearly established the existence of a new series of carbon-chain molecules, C_nS. In TMC1, the fractional abundances relative to molecular hydrogen of CCS and C_3S are 8 x 10^{-9} and 1 x 10^{-9}, respectively (2,5). The high abundances of CCS and C_3S may suggest that the sulfur chemistry is closely related to the carbon-chain chemistry. Therefore, CCS seems to be one of the important key molecules to understand the production mechanism of carbon-chain molecules in dark clouds.

Several chemical models have been proposed for interstellar production of CCS and C_3S (6-8). The S^+ ion does not react with molecular hydrogen, unlike O^+ (9), and hence, the S^+ ion is available to react with hydrocarbons. Smith et al. (7) considered the chemistry in a dark cloud without ultraviolet radiation, and proposed the reaction between S^+ and C_2H_2 as a principal route to produce $HCCS^+$, the precursor ion of CCS. Millar and Herbst (8) included the CCS chemistry in their reaction network, and calculated the CCS fractional abundance to be 1.4 x 10^{-9} at "early times" (3.16 x 10^5 yr). Suzuki et al. (6), on the other hand, extended the model of carbon-chain growth in relatively diffuse regions, where the carbon atoms are partially ionized by interstellar ultraviolet radiation (10). Since the ionization potential of S is lower than that of C, the S atom is also ionized in the partially ionized carbon region. The reactions of the S^+ ion with C_nH lead to the production of CCS and C_3S. These two models are complementary and both may actually work in interstellar clouds. Information on the distributions of CCS and other carbon-chain molecules is necessary to understand the chemistry of CCS more quantitatively.

In spite of the astrochemical significance, observational studies of CCS have still been limited. Recently, Fuente et al. (11) reported observations of CCS, C_3S, and $C^{34}S$ toward 18 astronomical sources, indicating that CCS is abundant in various dark clouds. In the present study, we conducted extensive survey observations of CCS, HC_3N, HC_5N, and NH_3 toward 49 dark cloud cores. We will discuss the results on the basis of the chemical model proposed.

OBSERVATIONS

The observations were made in Aug. 1987 and in Nov. 1987 by using the 45-m radiotelescope at Nobeyama Radio Observatory. The beam widths were 36 arcsec and 80 arcsec for the 45 GHz and 22 GHz observations, respectively. We observed eight molecular lines simultaneously by use of two receivers for the 45 GHz and 22 GHz regions; these were $HC_3N(J = 5 - 4)$, $HC_5N(J = 17 - 16, J = 9 - 8)$, CCS $(J_N = 4_3 - 3_2, J_N = 2_1 - 1_0)$, HC_7N $(J = 21 - 20, J = 20 - 19)$, and $NH_3(J, K = 1, 1)$.

We surveyed 49 dark cloud cores, most of which are in the Taurus and Ophiuchus regions. The positions of the cores are mainly taken from the papers of earlier survey observations of $C^{18}O$ (12) and HC_5N (13). The selected sources include cores with and without star formation. High-velocity CO gas has been reported in 16 cores, which indicates star formation. The integration time for each source was normally 40 min, and the rms noise obtained was 80 - 100 mK for both the 22 GHz and 45 GHz regions.

RESULTS AND DISCUSSION

1. Distribution and Abundance of CCS

Out of the 49 cores surveyed, the $J_N = 4_3 - 3_2$ and $J_N = 2_1 - 1_0$ lines of CCS were detected toward 27 and 16 cores, respectively. The detection rate is as high as 68 % in the Taurus region, which indicates wide distribution of CCS in this region. The lines of NH_3, HC_3N, and HC_5N were detected toward 39, 33, and 15 cores, respectively. The lines of HC_7N were detected only toward TMC1 in the present observations. An example of the spectra obtained toward TMC1 is shown in Figure 1.

The column density of CCS was derived from the observed antenna temperature by assuming LTE conditions. The dipole moment of CCS used is 2.81 D (14). The size of the CCS source is unknown for the most of the cores, and hence, we assumed that the source coupling factor is unity for both the 20 and 40 GHz transitions. We used the general formula including the effect of the optical depth, since the maximum value of τ is 1.5 and 3.5 for the $J_N = 2_1 - 1_0$ and $J_N = 4_3 - 3_2$ transitions, respectively, for TMC1. For several cores where the $J_N = 4_3 - 3_2$ and $J_N = 2_1 - 1_0$ lines were detected with a good S/N ratio, the excitation temperature is estimated to be 4.4 − 6.1 K on the basis of the intensity ratio between the two lines. For other cores, we assumed the excitation temperature to be be 5 K. A change in the excitation temperature by 1 K causes a change in column density of about 30 %. The column densities of HC_3N, HC_5N, and NH_3 were obtained in a similar way.

2. Abundance Correlations

The obtained column densities of CCS are plotted against those of HC_3N for various dark cloud cores, as shown in Figure 2. The column density of CCS is well correlated with that of HC_3N. A similar positive correlation is seen between the column densities of CCS and HC_5N. The average abundance ratio of [CCS]/[HC_3N] is 0.5, and that of [CCS]/[HC_5N] is 2. The positive correlations of the CCS column density with the HC_3N and HC_5N column densities indicate that the production chemistry of CCS is related to those of HC_3N and HC_5N. In Figure 2, star forming regions with high-velocity CO outflows are represented by open marks. The carbon-chain molecules such as CCS and HC_3N are generally deficient in the star forming regions, whereas they are abundant in quiescent cores without any sign of star formation.

On the other hand, the column density of CCS shows no positive correlation with that of NH_3(Figure 3). Although the beam size of the telescope is larger for the NH_3 line than those for the CCS $(J_N = 4_3 - 3_2)$ and HC_3N lines, NH_3 seems to be abundant in star-forming regions. Out of 13 sources with $N(NH_3)/N(CCS) > 100$, 8 sources (62 %) are star forming regions, as shown in Figure 3. Conversely, only 2 sources (20 %) out of 10 sources with $N(NH_3)/N(CCS) < 50$ have a sign of star formation. This suggests that the NH_3 chemistry takes place in physical and chemical conditions which are different from those suitable for CCS, and that star formation in a core may give favorable conditions for the production of NH_3. Later this point will be discussed in detail.

We also examined the correlation between $C^{18}O$ and CCS, where the data of $C^{18}O$ are taken from the paper by Benson and Myers (13). Although sources with higher column density of CCS tend to have higher column density of $C^{18}O$, sources with high $C^{18}O$ column density do not always have high CCS column density. The lack of an overall correlation between $C^{18}O$ and CCS indicates that the fractional abundance of CCS averaged over the line of sight varies remarkably from cloud to cloud. In other words, there are favorable and unfavorable physical and chemical conditions to produce CCS and carbon-chain molecules.

3. Possible Production Model of CCS

Suzuki (10) proposed a mechanism for gas-phase carbon-chain growth by reactions of C^+, which is formed by interstellar ultraviolet radiation. In this mechanism, a carbon chain grows up by single carbon units through the reaction of C^+. Since the carbon-chain growing processes compete with photodissociation, the most favorable region for this scheme is the intermediate region between diffuse and dense clouds where the carbon atoms are partially ionized. Dark cloud cores have experienced the diffuse stage during gravitational contraction, and, in such an evolutionary stage, carbon-chain molecules have been produced efficiently. The ionization potential of the sulfur atom is 10.360 eV, which is lower than that of the carbon atom (11.260 eV), so that the sulfur atom can also be ionized by the interstellar ultraviolet radiation in the partially ionized carbon region.

Suzuki (6) presented a possible production mechanism of C_nS, which combines the carbon-chain growth and the sulfur chemistry, as shown in Figure 4. In this scheme, reactions of C^+ with hydrocarbons such as CH, C_2H, and C_3H lead to the carbon chain growth, whereas the reactions of S^+ with the hydrocarbons lead to the production of C_nS. We assumed that the hydrogenation reactions of C_nS^+ proceed at a collisional rate. The HC_3N molecule is mainly produced by the reaction between atomic nitrogen and the $C_3H_3^+$ ion which is produced in the process of carbon-chain growth. Therefore, a global abundance correlation between CCS and HC_3N observed in various clouds is naturally expected.

In order to examine the model quantitatively, we made a numerical calculation for estimating the steady-state abundance of CCS. We treated 94 chemical species including H, C, O, N, S, He, and Mg with 380 reactions. The rate constants used were mostly taken from the papers by Leung, Herbst, and Huebner (15) and Prasad and Huntress (16). Figure 5 shows the fractional abundance of various species including CCS for a cloud with uniform density of 10^4 cm^{-3} and temperature of 10 K. No depletions for S and Mg (high metal abundances) were assumed in the calculation. Carbon-chain growth is most efficient in a region at $A_v = 0.2 - 1.5$, where the carbon atoms are partially ionized by the interstellar ultraviolet radiation. On the other hand, S^+ dominates C^+ in the region with $A_v > 1.4$, where CH, C_2H, and C_3H are converted into CS, CCS, and C_3S, respectively. The present model predicts the fractional abundance of CCS to be $(1-5) \times 10^{-8}$ in the region with $A_v = 0-2.2$. This abundance is slightly higher than that observed in TMC1. When we assume the depletion of the S and Mg atoms by two orders of magnitude (low metal abundances), the peak fractional abundance of CCS is only 10^{-9}, which is lower than the observed value. The abundance of CCS is generally low in the region with $A_v > 3$ for the high and low metal abundances.

The time scale for chemical equilibrium is not always shorter than that for a typical dynamical lifetime of a cloud (10^6-10^8 yr). In order to obtain the time scale for the chemical equilibrium of the present model, we carried out a preliminary time-dependent calculation under conditions of constant density (10^4 cm^{-3}) and constant temperature (10 K). The time scale for a cloud with $A_v > 2$ is comparable to or longer than the cloud lifetime, and hence, the chemical composition of an observed cloud is not always represented by the steady-state abundance. Suzuki (17) and Leung, Herbst, and Huebner (15) suggested that the abundances of carbon-chain molecules in a dark cloud with large A_v have a peak at several times 10^5 yr and become very low after 10^7 yr. This is because the carbon atoms are fixed to CO, as a cloud evolves, and

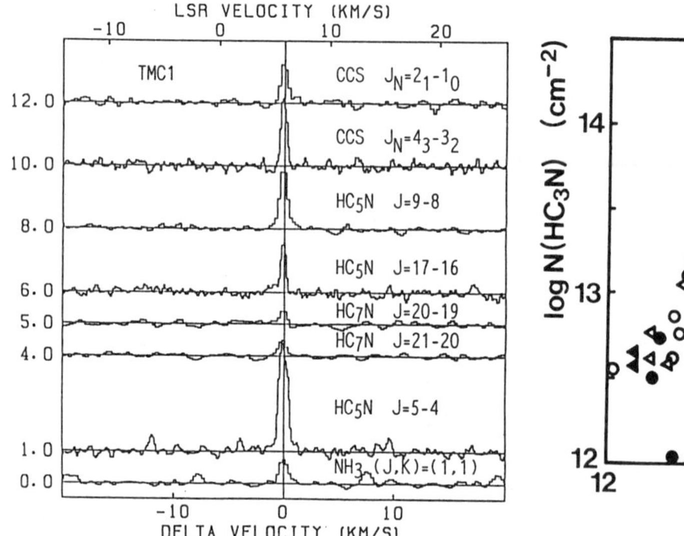

Fig. 1. An example of the spectra observed toward TMC1.

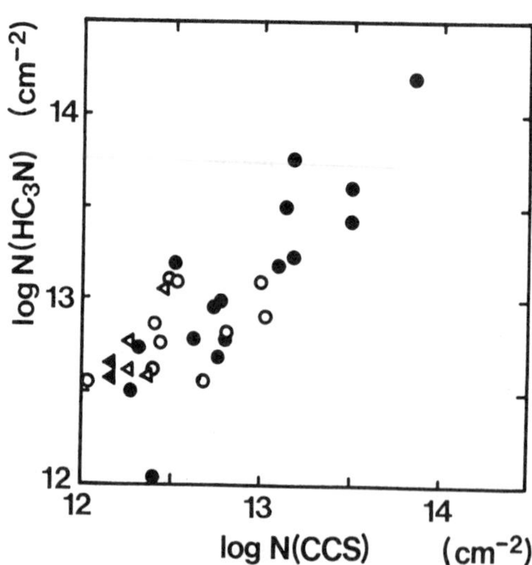

Fig. 2. Correlation between the column densities of CCS and HC_3N. Open marks represent cores with high-velocity CO outflows. Closed marks denote cores without any sign of star formation. Triangles represent upper limits.

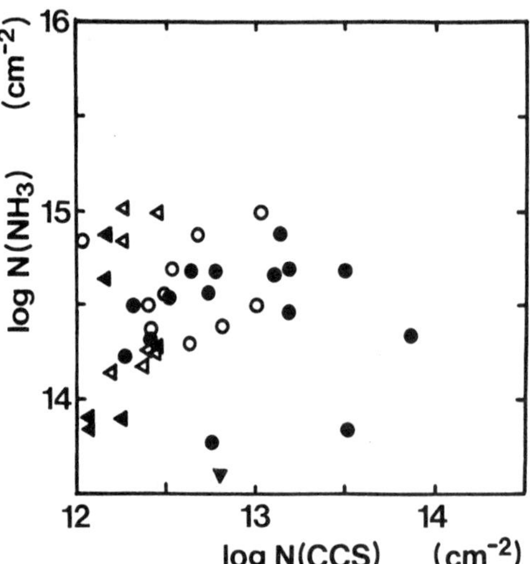

Fig. 3. Correlation between the column densities of CCS and NH_3. See caption of Figure 2.

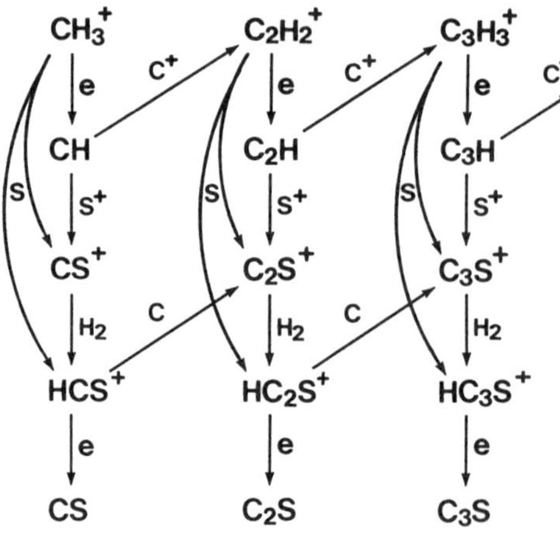

Fig. 4. A schematic illustration of the model for C_nS production (6).

the available number of carbon atoms is insufficient for the effective production of carbon-chain molecules in the steady state. The present time-dependent simulation of CCS shows a similar behavior; the abundance of CCS has a peak at the early stage of evolution. The peak fractional abundance of CCS is 6×10^{-9} for $A_v = 10$ even in the low metal case. This abundance is slightly larger than that reported by Millar and Herbst (8), and is close to the observed value in TMC1. The CCS radical and the other carbon-chain molecules are only abundant in a cloud with large A_v when the cloud is in the early stage of evolution. From the above considerations on the effects of the ultraviolet radiation and the cloud evolution, we can naturally see that the carbon-chain molecules such as CCS and HC_3N are abundant where and when the carbon ions and atoms are abundant.

4. Chemistry in the Low-Mass Star Forming Regions

In the present observations, the carbon-chain molecules were found to be deficient in star-forming regions, where NH_3 is relatively abundant. The effects of star formation on the chemical abundance have been extensively discussed for massive-star forming regions such as Ori KL (18-20). The present survey indicates that the chemistry of a dark cloud core also changes by the formation of a low-mass star.

There are several reasons for the deficiency of carbon-chain molecules in star-forming regions. (1) In general, a core with star formation is expected to be relatively older than that without star formation. Therefore, a certain portion of the carbon atoms have already been fixed to CO in the central part of the core, and the carbon-chain molecules are not efficiently produced because of the lack of available carbon atoms. (2) The density of the central part of the core would generally be larger than that of a core without star formation. The photoionization of C and S is restricted to the thin region near the cloud surface, which suppresses the production of the carbon-chain molecules in the Suzuki model. (3) A shock wave caused by the activities of a protostar would change the chemistry of the core entirely. Particularly, neutral-neutral reactions such as CCS + O would efficiently destroy the carbon-chain molecules.

On the other hand, NH_3 is not abundant in the early stage of cloud evolution. In our preliminary time dependent calculation, NH_3 is more abundant in the steady state than in the early stage of evolution, which is in contrast to the behavior of the carbon chain molecules. This trend is also seen in the results by Herbst and Leung (21). The gas phase production of NH_3 starts from the reaction; $He^+ + N_2 \rightarrow N^+ + N + He$. The N^+ ion thus produced has translational energy enough to overcome the small endothermicity of the next hydrogenation reaction; $N^+ + H_2 \rightarrow NH^+ + H$ (22). Further hydrogenations of NH^+ lead to NH_4^+, which is a precursor ion of NH_3. The formation time scale of N_2 is comparatively long, since it is mainly controled by neutral-neutral reactions. Therefore NH_3 is not abundant in the early stage. The grain surface chemistry is also considered to be important for the production of NH_3. NH_3 molecules formed on the surface of grains cannot return efficiently to the gas phase in the low temperature conditions. However, the energetic phenomena associated with star formation, such as molecular outflows, will provide energy sufficient for desorption of NH_3. Because of these principal reasons, NH_3 may be abundant in the cores with star formation.

Finally, consider cloud evolution from diffuse matter, where most of the carbon atoms are ionized by photoionization. As the cloud evolves and the density increases, the recombination processes become faster than the photoionization processes in the central core of the cloud. The carbon-chain molecules are efficiently produced in this early stage, since C^+ and C still remain throughout the cloud. As the cloud evolves further, the carbon atoms in the central core are converted to CO. The time scale for the conversion from C to CO is roughly comparable with the time scale of cloud contraction, and star formation will take place in some of these cores. The carbon-chain molecules would be deficient at this stage. On the other hand, the star formation provides conditions favorable to the NH_3 production. From these considerations, the carbon-chain molecules and NH_3 may represent different stages in chemical evolution; carbon chain molecules represent a relatively early stage of the cloud evolution, and NH_3 represents a relatively late stage mainly after star formation. Further observations of other related molecules toward a number of dark clouds will be important to establish this hypothesis.

The author is grateful to T.J. Millar, M. Ohishi, Y. Hirahara, N. Kaifu, S. Takano, and S. Saito for their valuable discussions. The author thanks W.M. Irvine for critical reading of the manuscript.

References

(1) Avery, L.W., Broten, N.W., MacLeod, J.M., Oka, T., and Kroto, H.W., 1976, *Ap. J. (Letters)*, **205**, L173.
(2) Saito, S., Kawaguchi, K., Yamamoto, S., Ohishi, M., Suzuki, H., Kaifu, N., 1987, *Ap. J. (Letters)*, **317**, L115.
(3) Suzuki, H., Kaifu, N., Miyaji, T., Morimoto, M., Ohishi, M., and Saito, S., 1984, *Ap. J.*, **282**, 197.
(4) Kaifu, N., Suzuki, H., Ohishi, M., Miyaji, T., Ishikawa, S., Kasuga, T., Morimoto, M., and Saito, S., 1987, *Ap. J. (Letters)*, **317**, L111.
(5) Yamamoto, S., Saito, S., Kawaguchi, K., Kaifu, N., Suzuki, H., Ohishi, M., 1987, *Ap. J. (Letters)*, **317**, L119.
(6) Suzuki, H., Ohishi, M., Kaifu, N., Kasuga, T., Ishikawa, S., and Miyaji, T., 1988, *Vistas in Astronomy*, **31**, 459.
(7) Smith, D., Adams, N.G., Giles, K., and Herbst, E., 1988, *Astron. Ap.*, **200**, 191.
(8) Millar, T.J. and Herbst, E., 1990, preprint.
(9) Prasad, S.S. and Huntress, W.T., Jr., 1982, *Ap. J.*, **260**, 590.
(10) Suzuki, H., 1983, *Ap. J.*, **272**, 579.
(11) Fuente, A., Cernicharo, J., Barcia, A., and Gomez-Gonzales, J., 1990, preprint.
(12) Myers, P.C., Linke, R.A., and Benson, P.J., 1983, *Ap. J.*, **264**, 517.
(13) Benson, P.J. and Myers, P.C., 1983, *Ap. J.*, **270**, 589.
(14) Murakami, A., 1990, preprint.
(15) Leung, C.M., Herbst, E., and Huebner, W.F., 1984, *Ap. J. (Suppl.)*, **56**, 231.
(16) Prasad, S.S. and Huntress, W.T., Jr., 1980, *Ap. J. (Suppl.)*, **43**, 1.
(17) Suzuki, H., 1979, *Prog. Theor. Phys.*, **62**, 936.
(18) Blake, G.A., Sutton, E.C., Masson, C.R., and Phillips, T.G., 1987, *Ap. J.*, **315**, 621.
(19) Brown, P.D., Charnley, S.B., and Millar, T.J., 1988, *Mon. Not. R. astr. Soc.*, **231**, 409.
(20) Jacq, T., Walmsley, C.M., Henkel, C., Baudry, A., Mauersberger, R., and Jewell, P.R., 1990, *Astron. Ap.*, **228**, 447.
(21) Herbst. E. and Leung. C.M., 1989, *Ap. J. (Suppl.)*, **69**, 271.
(22) Galloway, E.T. and Herbst, E., 1989, *Astron. Ap.*, **211**, 413.

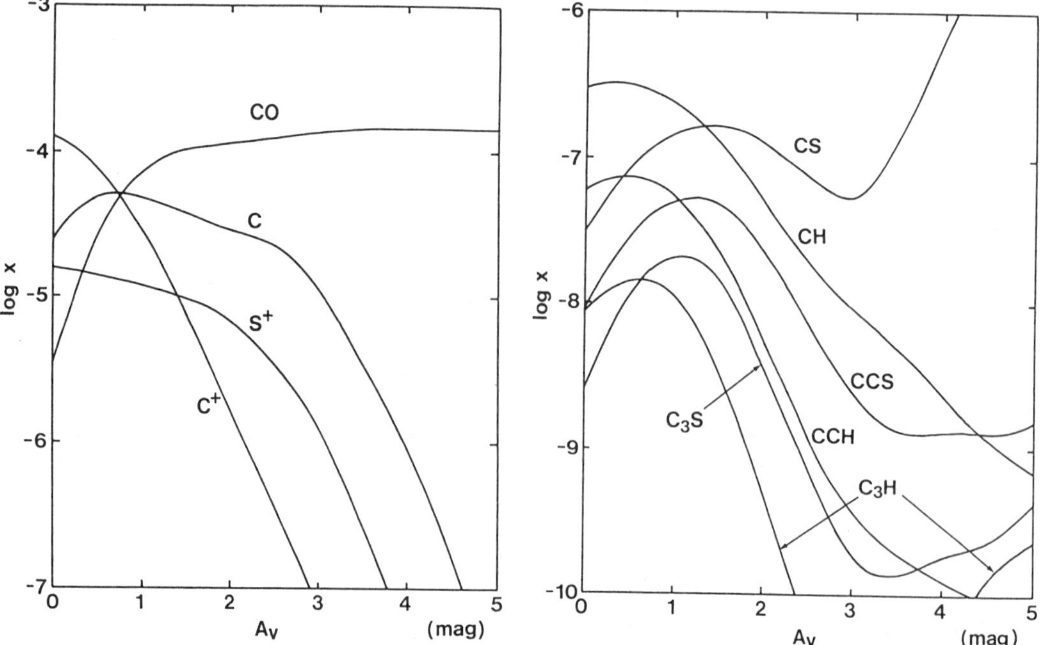

Fig. 5. Steady-state abundances of several species as a function of A_v. The H_2 density is 10^4 cm^{-3} and the temperature is 10 K.

Mapping Observations of Sulfur-Containing Carbon-Chain Molecules in the Taurus Molecular Cloud-1 (TMC-1)

Yasuhiro HIRAHARA[1], Akimasa MASUDA[1], Satoshi YAMAMOTO[2], Shuro TAKANO[2],
Masatoshi OHISHI[3], Hiroko SUZUKI[4], Shin-ichi ISHIKAWA[4], Kentarou KAWAGUCHI[4],
and Norio KAIFU[4]

[1] *Department of Chemistry, University of Tokyo, Bunkyo-ku, Tokyo 113, Japan*

[2] *Department of Astrophysics, Nagoya University, Chikusa-ku, Nagoya 464-01, Japan*

[3] *Department of Physics, Toyama University, 3190 Gofuku, Toyama 930, Japan*

[4] *Nobeyama Radio Observatory, Minamimaki, Minamisaku, Nagano 384-13, Japan*

Mapping observations of $C^{34}S$, CCS, C_3S, C_4H, HC_3N, and NH_3 were carried out toward Taurus Molecular Cloud-1 (TMC-1). The distributions of column densities of CCS and C_3S are similar to each other and peak at about 1' southeast of the cyanopolyyne peak. The distribution of NH_3 differs from those of CCS and C_3S. The results are discussed in connection with chemical evolution and the physical environment of TMC-1.

Introduction

The Taurus Molecular Cloud-1 (TMC-1) is known to be an outstanding source for a variety and high abundances of unsaturated carbon-chain molecules. Since Little et al. (1) first found the "anti-correlation" between distributions of HC_3N and NH_3 in TMC-1, the molecular distributions in this cloud have been extensively investigated (e.g. (2)). In 1987, the new carbon-chain molecules CCS and C_3S were detected (3-6). These molecules are expected to be suitable for the investigations of the dense cores of dark clouds because they emit intense but not opaque lines in the dark clouds. In addition, the comparison of the distributions of CCS and C_3S with those of CS and of other carbon-chain molecules would give information about the synthetic mechanism of carbon-chain molecules. In the present paper, we report mapping observations of TMC-1 with the lines of $C^{34}S$, CCS, C_3S, C_4H, HC_3N, and NH_3.

Observations

The observations were carried out in April 1988 and May 1989 by using the 45m radio telescope of Nobeyama Radio Observatory. Table I lists the observed lines, rest frequencies, and corresponding beam sizes (HPBW) of the telescope. We observed the transitions of CCS, C_3S, HC_3N, and NH_3 molecules toward 114 positions with a grid with 1 arcmin. spacing. The obtained rms noise was 0.2~0.3 K. The lines of $C^{34}S$ and C_4H were observed toward 20 positions with 1 arcmin. spacing along a NW-SE ridge of TMC-1, where the rms noise was 0.08~0.15 K.

Results

The integrated intensity maps of CCS $J_N=4_3-3_2$, HC_3N J=5-4, and NH_3 (J, K)=(1, 1) are shown in Fig. 1. The general distribution of CCS was found to be similar to that of HC_3N. On the other hand, the "anti-correlation" between CCS and NH_3 was observed clearly. From the map of CCS, several clumps are apparent in the ridge of TMC-1, which could not be found in the maps of other molecules. At the position of 2 arcmin. southwest of the ammonia peak ($4^h38^m19.0^s$(1950.0), 25°42'30"(1950.0)) we noticed the existence of a cloud component with an apparently different LSR velocity ($\sim$5.2 km·s^{-1}) from that of the ridge (5.85 km·s^{-1}). The NH_3 velocity distribution also indicates the existence of such a cloud component, which is consistent with the NH_3 observations of Little et al. (1) and Olano et al. (2).

Analysis

The column density distribution of CCS was derived from LTE calculations of $J_N = 4_3 - 3_2$ and $2_1 - 1_0$. The excitation temperatures were about 6 K at most of the observed positions. The result is shown in Fig. 2. We also obtained the column density distribution of C_3S with the LTE calculations of the J=7-6 and J=4-3 transitions. The column density distribution of C_4H was derived from assumptions that the line of C_4H is optically thin, that T_{ex} = 6 K, and that the beam-filling factor is unity. The column density distributions of CCS, C_3S, and C_4H along the NW-SE ridge of TMC-1 are shown in Fig. 3. The maximum value of the CCS column density is $(1.0 \pm 0.3) \times 10^{14}$ cm^{-2} at the position about 1~2 arcmin. southeast of the cyanopolyyne peak $(4^h 38^m 38.6^s (1950.0), 25^\circ 35' 45'' (1950.0))$ along the NW-SE ridge. This position also gives the maximum of the column densities of $C^{34}S$, C_3S, and C_4H. Olano et al. (2) reported a similar deviation from their mapping observation of C_4H in the 20 GHz region. Moreover, the abundant region of CCS is compact and has a FWHM radius of ~0.05 pc., assuming a distance to TMC-1 of 140 pc..

Since the excitation of $C^{34}S$ was not explained well with the LTE approximation, a LVG model at T_k = 10 K was adopted to derive the column density distribution of $C^{34}S$ as well as the number density distribution of H_2. As a consequence, $n(H_2)$ was derived to be $(4.0^{+3.1}_{-2.7}) \times 10^4$ cm^{-3} at the cyanopolyyne peak, and $(4.0 \pm 2.0) \times 10^5$ cm^{-3} at the ammonia peak. The northwest part of the TMC-1 ridge has somewhat higher density than the southeast part. The $C^{34}S$ column density is $(1.0 \pm 0.3) \times 10^{13}$ cm^{-2} at the cyanopolyyne peak. From this value, the fractional abundance of $C^{32}S$ is determined to be $(2.3 \pm 0.7) \times 10^{-8}$ by assuming the isotope ratio $C^{32}S/C^{34}S$ = 22.6, and the hydrogen column density, $N(H_2) = 10^{22}$ cm^{-2}. The [CCS]/[CS] abundance ratio at the CCS peak position is estimated to be 0.3 ± 0.1, and the [C_3S]/[CCS] ratio to be 0.3 ± 0.15. These two ratios, [CCS]/[CS] and [C_3S]/[CCS] have almost the same value at the positions around the cyanopolyyne peak. On the other hand, both ratios are lower at the other regions. These results show the existence of clumps with significantly different chemistry for the growth of the C_nS carbon-chain series in TMC-1.

Discussion

Two models have been postulated so far for the synthesis of C_nS. Smith et al. (7) proposed one model of synthesis of C_nS in dense clouds, where CCS is produced via a reaction between S^+ and C_2H_2, and C_3S via a reaction between S^+ and cyclic-C_3H_2. Suzuki et al. (8) proposed another model with a coupled mechanism for C_nS and C_nH production from C^+ and S^+ in UV-ionized (C^+-S^+) regions.

We made a time dependent calculation of the formation of C_nS and C_nH based on Suzuki et al's model considering the UV-ionization of C and S in the cloud. From the preliminary result, these carbon-chain molecules can be produced effectively at a relatively early stage ($\sim 10^5$ year) starting from the diffuse environment where most of the carbon and sulfur atoms are photo-ionized. Millar et al.(9) calculated the abundance of C_nS based on Smith et al's model, and also showed that these species are produced at the age of $\sim 10^{5.5}$ year. Therefore, the carbon-chain enhanced region in TMC-1 might be in a relatively early stage of the dynamical evolution of a clump, where carbon-chain molecules have neither attached to the dust grains, nor been converted to CO in gas phase. Because the CCS abundant region in TMC-1 is found to be compact in size and has a moderate density of H_2, $(6 \pm 2) \times 10^4$ cm^{-3}, the UV-radiation can reach deeper inside the cloud core, and the carbon-chain molecules might have been synthesized at a more diffuse stage.

In contrast, the northern part of TMC-1, where such carbon-chain molecules are deficient and NH_3 is prominent, can be interpreted to be in relatively advanced stage of cloud evolution. Carbon-chain molecules which had been produced in an earlier stage of the gas-phase reactions might have already been destroyed in this region. In addition, NH_3 might have been enhanced by overcoming the slight endothermicity in its gas-phase reaction because of some heat-up mechanism resulted from the process of gravitational contraction and/or energy input from neighboring active sources. The physical characteristics of this region, such as the larger size and more complicated velocity structures compared with the CCS core, suggest that these region had been influenced by star-formation. The IRAS point source (04381+2540) at ~3 arcmin. southwest of the ammonia peak, which is a candidate for a proto-star, might be the energy source, although the distance between the IRAS source and the ammonia peak is unknown. The above interpretation of the chemical differences between the carbon-chain peak and the ammonia peak is consistent with the observational facts that the NH_3 cores in dense cloud statistically coexist with star-formation phenomena, while the CCS cores do not (10). More detailed observations of the molecules in TMC-1 as well as the chemical model calculations of contracting and UV-ionized clouds are required for the understanding of the chemical and physical evolutions of dark cloud cores.

<u>References</u>

(1)Little, L. T., Macdonald, G. H., Riley, P. W., and Matheson, D. N. 1979, <u>M. N. R. A. S.</u> **189**, 539.
(2)Olano, C. A., Walmsley, C. M., and Wilson, T. L. 1988, <u>Astr. Ap.</u> **196**, 194.
(3)Suzuki, H., Kaifu, N., Miyaji, T., Morimoto, M., Ohishi, M. and Saito, S. 1984, <u>Ap. J.</u> **282**, 197.
(4)Kaifu, N., Suzuki, H., Ohishi, M., Miyaji, Ishikawa, T. S., Kasuga T., Morimoto, M., and Saito, S. 1987, <u>Ap. J. (Letters)</u> **317**, L111.
(5)Saito, S., Kawaguchi, K., Yamamoto, S., Ohishi, M., Suzuki, H., Kaifu, N., 1987, <u>Ap. J. (Letters)</u> **317**, L115.
(6)Yamamoto. S., Saito, S., Kawaguchi, K., Kaifu, N., Suzuki, H., and Ohishi, M. 1987, <u>Ap. J. (Letters)</u> **317**, L119.
(7)Smith, D., Adams, N. G., Giles K., and Herbst, E. 1988, <u>Astr. Ap.</u> **200**, 191.
(8)Suzuki, H. Ohishi M., Kaifu N., Kasuga, T. Ishikawa, S., and Miyaji T. 1988, <u>Vistas in Astronomy</u> **31**, 459.
(9)Millar, T. J. and Herbst, E., 1990, <u>Astr. Ap.,</u> **231**, 466.
(10)Yamamoto. S., 1989, <u>"The 1989 international Chemical Congress of Pacific Basin Societies"</u>

Table I. Observed transitions in TMC-1.

Molecule	Transition	Frequency(MHz)	Beam size(HPBW)
$C^{34}S$	$J=1-0$	48206.956	34"
$C^{34}S$	$J=2-1$	96412.981	17"
CCS	$J_N=2_1-1_0$	22344.032	74"
CCS	$J_N=4_3-3_2$	45379.039	36"
C_3S	$J=4-3$	23122.985	71"
C_3S	$J=7-6$	40465.017	41"
HC_3N	$J=5-4$	45490.314	36"
C_4H	$N=5-4,J=11/2-9/2$	47566.800	34"
C_4H	$N=5-4,J=9/2-7/2$	47605.515	34"
NH_3	$(J, K)=(1, 1)$	23694.495	71"

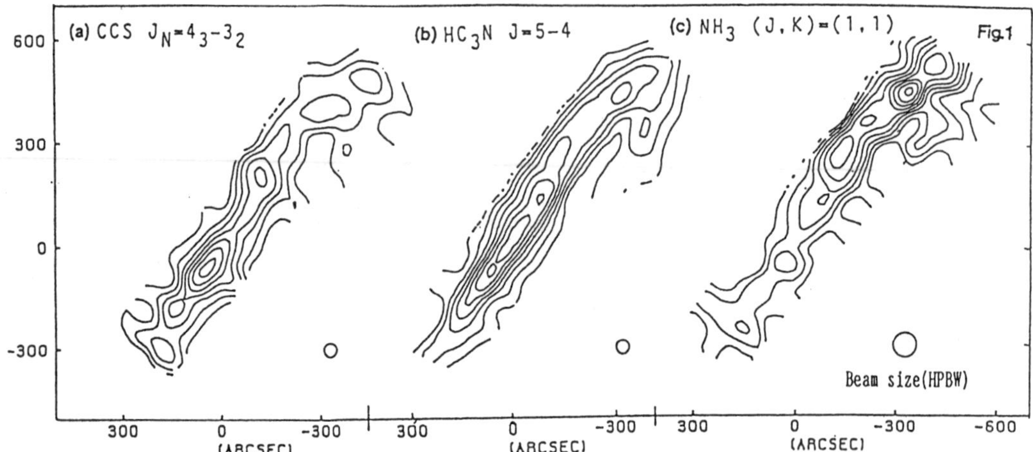

Fig. 1.
Maps of integrated intensities of (a) CCS $J_N=4_3-3_2$, (b) HC$_3$N J=5-4, and (c) NH$_3$ (J, K)=(1, 1). The interval between contours is 0.2 K*km/s for (a),(c), and 0.4 K*km/s for (b).
The map origin is the cyanopolyyne peak.

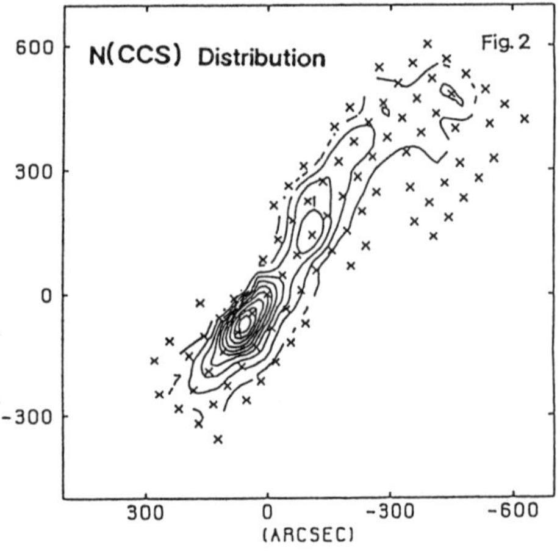

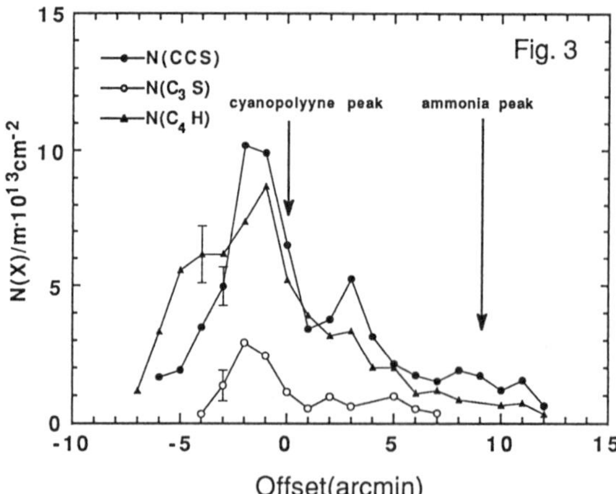

Fig. 2.
Column density distribution of CCS. The minimum contour level and the contour interval is 10^{13} cm^{-2}. The map origin is the cyanopolyyne peak.

Fig. 3.
Column density distributions of CCS, C$_3$S, and C$_4$H along NW-SE ridge of TMC-1. The absissa represents the distance from the cyanopolyyne peak. The ordinate represents the column density in m x 10^{13} cm^{-2} unit, where m=1 for CCS and C$_3$S, and m=20 for C$_4$H.

Observation of the ^{13}C Isotopic Species of HC$_3$N in TMC-1: Evidence for Isotopic Fractionation

Shuro Takano[1], Akimasa Masuda[2], Yasuhiro Hirahara[2], Hiroko Suzuki[3],
Shin-ichi Ishikawa[3], Norio Kaifu[3], and Masatoshi Ohishi[4]

[1] Department of Astrophysics, Faculty of Science, Nagoya University, Chikusa-ku, Nagoya 464-01,
Japan
[2] Department of Chemistry, Faculty of Science, University of Tokyo, Hongo, Bunkyo-ku, Tokyo
113, Japan
[3] Nobeyama Radio Obeservatory, Minamimaki, Minamisaku, Nagano 384-13, Japan
[4] Department of Physics, Faculty of Science, Toyama University, Gofuku, Toyama 930, Japan

The J=5-4 rotational transitions of the three ^{13}C isotopic species of HC$_3$N
were observed at three positions toward TMC-1. A significant difference in
the intensities was detected among the spectral lines of the three isotopic species.
This difference is considered to be due to isotopic fractionation. The isotopic
fractionation mechanism of the ^{13}C isotope between the three carbon sites in HC$_3$N
is discussed.

INTRODUCTION The carbon-chain molecules, H-(C≡C)$_n$-C≡N (n=1~5) and so on, are abundant
in cold dark clouds and carbon-rich circumstellar envelopes. However, the carbon-chain growing
mechanism is not well established. A detailed analysis of isotopic fractionation of the ^{13}C isotope
in the carbon-chain molecules may be one of the methods to obtain information on the mechanism.
Among the carbon-chain molecules, HC$_3$N is considered to be the best one suitable for observations
of its molecular species with single substitution by the ^{13}C isotope, because of its relatively strong
intensity. Until now, the three ^{13}C isotopic species, H$^{13}CCCN$, HC^{13}CCN and HCC^{13}CN, were
observed in Sgr B2 (1~5), Ori A (3,5 and 6), IRC+10216 (6~8) and TMC-1 (9). In Sgr B2, Ori A
and IRC+10216, no apparent difference in the intensities was found among the spectral lines of the
three ^{13}C isotopic species. In TMC-1, the signal-to-noise ratio of the obtained spectra (J=1-0)
of the ^{13}C isotopic species was not good enough to give the meaningful difference in the intensities
suggesting isotopic fractionation. We studied the three ^{13}C isotopic species of HC$_3$N by the J=5-4
rotational transition toward TMC-1, aiming to obtain clear evidence for isotopic fractionation.

OBSERVATION The observation was carried out in March 1988 using the 45m radiotelescope
at Nobeyama Radio Observatory (NRO). The observed species, transitions, frequencies, and
receivers used are listed in Table I. The J=5-4 transitions of HC$_3$N, HC^{13}CCN and HCC^{13}CN
were observed with an SIS mixer receiver, and simultaneously the J=5-4 transition of H$^{13}CCCN$
was observed with a cooled Schottky barrier diode mixer receiver. The three positions in TMC-1
were studied: a point near the ammonia peak (R.A.(1950)= $4^h38^m16.^s4$, Dec.(1950)=25°41'45"),
the cyanopolyyne peak ($4^h38^m38.^s6$, 25°35'45") and a southeast point of the cyanopolyyne peak (
$4^h38^m47.^s5$, 25°33'45").

RESULTS The observed spectra of HC^{13}CCN and HCC^{13}CN for the three positions are shown
in Figure 1. The intensity of the spectral line of HCC^{13}CN is stronger than that of HC^{13}CCN
at all of the positions. In particular, the difference in the intensities is larger than three times
the rms noise at the cyanopolyyne peak. The intensity of the other isotopic species, H$^{13}CCCN$,
is weaker than that of HC^{13}CCN. The integrated intensities of the spectral lines of the three
^{13}C isotopic species at the cyanopolyyne peak are 0.21±0.04, 0.30±0.04 and 0.42±0.04 in K·km/s
for H$^{13}CCCN$, HC^{13}CCN and HCC^{13}CN, respectively. The errors correspond to three times
the standard deviation. The relative ratio of the integrated intensities is about 0.7 : 1.0 : 1.4.
However, the intensity for H$^{13}CCCN$ may suffer from some systematic errors, because H$^{13}CCCN$
was observed with a different receiver (Table I).

DISCUSSION The significant difference in the intensities may be caused by an anomalous rotational excitation or by a difference in the abundances among the isotopic species. In order
to discriminate between these two causes, the intensities of the spectral lines of HC^{13}CCN and
HCC^{13}CN were compared by using the spectra of the J=4-3 transition, which was observed by

the line survey project at NRO.   The intensity of the spectral line of $HCC^{13}CN$ was found to be also stronger than that of $HC^{13}CCN$ in the J=4-3 transition.   Therefore we conclude that the abundance of the isotopic species is directly reflected by the intensity.   The difference in the abundances is considered to be caused by the chemical reaction which is related to HC_3N, namely, by chemical fractionation.

Chemical fractionation usually occurs as a result of isotope exchange reactions at low temperatures.   In the case of HC_3N, the following isotope exchange reactions may be responsible for fractionation; $HC_3N + {}^{13}C^+ \Longleftrightarrow (H^{13}CCCN$ or $HC^{13}CCN$ or $HCC^{13}CN) + {}^{12}C^+$.   However, these isotope exchange reactions must be excluded as a main fractionating process for the following reasons.   [1] Generally, the heavier isotope concentrates in molecules by the isotope exchange reaction in the cold environment of the dark clouds as a consequence of the small exothermicity caused by the difference in the zero-point energy.   In order to estimate this effect on the above isotope exchange reactions, the differences in the zero-point energy were calculated among the normal HC_3N and the three ${}^{13}C$ isotopic species.   The result is shown in Figure 2.   The differences in the zero-point energy among the three ${}^{13}C$ isotopic species were already calculated by Wolfsberg et al.(11).   According to the calculation, the difference between the normal HC_3N and $H^{13}CCCN$ is the largest of the differences among the three isotopic species.   Therefore, if an isotope exchange reaction is responsible for fractionation, the ${}^{13}C$ isotope must concentrate in all of the three carbon sites in HC_3N, and the carbon isotope ratios $({}^{12}C/{}^{13}C)$ calculated by the abundances of HC_3N and the three ${}^{13}C$ isotopic species must become much lower than the solar value $({}^{12}C/{}^{13}C = 89)$. However, the obtained carbon isotope ratios, shown in Table II, indicate that the ${}^{13}C$ isotope in $H^{13}CCCN$ and $HC^{13}CCN$ is not concentrated.   [2] The laboratory experiment of the ion-molecule reaction between HC_3N and C^+ shows that the main products are C_4N^+ and C_3H^+ (12,13).

These reasons indicate that the former isotope exchange reactions are not responsible for fractionation, and it is concluded that the ${}^{13}C$ isotope must be fractionated in the precursor in the formation process of HC_3N.   Therefore information on the precursor can be obtained from fractionation in HC_3N.   Fractionation among the three carbon atoms in HC_3N entails that the precursor should have nonequivalent carbon atoms.   For example, the bare carbon-chain molecule, C_3, which has two equivalent carbons is excluded from the main precursors of HC_3N in cold dark clouds.   In addition, cyclic-$C_3H_3^+$ which has three equivalent carbons should not be a main precursor.   Hydrogen or nitrogen must bind carbon to introduce asymmetry in the skeletal structure of the precursor in the early stage of the HC_3N formation.   The consideration given so far claims that the fractionation of the ${}^{13}C$ isotope in the possible precursors CN, HCN, HNC and CH must be investigated to understand the mechanism of fractionation in HC_3N.   The study of fractionation of the ${}^{13}C$ isotope in a related molecule HC_5N will also supply more information.   The ${}^{13}C$ isotopic species of HC_5N already have been detected in TMC-1 (14).   The study of the mechanism of fractionation in HC_3N will contribute to the understanding of the formation process of the carbon-chain molecules.

ACKNOWLEDGEMENT   We are grateful to S. Yamamoto for discussions and calculations of the zero-point energies.

REFERENCES
(1) Gardner, F.F. and Winnewisser, G. 1975, *Ap.J.(Letters)*, **197**, L73.
(2) Churchwell, E., Walmsley, C.M. and Winnewisser, G. 1977, *Astr.Ap.*, **54**, 925.
(3) Wannier, P.G. and Linke R.A. 1978, *Ap.J.*, **226**, 817.
(4) Cummins, S.E., Linke, R.A. and Thaddeus, P. 1986, *Ap.J.Suppl.*, **60**, 819.
(5) Turner, B.E. 1989, *Ap.J.Suppl.*, **70**, 539.
(6) Johansson, L.E.B., Andersson, C., Elldér, J., Friberg, P., Hjalmarson, Å., Höglund, B., Irvine, W.H., Olofsson, H. and Rydbeck, G. 1985, *Astr.Ap.Suppl.*, **60**, 135.
(7) Olofsson, H., Johansson, L.E.B., Hjalmarson, Å. and Nguyen-Quang-Rieu 1982, *Astr.Ap.*, **107**, 128.
(8) Kahane, C., Gomez-Gonzalez, J., Cernicharo, J. and Guélin, M. 1988, *Astr.Ap.*, **190**, 167.
(9) Mangum, J.G., Rood, R.T., Wadiak, E.J. and Wilson T.L. 1988, *Ap.J.*, **334**, 182.
(10) Creswell, R.A., Winnewisser, G. and Gerry, M.C.L. 1977, *J.Mol.Spectrosc.*, **65**, 420.
(11) Wolfsberg, M., Bopp, P., Heinzinger, K. and Mallinson, P.D. 1979, *Astr.Ap.*, **74**, 369.
(12) Raksit, A.B. and Bohme, D.K. 1985, *Can.J.Chem.*, **63**, 854.
(13) Knight, J.S., Freeman, C.G., McEwan, M.J., Smith, S.C., Adams, N.G. and Smith, D. 1986, *Mon.Not.R.Astr.Soc.*, **219**, 89.
(14) Takano, S., Suzuki, H., Ohishi, M., Ishikawa, S., Kaifu, N., Hirahara, Y. and Masuda, A. 1990, *Ap.J.(Letters)*, **361**, L15.

TABLE I   Observed species, frequencies, and receivers used.

SPECIES	TRANSITION	FREQUENCY [a]	RECEIVER(T_{SYS})
H^{13}CCCN	J=5-4	44084.170(MHz)	cooled Schottky barrier diode mixer (600 K)
HC^{13}CCN	5-4	45297.345	SIS mixer (300 K)
HCC^{13}CN	5-4	45301.708	SIS mixer
HC$_3$N	5-4	45490.307	SIS mixer
HC$_3$N	10-9	90979.023	cooled Schottky barrier diode mixer (450 K)

[a] Reference 10.

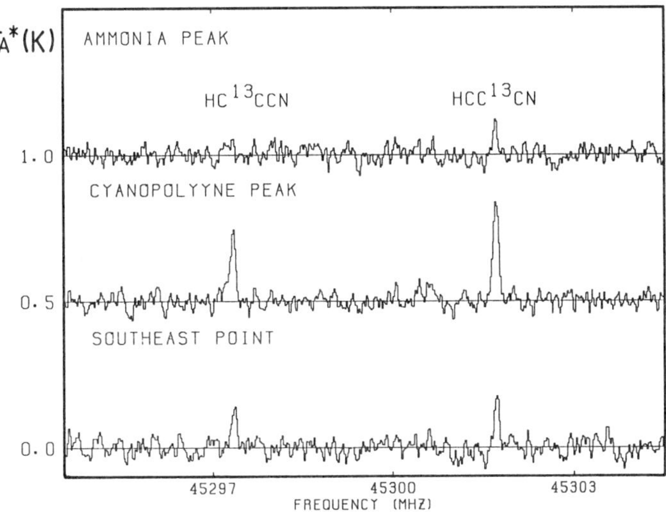

Fig.1   The observed spectra of HC^{13}CCN and HCC^{13}CN at three positions in TMC1.

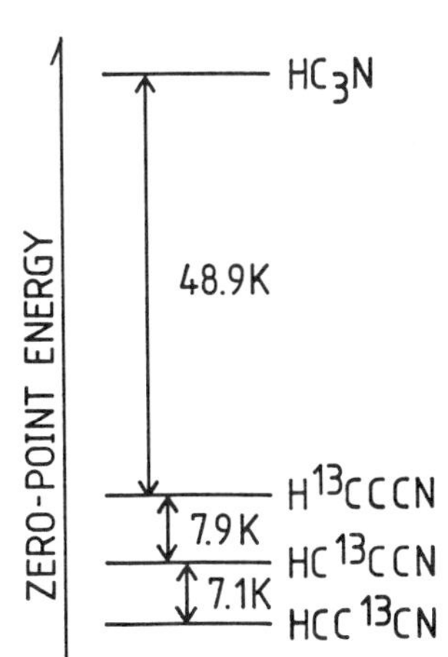

TABLE II   The carbon isotopic ratio (^{12}C/^{13}C) obtained by each ^{13}C isotopic species at the cyanopolyyne peak[a].

H^{13}CCCN	HC^{13}CCN	HCC^{13}CN
120(102,140)	87(75,99)	63(55,71)

[a] The values in parentheses indicate the lower and upper limit.

Fig.2   The differences in the zero-point energy between the normal HC$_3$N and the three ^{13}C isotopic species.

Molecular Line Survey toward Sgr B2 with the Nobeyama 45m Radio Telescope

Masatoshi OHISHI*

Department of Physics, Toyama University, 3190 Gofuku, Toyama, 930, Japan

** Present address: Nobeyama Radio Observatory, National Astronomical Observatory of Japan,
Nobeyama, Minamimaki, Minamisaku, Nagano, 384-13, Japan*

Recent new results from the Nobeyama Molecular Line survey are
presented. These results were obtained by installing four new
very low-noise receivers (SIS40, SIS100, and two HEMT amplifiers).
In the long-mm wave region and the cm wave region ($\nu \le 50$ GHz), most
observable frequency regions were surveyed for Sgr B2 (M). A
new "Local Peak" of molecular abundance was found away from Sgr
B2 (M) through 3-point line survey in Sgr B2. The chemical significance of the absorption features of SiO is discussed.

INTRODUCTION

More than 80 interstellar molecules have been detected so far. Many of
them were detected in the early 70's by mm-wave radio telescopes. In the late
80's, we experienced "the second Gold-rush" of detecting new interstellar
molecules with larger, more sensitive radio telescopes such as the 45m telescope
of Nobeyama, Japan, and the 30m telescope of IRAM in Spain. The second Gold-
rush of detecting new interstellar molecules can be said to have resulted from
the wideband Molecular Line Survey performed in these observatories.
The first spectral line survey was made by the Onsala 20m telescope in
Sweden toward the massive star-forming region, Orion KL, and the circumstellar
molecular envelope of the super-redgiant star, IRC+10216 (1). Following this
this work, Sutton et al. (2) and   Blake et al. (3) made a survey toward Orion
KL in a frequency region of 215 - 263 GHz using the 10m dish of Caltech, and
Cummins et al. (4) made a survey toward the huge molecular cloud / HII region
complex in the center of our Galaxy, Sgr B2, with the 7m telescope of Bell
laboratory. Recently Turner (5) published results of his survey for Orion KL
and Sgr B2 with the mm-wave telescope of NRAO at Kitt Peak. Two unpublished
molecular line surveys exist : the Nobeyama Molecular Line Survey toward Orion
KL, the cold and quiescent cloud TMC-1, Sgr B2, and IRC+10216, and the IRAM
survey toward IRC+10216 in the short mm-wave region.
The Nobeyama Molecular Line Survey was started in 1982. The survey staff
are N. Kaifu, H. Suzuki, T. Miyaji, and I. Later S. Ishikawa, S. Saito, S.
Yamamoto, K. Kawaguchi, and Y. Hirahara joined the group. At first, we made a
survey toward Orion KL in the 86 - 91.5 GHz region, then in the 34.5 - 50 GHz
region (6). This survey showed higher sensivity of the data aquisition compared
with other survey works performed before. After the work toward Orion KL, we
started to survey a famous cold, dark cloud TMC-1 in the 34 - 50 GHz region. It
is well known that this cloud has no sign of current star-formation, and thus
seems to show simple and basic chemistry. Especially, the molecular abundances
of linear carbon-chain molecules are quite high in TMC-1. Hiroko Suzuki was
investigating the formation mechanism of such linear carbon-chain molecules, and
therefore the survey toward TMC-1 was a very important project for her. Most of
the results of the Molecular Line Survey toward TMC-1 are summarized by Kaifu
(7) in this issue.
Following by the survey toward TMC-1, we began to survey Sgr B2 and

"

IRC+10216.   In this paper, we report some topics from our results of the survey
toward Sgr B2 in the cm- and long mm-wave region ($v \leq 50$ GHz).

OBSERVATIONS

The observations were made using the 45m radiotelescope of Nobeyama Radio
Observatory (NRO) since August 1982.   The frontends used for the survey are
summarized in Table I in the order of frequency.   The backend is a bank of eight
acousto-optical radiospectrometers (AOS), each of which has 250 kHz resolution,
250 MHz bandwidth, and 2000 ch outputs.

Table I.   Frontends at Nobeyama Radio Observatory

RX name	frequency range (GHz)	bandwidth (MHz)	Tsys (SSB) (K)
A10	8.8 ~ 10.8	2000	~ 100
H15	14.0 ~ 16.0	2000	~ 100
H22	18.0 ~ 26.0	2000	~ 150
A40	34.0 ~ 50.0	2000	600 ~ 1000
S40	33.0 ~ 50.0	700	250 ~ 600
A80	70.0 ~ 90.0	2000	600 ~ 800
M100	84.5 ~ 115.6	500	350 ~ 700
S100	82.0 ~ 115.6	500	250 ~ 700

First characters of RX name mean the following :
  H -- HEMT amplifier receiver
  S -- SIS mixer receiver
  A -- cooled Schottky diode mixer receiver
  M -- cooled Schottky diode mixer receiver provided by Millitech

The position-switching mode was used for the observations 0.5 degrees in
azimuth in a period of one minute.   The telescope pointing was checked by the
SiO maser from nearby stars (for Sgr B2, we selected VX Sgr), and the pointing
accuracy is estimated to be better than 5" rms.

The survey was made toward three positions in the Sgr B2 molecular cloud :
the Sgr B2 (M) (α(1950.0) = $17^h 44^m 10.^s6$, δ(1950.0) = -28˚ 22'05", the reference
position), the Sgr B2 (N) (($\Delta\alpha,\Delta\delta$) = (+0", +60")), and a "Local Peak" (($\Delta\alpha,\Delta\delta$) =
(-60", +45")).   The "Local Peak" was found through our mapping observations of
the HC_3N and the HNCO molecules. It is a local peak of their column densities,
and contains no continuum sources.   We note that we have NOT observed the
usually observed position, Sgr B2 (OH).

COMPARISON of SPECTRA from the THREE POSITIONS in SGR B2

Sample spectra of our survey for the three positions of Sgr B2 are shown in
Figures 1 and 2.   Now we will compare them to investigate similarities and
differences among them.

(1) Sgr B2 (M)

There is a powerful ultracompact HII region, MD5, at this position (8), and
the existence of a molecular outflow has been suggested (9).   These facts
indicate that very active OB star-formation takes place at this position.   We
can observe most of the interstellar molecules detected toward this position
except for long carbon-chain molecules and some molecules characteristic of the
circumstellar envelope of IRC+10216.   Especially, antenna temperatures of SO and
SO_2 peak at this position as was reported by Goldsmith et al. (10).   Preliminary
excitation analyses for these molecules show that the column densities of SO and

SO$_2$ also peak here. Line profiles of the SO and SO$_2$ spectra show no signs of absorption, indicating that these molecules distribute only in the dense and hot core region. The situation is very similar to that in Orion KL core, which strongly suggests that the formation of SO and SO$_2$ molecules is related to active OB star-formation.

On the other hand, long carbon-chain molecules, e.g. HC$_{2n+1}$N, are relatively less abundant here. We can compare the distribution of HC$_7$N J = 38-37 transition in Figure 1. There is no hint of this transition in Sgr B2 (M), while we can clearly see it in the "Local Peak" and possibly in Sgr B2 (N). Such characteristics can be seen in other linear cyanopolyyne molecules. We could observe HC$_{2n+1}$N (n = 0 ~ 3), but no features for HC$_9$N nor HC$_{11}$N.

Some molecules detected in our survey were observed with absorption in Sgr B2 (M) : CS, CCS, SiO, CH$_3$CHO, CH$_3$CN, H$_2$CO, NH$_3$, C$_3$H$_2$, although our survey has not been completed. The absorption feature of the SiO molecule is especially worthy of note. As you can see in Figure 2, the deep absorption feature at V$_{LSR}$ ~ 62 km s^{-1} can be seen in all positions, suggesting that the absorbing material is extended. We will describe the SiO spectra later in detail.

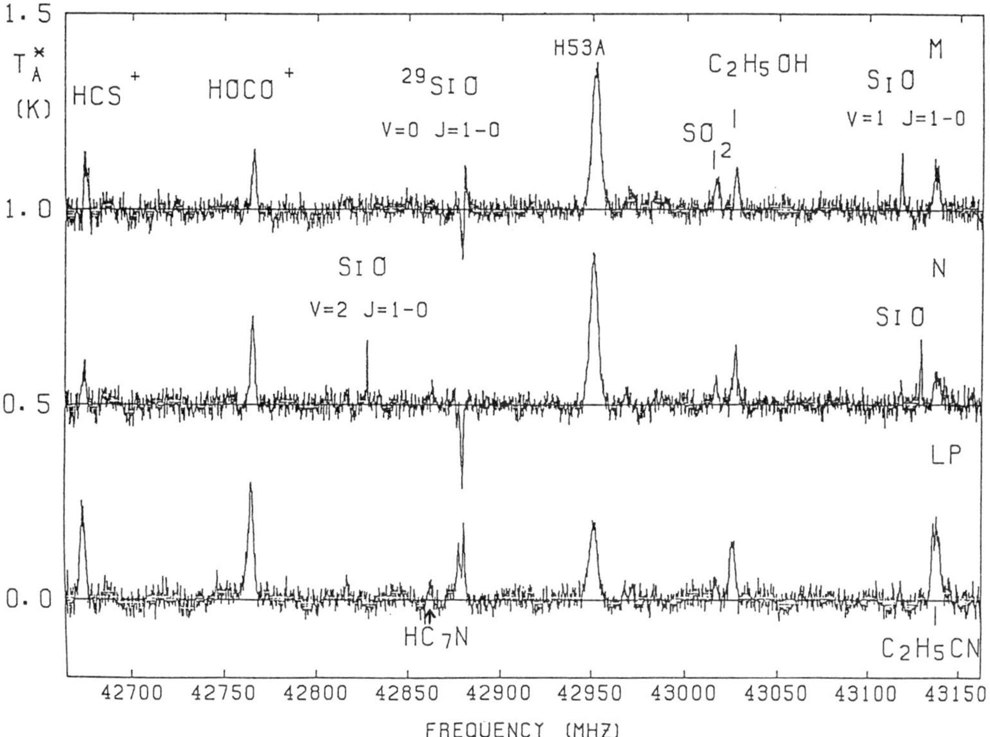

Fig. 1.   Sample spectra of the Sgr B2 line survey in the frequency region of   42663 to 43163 MHz. The top spectrum was taken at Sgr B2 (M), the middle   at Sgr B2 (N), and the bottom at the "Local Peak", respectively.

(2) Sgr B2 (N)

There is another very ative ultracompact HII region, MD4, in Sgr B2 (N) (8). It is well known that a lot of H$_2$O maser spots distribute along the SW edge of MD4. Furthermore, vibrationally excited HC$_3$N is quite strong at this position (10). The dust temperature is higher than that in Sgr B2 (M) (11). From these facts, Sgr B2 (N) is interpreted as a younger OB star-forming region than Sgr B2 (M).

Molecular species that we have observed in Sgr B2 (N) are very similar to those in Sgr B2 (M). Line intensities are generally lower than those in Sgr B2 (M) except for transitions from high energy-levels. SO and SO_2 also peak in Sgr B2 (N). It should be noted that there is a possible feature for the HC_7N J = 38 - 37 transition (see Figure 1). This is consistent with the fact that the column density of HC_3N in Sgr B2 (N) is larger that that in Sgr B2 (M) (12), which suggests that the HII regions of Sgr B2 (N) have just begun to photoionize its surrounding molecular gas.

Furthermore, we can clearly see the v=2 J=1-0 maser feature of SiO as well as the v=1 J=1-0 maser (Figure 1). The apparent linewidth of the v=2 J=1-0 maser is 6 km s^{-1} and the radial velocity is 106.4 km s^{-1} which is consistent with that of one of the H_2O maser spots in Sgr B2 (N).

(3) The "Local Peak"

This "Local Peak" was found through our mapping observations of HC_3N and HNCO. The column densities of these molecules peak at $(\Delta\alpha,\Delta\delta)$ = (-60", +45") from Sgr B2 (M). HNCO also peaks at (0", +90") where HOCO$^+$ has a local maximum (13). But most of other molecules are not so abundant at this position.

On the other hand, our "Local Peak" is very rich in observable molecular species. Most of the emission lines have the highest intensities in this position. Some molecules, for example the SiO molecule, show clear evidence of self-absorption (see figure 2). This fact might indicate that most of the region of the Sgr B2 molecular cloud is cold and very abundant in molecules, and several active star-forming regions have just been created in the core region.

SO and SO_2 are very weak at this position, which, combined with the results in Sgr B2 (M) and (N), suggests that the formation of these sulfur-bearing molecules is related with the activity associated with OB star-formation.

We believe that the molecular abundances determined in our "Local Peak" represent those typical in the Sgr B2 molecular cloud.

The SiO Molecule

The first detection of interstellar SiO was made in 1971 (14). Followed by this work, the SiO molecule was detected in dense and hot regions, and in the circumstellar shells around red-giant stars. Although this molecule is considered to be a very important constituent of interstellar dust, the SiO molecule has not been detected in cold and dense molecular clouds despite of an extensive search (15).

It is well known that the SiO spectra in Sgr B2 have a very complex line-shape. The v=0 J=2-1 line has an absorption feature at $V_{LSR} \sim$ 62 km s^{-1} (16). The distribution of the absorption features was investigated by mapping the 80" x 80" region of the Sgr B2 core in the v=0 J=1-0 line (17), and was revealed to coincide with that of the continuum.

We have obtained very sensitive spectra for SiO as are shown in Figure 2. We can see several velocity components in the spectra from Sgr B2 (M) and (N). The most prominent absorption feature at $V_{LSR} \sim$ 62 km s^{-1} is seen in all positions, and this absorption feature is flat-bottomed. Furthermore, the 62 km s^{-1} component can be seen even in the ^{29}SiO spectra (figure 1). These facts suggest that this velocity component is optically very thick. It should be stressed that this velocity component can be seen in the "Local Peak" where no background continuum emission from HII region exists. This fact means that the 62 km s^{-1} foreground cold, absorbing material contains enough SiO molecules in the gas-phase, and is extended. The characteristics of the 62 km s^{-1} component should be studied more.

The 7 km s^{-1} component, which is seen toward Sgr B2 (M) and (N), has identical radial velocity with the deep absorption feature of HI in the spiral arms between the Sgr B2 molecular cloud and us (18). No prominent emission lines except for CO have been observed at $V_{LSR} \sim$ 7 km s^{-1} (16). This fact

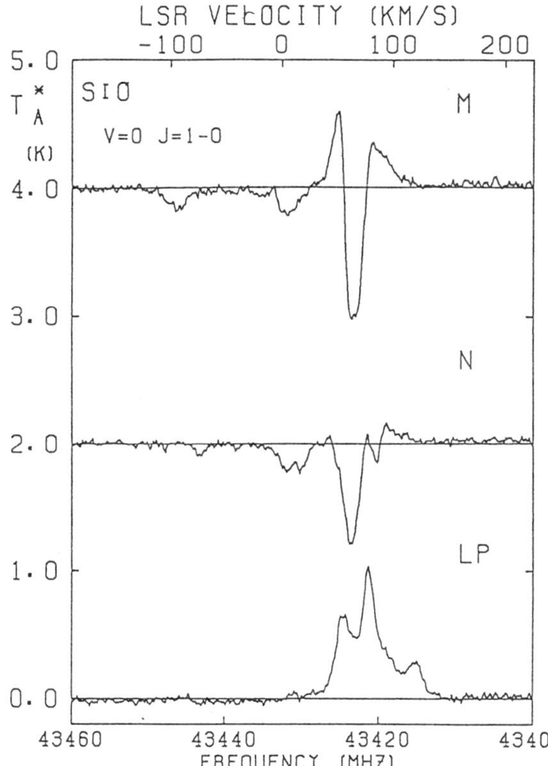

Fig. 2. Comparison of the absorption feature of SiO v=0 J=1-0 transition toward Sgr B2 (M) (top), Sgr B2 (N) (middle), and at the "Local Peak" (bottom), respectively.

indicates that molecular clouds producing absorption of SiO are not so dense (probably $n(H_2) < 10^4$ cm^{-3} judging from the critical density of CS J=1-0). The electric permanent dipole monent of SiO is 3.1 Debye, and the excitation temperature of the SiO molecule under such density will be very low. Indeed, an LVG calculation for SiO gave Tex $\sim$ 2.7 K if the line was optically thin. Therefore we have calculated the column density of the absorbing SiO molecule with the following assumptions : Tex = 2.7 K, the line is optically thin, and the beam-filling factor is unity. The result is N(SiO) $\sim$ 3 x 10^{12} cm^{-2} in the direction of Sgr B2 (M), which is comparable to those obtained in NGC7538S and IC443G (15). Furthermore, the column density of SiO in B1 which has recently been detected is 1.7 x 10^{12} cm^{-2} (19).

The fractional abundance of SiO relative to H_2 in IC443G and B1 are 8.5 x 10^{-10} (15) and (2-3) x 10^{-9} (19). It is difficult to estimate the fractional abundance of SiO for the 7 km s^{-1} absorbing feature because there is no estimate of the column density of H_2 for this velocity component. If we compare the source sizes of IC443G and B1 (< 1 pc) and the distance between Sgr B2 and us (8.5 kpc), we believe that the fractional abundance of SiO for the 7 km s^{-1} component is smaller by at least a factor of 100 than those for IC443G and B1. Then the fractional abundance of SiO for the 7 km s^{-1} component would be <10^{-11}, which is comparable to the upper limits of the fractional abundance of SiO in several cold and dense cloud cores (15), and is not inconsistent with the result of theoretical calculations (20).

The silicon chemistry has not been understood so well. Therefore investigations of the absorption features of SiO in many directions would be very important for the silicon chemistry.

<u>REFERENCES</u>

(1)   Johansson, L.E.B., Andersson, C., Elldér, J., Friberg, P., Hjalmarson, Å,
       Irvine, W.M., Olofsson, H., and Rydbeck, G. Astron. Astrophys., 130, 227
       (1984).
(2)   Sutton, E.C., Blake, G.A., Masson, C.R., and Phillips, T.G. Astrophys. J.
       Suppl., 58, 341 (1985).
(3)   Blake, G.A., Sutton, E.C., Masson, C.R., and Phillips, T.G. Astrophys. J.
       Suppl., 60, 357 (1986).
(4)   Cummins, S.E., Linke, R.A., and Thaddeus, P. Astrophys. J. Suppl., 60, 819
       (1986).
(5)   Turner, B. Astrophys. J. Suppl., 70, 539 (1989).
(6)   Ohishi, M. Ph.D. thesis, University of Tokyo (1984).
(7)   Kaifu, N. this volume (1990).
(8)   Benson, J.M., and Johnston, K.L. Astrophys. J., 277, 181 (1984).
(9)   Vogel, S.N., Genzel, R., and Palmer, P. Astrophys. J., 316, 243 (1987).
(10)   Goldsmith, P.F., Snell, R.L., Hasegawa, T., and Ukita, N. Astrophys. J.,
       314, 525 (1987).
(11)   Lis, D.C., and Goldsmith, P.F. Astrophys. J., 337, 704 (1989).
(12)   Chung, S.H., Ohishi, M., and Morimoto, M. submitted to Publ. Astron. Soc.
       Japan (1990).
(13)   Minh, Y.C., Irvine, W.M., and Ziurys, L.M. Astrophys. J., 334, 175 (1988).
(14)   Wilson, R.W., Penzias, A.A., Jeffers, K.B., Kutner, M.L., and Thaddeus, P.
       Astrophys. J. (Letters), 167, L97 (1971).
(15)   Ziurys, L.M., Friberg, P., and Irvine, W.M., Astrophys. J., 343, 201
       (1989).
(16)   Scoville, N.Z., Solomon, P.M., and Penzias, A.A. Astrophys. J., 201, 352
       (1975).
(17)   Haschick, A.D., and Ho. P.T.P. Astrphys. J., 352, 630 (1990).
(18)   Burton, W.B. Astron. Astrophys., 10, 76 (1971).
(19)   Bachiller, R., Menten, K.M., and Rio-Alvarez, D.D. IRAM preprint (1990).
(20)   Langer, W.D., and Glassgold, A.E. Astrophys. J., 352, 123 (1990).

Observational Studies of Silicon and Metal Chemistry in Dense Clouds

Lucy M. ZIURYS

Department of Chemistry, Arizona State University, Tempe, AZ 85287-1604, U. S. A.

Recent observations of silicon and metal-containing molecules towards dense interstellar clouds are summarized. Work on SiO and SiS are discussed in particular. The observations of SiO suggest that this species is produced by a process within an activation energy of ~90 K. Studies of SiS also indicate that the synthesis of this molecule is favored in warmer gas. The chemistry of silicon is likely to be linked to elevated temperatures. Formation of metal-containing molecules may also involve hot regions, but such species have yet to be detected in dense clouds.

I. Introduction

Studies of interstellar molecules containing silicon and metals (in the chemist's sense) are of interest for a variety of reasons. First of all, many of these elements have relatively large cosmic abundances. For example, both silicon and magnesium have abundances only one order of magnitude smaller than that of carbon, (i.e. C/H ~ 3 x 10(-5), while Si or Mg/H ~ 3 x 10(-6) (1).) Iron is also relatively common, with Fe/H ~ 8 x 10(-6). Calcium, sodium and aluminum have abundances a little over a factor of 100 less than carbon. This is in contrast to phosphorus, which has an abundance, relative to hydrogen, of 2 x 10(-7). Since at least one molecule containing phosphorus has been observed in dense interstellar clouds (2), it would seem likely that species with silicon or a metal might also be detectable.

Silicon and metals, being refractory elements, are also central to questions of depletion in interstellar clouds. Towards diffuse clouds, these elements have been measured to be highly depleted, as much as 99% (3), presumably condensed onto surfaces of dust grains. Such depletions are measured in diffuse gas by observing atomic transitions of the elements in absorption against background stars. Unfortunately, this technique cannot be readily applied to denser material. Thus, depletions of the refractory elements in dense clouds are virtually unknown. Observational studies of molecules containing these elements might be one method of obtaining depletion factors. Also, since dense clouds are often sites of star formation, they may possess energetic regions involving shock waves and outflows. A certain amount of grain destruction is likely to occur in these sources. Therefore, observation of refractory molecules towards star-forming clouds might be useful in evaluating the degree of grain recycling in hot, shocked material.

From the viewpoint of interstellar chemistry, these species are exciting as well. Several molecules of this type are predicted by models of "shock" or high temperature chemistry to have enhanced abundances in hot gas, for example SiO (4). Thus, studies of these molecules in sources where elevated temperatures are present can test shock models. Also, examining the chemistry of species with refractory elements may give useful clues to the relative roles of high

Table I

Element	X/H	H	O	C	S	N	C_2
Si	3×10^{-5}	-	L,A	L,A	L,A	L	L,A
Mg	3×10^{-5}	L	L	-	L	-	-
Fe	8×10^{-6}	-	L	-	-	-	-
Na	2×10^{-6}	L	L	-	-	-	-
Ca	2×10^{-6}	-	-	-	L	-	-
Al	2×10^{-6}	-	L	-	-	-	-

L - Species for which rest frequencies have been measured
A - Species detected in interstellar gas

temperature gas phase reactions vs. dust grain destruction in molecule formation.

Studies of these molecules have an additional intriguing chemical aspect. Simple compounds with individual bonds to silicon or metals are rare in the terrestrial laboratory. These species may therefore be termed "non-terrestrial". Since bonding in these molecules often involves d and f electrons, they are usually open-shell (i.e., contain unpaired electrons) and have large electron spins. They are thus fascinating from the aspect of quantum mechanics.

In order to observe these molecules in dense interstellar gas, however, their rotational spectra, and corresponding rest frequencies, must be accurately known. Table I gives a summary of simple compounds composed of silicon and metals, with carbon, nitrogen, oxygen, sulfur, and hydrogen, that could conceivably be present in interstellar gas. The table also indicates which molecules have measured high resolution spectra, as well as which have been detected in the ISM. As the table illustrates, for many species, accurate rest frequencies are not available. Also, it shows that only four molecules with well-known rest frequencies have been observed in interstellar gas, SiO, SiS, SiC, and SiC_2. In addition, although all four compounds have been detected in circumstellar shells of late-type stars, in particular IRC + 10216 (5,6), only SiO and SiS have been observed in molecular clouds.   A few other metal-containing molecules not listed in the table have also been seen only in circumstellar envelopes: NaCl, KCl, AlCl, and, tentatively AlF (7). Consequently, studies of silicon and metal chemistry in dense clouds are limited at present to examining only two known species, as well as continued searches for new molecules.

II. Studies of Silicon-Containing Molecules in Dense Clouds

Certain silicon species are not obviously present in molecular clouds, including SiC, SiC_2, and SiN. SiO, in contrast, is a rather abundant and common interstellar molecule towards active star-forming regions, as initially found by Downes et al. (8). These authors surveyed outflow sources in the $J = 2\rightarrow1$ transition of this species and detected it in many clouds, including W51, W49, NGC7538, DR21(OH), and W3. They also noticed that the SiO profiles almost always showed evidence of line wings, indicating outflows. The presence of SiO in outflow regions was clearly illustrated by Hat Creek interferometer maps of Orion-KL as well, where the species was shown to exclusively trace hot, outflowing gas (9). SiS, on the other hand, until the present time had only been observed in one cloud, SgrB(OH) (10).

Recently Ziurys, Friberg, and Irvine (11) did a new survey of SiO, observing the $J = 2\rightarrow1$ transition towards a variety of sources including cold, dark clouds, where temperatures are in the order of $T_K = 10K$, as well as other shocked regions. The results they obtained were quite interesting. SiO was readily observed in all dense clouds where $T_K > 30K$, and was most abundant in the hottest regions. In contrast, the species was not detected in any of the cold clouds, including TMC-1, L134N and B335, down to limits of $[SiO]/[H_2] \lesssim 10(-12)$.

These measurements suggest that the formation of SiO depends critically on
gas kinetic temperature. In fact, if a plot is made of the natural log of the
abundance of SiO, measured towards the surveyed clouds, versus the inverse of
their kinetic temperature, the result is an apparent linear relationship shown in
Fig. 1. This relationship implies that the process leading to the formation of
SiO has an exponential temperature dependence exp($-\Delta E/KT$), with ΔE being an
activation energy. The slope of the line in Fig. 1 gives a ΔE ~90K. This
barrier naturally favors SiO production in hot gas. The high temperature
character of this species is clearly illustrated in Fig. 2, which is a map of
integrated intensity of J = 2→1 line of SiO made toward OMC-1 (12). As the map
shows, the SiO emission is clearly confined to two regions in the cloud: a
northern source, which is centered on the KL/IRCc2 outflow, and the region
located at (0,0), known as Orion-S. The Orion-S region has recently been found
to contain a separate outflow. SiO clearly traces the hotter material of the two
outflow sources, but fails to appear in the extended material that characterizes
the cool, quiescent gas.

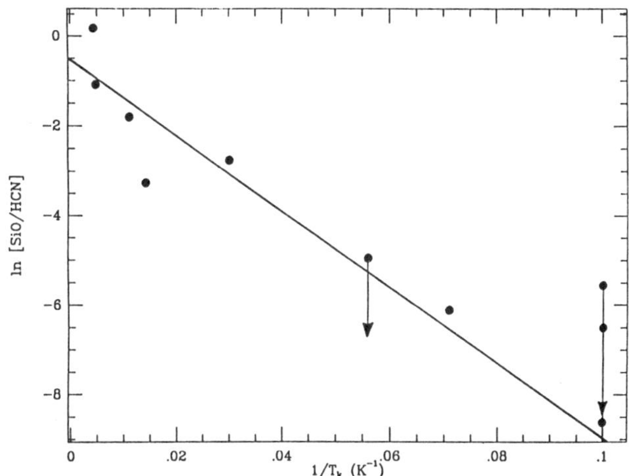

Figure 1. Plot of the natural log of the SiO abundance, observed towards dense
clouds, vs. the inverse of kinetic temperature. The apparent linear relationship
suggests that SiO is formed by a process involving an activation energy near 90K.

The SiO temperature dependence may be a result of gas phase high temperature
reactions, or a result of silicate dust grain destruction, or a combination.
Langer and Glassgold (13) have suggested an explanation for the temperature
dependence of SiO using gas phase reactions, based on the work of Graff (14).
Graff postulated that neutral-neutral reactions involving an atom as a reactant
occur while the atom is an excited fine structure state. Langer and Glassgold
applied this idea to silicon, whose first excited fine structure state lies
~111 K above ground state. They postulated that Si must be in this first excited
fine structure state in order to form SiO. Such a formation scheme results in
the SiO abundance being proportional to exp($-\Delta E/KT$), where ΔE, the activation
energy, is 111 K. In addition, such a production scheme for SiO requires gas
densities n(H$_2$)>10(5) cm(-3), necessary to collisionally-excite the fine
structure line. These densities are clearly present in the sources where SiO has
been detected. Overall, the theoretical predictions of Langer and Glassgold
agree quite well with the observations.
 Are silicon compounds generally associated with high temperatures? The next
molecule to study in this regard is SiS. Ziurys has recently done observations
of this species in dense clouds containing outflows (15,16). Previous work had
suggested that SiS was present in the Orion-KL region (17). Ziurys firmly

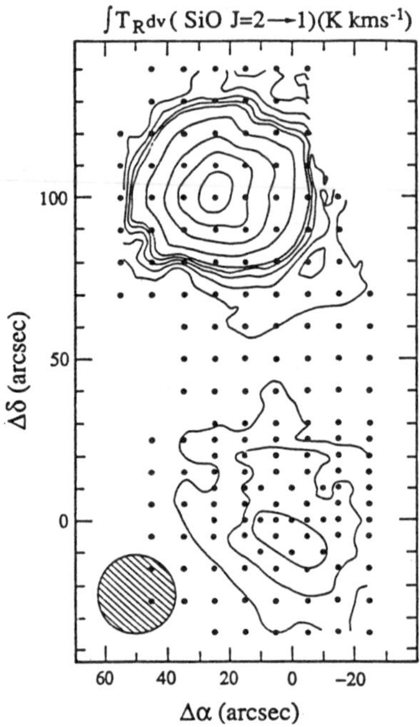

Figure 2. A map of the integrated intensity of the J = 2→1 transition of SiO towards OMC-1, made with the IRAM 30 telescope with 25" spatial resolution (13). The (0,0) position marks the center of the Orion-S outflow. The KL/Rc2 region is located at $\Delta\alpha$ = 25", $\Delta\delta$ = 100". The map shows that SiO clearly traces the hot, outflow sources and not the quiescent gas.

established the existence of this species in Orion-KL by detecting the J = 5→4 and J = 6→5 transitions of SiS at 3mm towards this object. Although rather weak (T_A^* ~ 40-50 mk), the SiS spectra appeared to resemble the optically thin, thermal transitions of SiO, having broad parabolic line shapes, indicating an origin in the Orion outflow. The line profiles showed no evidence of the cool, quiescent extended ridge emission. It would therefore appear that SiS and SiO arise from the same region in Orion, which is thought to have a kinetic temperature near 200 K (9).

The high temperature nature of SiS in Orion-KL is further evidenced by the recent detection of the J = 15→14 transition of this species at 272 GHz towards this source, as shown in Fig. 3 (16). This spectrum has a broad, almost flat-topped profile, which again indicates an outflow. In fact, the line shape closely resembles that of the J = 1→0 transition of SiO in Orion (18). Although the J = 15→14 spectrum is more flat-topped than the 3mm SiS lines, all transitions have $\Delta V_{1/2}$ ~ 20 km/s and V_{LSR} ~ 8 km/s. Thus, they are likely to come from the same region.

The J = 15→14 line of SiS lies ~105 K above ground state; yet, its measured antenna temperature is T_R^* ~ 0.3K. A five point map of this transition showed that it is confined to a region <30 arcsec in extent (16). In fact, if SiS and SiO do arise from the same material, then the likely size of the SiS source is on the order of 10 arcsec. If the J = 15→14 line is optically thin and confined to a 10 arcsec region, the excitation temperature for the transition is >100 K. This agrees with temperatures derived from SiO (9).

In addition to Orion-KL, the J = 6→5 and J = 5→4 transitions of SiS have been recently detected in a few other active star forming clouds where SiO is abundant (16). These include W51, Orion-S, and tentatively, W49. Representative spectra are given in Fig. 4. As these spectra demonstrate, SiS emission in these clouds is typically quite weak (T_R^* ~ 10mk), corresponding to column densities of N_{tot} ~ 1-2 x 10(12) cm(-2), for a source uniformly filling the telescope beam. In terms of linewidth and LSR velocity, the line profiles of SiS in these sources do resemble those of SiO, as best can be determined (8,12). This certainly is the case for Orion-S. It is therefore likely that SiO and SiS arise from the same material in these sources. A meaningful comparison can then be made between

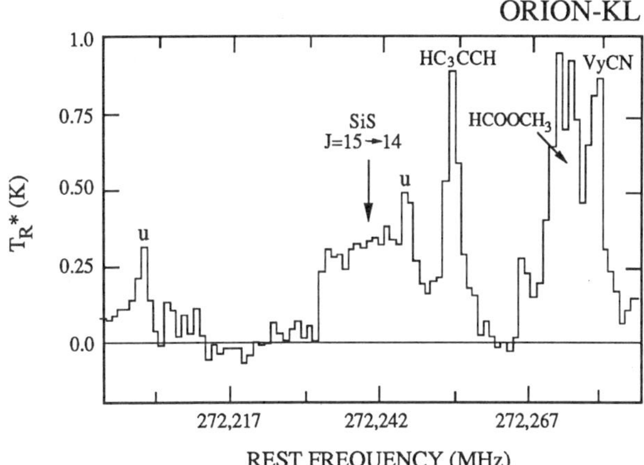

Figure 3. Spectrum of the J = 15→14 line of SiS at 272 GHz observed toward
Orion-KL with the NRAO 12m antenna (16). The broad, flat-topped profile of the
line suggests it arises from the hot outflow region in Orion.

the abundances of the two species. The ratio of SiO to SiS in these sources,
including Orion-KL, is typically SiO/SiS ~ 40-60.

Thus, as is the case for Orion-KL, SiS appears to be preferentially formed in
warm gas where SiO is produced. Indeed, searches for the J = 2→1 transition of
SiS towards the cold, dark cloud TMC-1, performed at Nobeyama, have proved
negative, down to antenna temperature limits of T_A^* ~ 30 mk (19). It would be
interesting to see if this apparent high temperature trend applies to other
silicon containing species. This, unfortunately, must wait until further
detections are made.

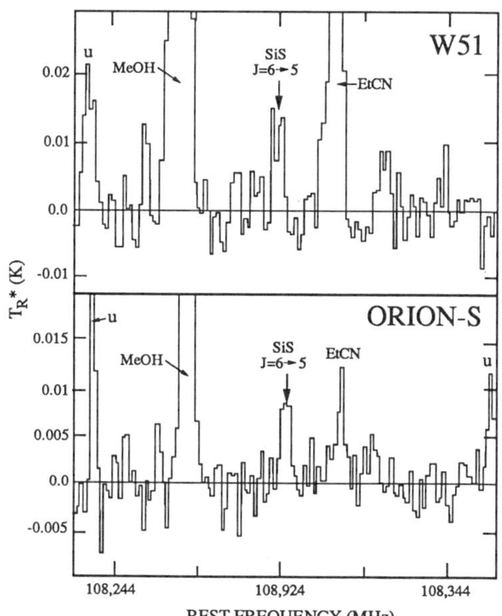

Figure 4. Spectra of the J = 6→5
transition of SiS at 109 GHz,
observed towards W51 and Orion-S,
using the NRAO 12m telescope. The
lines, although weak, probably arise
from the same gas as SiO.

III. Metal-Containing Molecules in Dense Clouds: Where Are They?

No metal-containing molecules have been detected in dense interstellar
clouds, although several have been searched for. These include species such as
MgO, FeO, and AlO, i.e., chiefly metal oxides. This situation is somewhat
peculiar, since SiO is quite abundant in dense clouds. The apparent absence of
MgO (20) is particularly puzzling, since magnesium and silicon have identical
cosmic abundances. Also, both are important constituents of silicates, which are
thought to partly compose interstellar grains (21). If SiO is actually formed by
grain destruction, then why not also produce MgO? On the other hand, reactions
of metal atoms to form molecules generally possess activation energy barriers.
However, in the hot regions where SiO is abundant, such barriers should be
readily overcome. It could be that carbides and nitrides are more likely forms
of the metals, although searches for SiC towards star-forming regions have thus
far proved negative (22). Alternately, the metals may remain in atomic form.
Certainly additional metal-containing species must be searched for, however,
before that conclusion can be reached.

IV. Conclusion

Studies of molecules containing silicon, and metals such as magnesium, iron,
and aluminum are relevant to many aspects of astrophysics. Observations of SiO
and SiS in dense molecular clouds indicate that these species are likely to be
formed by high temperature gas phase reactions. Much more observational and
laboratory spectroscopy work must be done, however, before silicon and metal
chemistry can really be understood.

V. References

1.   Allen, C.W. 1973, _Astrophysical Quantities_ (London: Athlone Press), 3rd ed.
2.   Ziurys, L.M. 1987, Ap.J. (Letters), _321_, L81.
3.   Morton, D.C. 1974, Ap.J. (Letters), _193_, L35.
4.   Hartquist, T.W., Oppenheimer, and Dalgarno, A. 1980. Ap. J., _236_, 182.
5.   Thaddeus, P., Cummins, S., and Linke, R.A. 1984, Ap.J. (Letters), _283_, L45.
6.   Cernicharo, J., Gottlieb, C., Guélin, M., Thaddeus, P., and Vrtilek, J.
      1989, Ap.J. (Letters), _341_, L25.
7.   Cernicharo, J., and Guélin, M., 1987, Astr. Ap. _183_, L10.
8.   Downes, D., Genzel, R., Hjalmanson, A., Nyman, L.A., Ronnang, B., 1982,
      Ap.J. (Letters), _252_, L29.
9.   Wright, M.H.C., Plambeck, R.L., Vogel, S.N., Ho, P.T.P., and Welch, W.J.
      1983, Ap.J. (Letters), _267_, L41.
10.   Morris, M., Gilmore, W., Palmer, P., Turner, B.E., and Zuckerman, B. 1975,
      Ap. J. (Letters), _199_, L47.
11.   Ziurys, L.M., Friberg, P., and Irvine, W.M. 1989, Ap.J., _343_, 201.
12.   Ziurys, L.M., Wilson, T.L., and Mauersberger, R. 1990, Ap.J. (Letters), in
      press.
13.   Langer, W.D., and Glassgold, A.E. 1990, Ap.J., in press.
14.   Graff, M.M. 1989, Ap.J., _339_, 239.
15.   Ziurys, L.M. 1988, Ap.J., _324_, 544.
16.   Ziurys, L.M. 1990, in preparation.
17.   Dickinson, D.F., and Rodriguez-Kniper, E. 1981 Ap.J., _247_, 112.
18.   Genzel, R., Downes, D., Schwartz, P.R., Spencer, J.H., Pankonin, V., and
      Baars, J.W.M. 1980, Ap.J., _239_, 519.
19.   Kaifu, N. Private communication.
20.   Turner, B.E., and Steimle, T.C. 1985, Ap.J., _299_, 956.
21.   Knacke, R. 1988, in _Molecular Clouds in the Milky Way and External Galaxies_
      ed. R. Dickman, R. Snell, and J.S. Young (New York-Springer-Verlag) p. 151.
22.   Ziurys, L.M. 1990, in preparation.

Optical Observations of Molecules in Diffuse Interstellar Clouds

S. R. Federman

Department of Physics and Astronomy, The University of Toledo, Toledo, OH 43606, U. S. A.

With the advent of silicon-based photodiode detectors, observations
at visible wavelengths of moderately reddened stars behind molecular
interstellar clouds are now possible. The most recent measurements
of molecular absorption lines sample regions corresponding to visual
extinctions of 1 to 4 mag. Results for CH, C_2, and CN are stressed
in this review. Because of the linearity of these photodiode
detectors, weak lines are measurable with precision. Recent work
on the $^{12}C/^{13}C$ ratio in CH^+ and CN, on the excitation temperature of
CN, and on upper limits for several chemically interesting species
is also discussed.

INTRODUCTION

Much of the recent work on molecular absorption at optical wavelengths
focuses on the acquisition of the data with silicon-based photodiode detectors, as
was anticipated in the review for IAU Symposium No. 120, Astrochemistry, (11).
The great sensitivity and a dynamic range that is linear over many decades are two
important qualities of these detectors and are especially appropriate in describing
charged-coupled devices (CCDs). The enhanced sensitivity allows observers to
acquire data with signal-to-noise ratios of ~100 on 9th magnitude stars. The
linear dynamic range provides the means to measure weak lines (~0.1 mÅ) with high
precision. This review stresses recent measurements made possible because of these
two properties of CCDs. Surveys of absorption toward moderately reddened stars,
measurements of weak lines, and the outlook for the future are discussed in the
following sections.

SURVEYS

The surveys take two forms. One involves data of CH, C_2, and CN absorption
toward many random lines of sight; the other focuses on measurements toward stars
associated with the same molecular cloud complex. Surveys of the former type
include the work of Federman and Lambert (15), who acquired data on CH, C_2, and
the Violet System of CN, and the work of van Dishoeck and Black (34), who studied
absorption from C_2 and the Red System of CN. The results for the Red System of CN
at 7900 Å and 9200 Å are important because for highly reddened stars, the low flux
anticipated at 3880 Å may make measurements all but impossible.

Compilations of all the available data may provide clues to the underlying
chemistry. One way of analyzing the data is to seek correlations between species.
Because of problems associated with seeking relationships of the form,
$$N(A) \propto a\, N(B)^b,$$
(see 14), comparisons involving logarithmic relationships,
$$\log N(A) \propto b \log N(B),$$

are used.

Figure 1 shows the correspondence between log $N(C_2)$ and log $N(CH)$. This is equivalent to a comparison of C_2 and H_2 because of the linear relationship found for CH and H_2 (2, 23). These data are for stars that have 1 to 4 mag of foreground extinction. A moderately strong correlation is present (r ~ 0.8) with a slope of

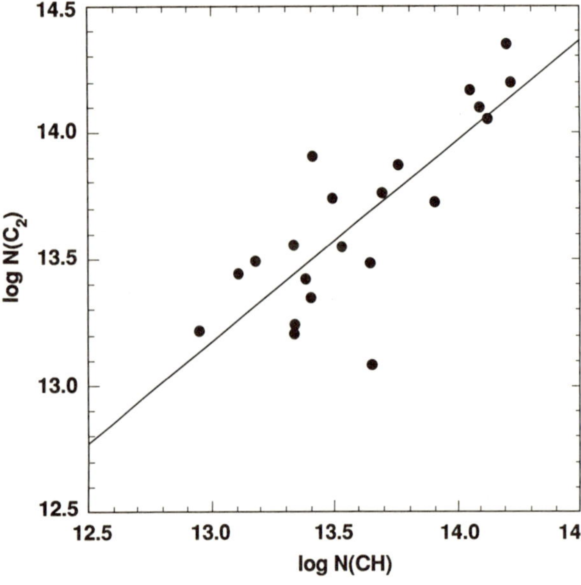

Figure 1 – A comparison between column density of C_2 and that of CH. Because of the observed linear relationship between CH and H_2 (see text), this graph is also a comparison between C_2 and H_2. A least-squares fit to the data is indicated.

0.8. The dispersion among the data points probably arises from the variation in physical conditions (density, temperature, radiation field, etc.) from one cloud to another. The observed trend may aid in our understanding of the transition from a chemistry dominated by photochemical processes when $A_V \leq 2$ mag to a chemistry dominated by gas phase reactions at greater extinction (15).

Several groups studied individual molecular cloud complexes in order to discern variations in chemistry over small distances and from one cloud complex to another. Of course, these studies form a subset of the data acquired from random directions. One of the first examples of this type of investigation is the work of Chaffee and Dunham (3), who measured molecular absorption toward stars in the Cepheus OB2 association. More recently, Cardelli and Wallerstein (2) observed reddened stars in the Scorpius OB2 association. The ρ Ophiuchi Molecular Cloud lies within this association. Crawford (8) examined the absorption from stars in the Sco OB1 association. He found that one velocity component is seen toward the stellar cluster NGC 6231, but common velocity components are not present for stars beyond the tight group representing NGC 6231.

A program that combines the efforts of several investigators is nearing completion (16). Cardelli and Smith are contributing data on stars in the Cep OB3 association, and Federman and Lambert are contributing data for stars in the vicinity of TMC-1 and TMC-2 and for stars in the Sco OB2 association. Federman and Lambert measured C_2 toward HD147701, a star previously observed by Cardelli and Wallerstein (2), and measured molecular absorption toward several other reddened stars in the association. One aim of the program is to quantify the relationship between stellar ultraviolet field and chemistry. Cep OB3 has massive O stars, Sco OB2 has mostly early B stars, and the Taurus region has late B stars. As observations probe gas in front of stars with spectral types later than B5 - B7, a

problem arises: Narrow stellar features are no longer discernible from interstellar features. Stellar metal lines can totally mask the region where interstellar lines are expected (see 36), especially for measurements acquired with
moderate ($\sim$ 25,000) spectral resolution.

Such studies allow a direct comparison between data from optical absorption
lines and data from measurements of radio emission. One example of a coordinated
investigation involves observations of an isolated cloud in front of HD 169454 (20).
Other examples include searches for optical absorption from high-latitude molecular
clouds, first detected by their CO radio emission (21). For example, Welty et al.
(36) sought absorption from MBM 40, 53, 54, and 55. In a similar vein, de Vries
and van Dishoeck (10) found CH absorption toward HD 210121, which lies behind a
cloud associated with infrared cirrus. Their upper limit on CH^+ absorption places
stringent limits on the importance of shock chemistry in these clouds.

The surveys are useful in probing details of the chemistry through comparisons
involving observed and calculated column densities. A recent investigation into
the chemistry of diatomic carbon-bearing species, with particular emphasis on C_2,
revealed the need for several modifications to the chemical schemes (13). The
analysis is based on an analytical expression for steady-state chemistry in terms
of column densities. Steady-state is appropriate in the outer portions of molecular
clouds where chemical equilibrium is reached (30). Because the local fractional
abundances are increasing smoothly and monotonically in this portion of the cloud
(1), the use of rate equations in terms of column densities appears to be a
reasonable approximation.

For C_2 the major production channel involves reactions between C^+ and CH,
while photodissociation and reactions with oxygen and nitrogen are the important
destruction terms. The best correspondence between observational and analytical
results required two changes to the chemical scheme. First, the fractional abundance of C^+ has to decrease by a factor of $\sim$ 15 from A_V of 2 mag to A_V of 4 mag.
Although enhanced depletion onto grain mantles in opaque portions of a cloud is the
cause for some of the decrease, most of the decrease results from the chemical processing of C^+ into CO (14). Second, the initiating step - $C^+ + CH \rightarrow C_2H^+ + H$ -
has a rate coefficient of $\sim$ 3 x 10^{-10} cm^3 s^{-1}, a value that is about a tenth of the
collisional rate coefficient. The presence of competing channels lowers the
efficiency for the reaction of interest.

These changes are also consistent with the analysis of Federman and Huntress
(13) for the observations of CH and CO when the rate coefficient for the reaction
$$C^+ + H_2 \rightarrow CH_2^+ + h\nu$$
is reduced to 1 x 10^{-16} cm^3 s^{-1}. Although somewhat lower than the value used by
van Dishoeck and Black (33) in their modeling of several diffuse clouds, this
value is consistent with the value adopted for models of dark clouds (1).

WEAK LINES

Measurements of weak lines with $W_\lambda < 0.5$ mÅ provide information on stellar
nucleosynthesis, on the temperature of the cosmic background radiation, and of
course on interstellar chemistry. Column densities for $^{13}CH^+$ and ^{13}CN, when compared to column densities for $^{12}CH^+$ and ^{12}CN, yield information on the history of
stellar nucleosynthesis. The relative populations of the J = 1 and J = 2 levels
of the ground state of CN allow a determination of the cosmic background temperature at 2.6 and 1.3 mm. Limits recently obtained on absorption from NaH, MgH, SH^+,
OH^+, and C_3 help place constraints on chemical models.

The solar value for the ratio of ^{12}C to ^{13}C is 89. Chemical fractionation
is not expected to be significant in diffuse interstellar clouds because of the
moderate temperatures ($\sim$ 50 K) present (35). As a result, the amount of ^{13}C incorporated into molecules gives us information on the nucleosynthetic history of

the Galaxy. Recent results include measurements involving CH^+ and CN. The unfortunate situation is that systematic effects appear to mask the true ratios.

Hawkins, Jura, and Meyer (18), Hawkins and Jura (17), Hawkins and Meyer (19) derived $^{12}C/^{13}C$ ratios toward several stars within 2 kpc of the Sun and found a constant value of ~ 45. This value indicates chemical processing occurred in the interstellar medium since the formation of the Solar System. On the other hand, Stahl et al. (29) measured a ratio from CH^+ data of 77, a value previously derived by Vanden Bout (31) and Vanden Bout and Snell (32). This value is very similar to the solar value. Stahl et al. used a technique first discussed by Crane and Hegyi (5) whereby an equivalent width for each exposure is obtained and the distribution of W_λ is used to find a mean value. My expectation is that this technique should provide accurate equivalent widths. The reason for differences with the work of Hawkins and collaborators remains unknown, however.

The situation is no better for ^{13}CN. Crane and Hegyi (5) determined a $^{12}C/^{13}C$ ratio of 47 toward ζ Ophiuchi, while Meyer, Roth, and Hawkins (25) obtained a value of 122 toward HD 21483. The results for ζ Oph agree with the results of Hawkins, Jura, and Meyer (18), but not with the results of Stahl et al. (29). As mentioned by the latter authors, a similar value is not expected in CH^+ and CN when chemical fractionation is important because these molecules exist under different physical conditions. If similar values for the ratio are appropriate toward ζ Oph because fractionation is not important, then why are the CN results toward HD 21483 larger than the solar value when Hawkins and Jura (17) found a constant ratio in CH^+ in the solar neighborhood? A critical evaluation of systematic effects may hold the clue for obtaining a self-consistent picture.

Although the systematic effects in measured equivalent widths appear in the determinations of the cosmic background temperature from the excitation temperature of CN, the resulting temperatures agree extremely well. A value of ~ 2.8 K is derived in all cases (6,7,25) and is consistent with bolometric determinations of the cosmic background temperature. Through measurements by Meyer and Roth (24) of absorption from the (1,0) vibrational band of CN, the results toward ζ Oph are more firm.

New upper limits are available for several molecular species. Czarny, Felenbok, and Roueff (9) searched for absorption from NaH and MgH. While predictions involving gas phase synthesis are comfortably below the measurements, their results place constraints on the chemistry involving grain surfaces. The constraint to synthesis on grain surfaces is that these molecules cannot be produced with high efficiencies on grain surfaces. The upper limits on SH^+ (22, 26) place weak constraints on the production route in MHD shocks (27) because the upper limits are within the uncertainties of the modeling parameters. A similar situation arises with the upper limit for OH^+ (12) in that quiescent gas phase schemes predict comparable column density for OH^+. Snow, Seab, and Joseph (28) improved upon the upper limits for C_3 first measured by (4); the value is well above the column density predicted by gas phase chemistry.

FUTURE WORK

The future for absorption-line studies of interstellar molecules is bright. With the launch of the Hubble Space Telescope this year, high resolution spectroscopy at ultraviolet wavelengths will be possible. With the advances being made in detector technology for the near infrared, high spectral resolution, high signal-to-noise data will be readily obtained. Detections of important species, such as CH_2 and C_2H, are very likely with the new instrumentation. In the traditional wavelength region for optical spectroscopy, the visible portion of the spectrum, new measurements of dense gas will provide data that can be directly compared to observations at radio wavelengths. Once systematic effects are better understood, the ability to probe the nucleosynthetic history of the Galaxy through measurements of $^{13}CH^+$ and ^{13}CN will become routine.

REFERENCES

 (1) Boland, W., and de Jong, T. 1984, Astr. Ap., 134, 87.
 (2) Cardelli, J.A., and Wallerstein, G. 1986, Ap. J., 302, 492.
 (3) Chaffee, F.H., and Dunham, T. 1979, Ap. J., 233, 568.
 (4) Clegg, R.E.S., and Lambert, D.L. 1981, M.N.R.A.S., 201, 723.
 (5) Crane, P., and Hegyi, D.J. 1988, Ap. J. (Letters), 326, L35.
 (6) Crane, P., Hegyi, D.J., Kutner, M.L., and Mandolesi, N. 1989, Ap. J., 346, 136.
 (7) Crane, P., Hegyi, D.J., Mandolesi, N., and Danks, A.C. 1986, Ap. J., 309, 822.
 (8) Crawford, I.A. 1989, M.N.R.A.S., 241, 575.
 (9) Czarny, J., Felenbok, P., and Roueff, E. 1987, Astr. Ap., 188, 155.
(10) de Vries, C.P., and van Dishoeck, E.F. 1988, Astr. Ap., 203, L23.
(11) Federman, S.R. 1987, IAU Symposium No. 120, Astrochemistry, ed. M.S. Vardya
     and S.P. Tarafdar [Dordrecht: Reidel], p. 123.
(12) Federman, S.R., Curtis, L.J., Trauger, J.T., and Meyer, D.M. 1990, submitted
     to Ap. J.
(13) Federman, S.R., and Huntress, W.T. 1989, Ap. J., 338, 140.
(14) Federman, S.R., Huntress, W.T., and Prasad, S.S. 1990, Ap. J., in press.
(15) Federman, S.R., and Lambert, D.L. 1988, Ap. J., 328, 777.
(16) Federman, S.R., Lambert, D.L., Cardelli, J.A., Smith, V.V., and Joseph, C.L.
     1990, in preparation.
(17) Hawkins, I., and Jura, M. 1987, Ap. J., 317, 926.
(18) Hawkins, I., Jura, M., and Meyer, D.M. 1985, Ap. J. (Letters), 294, L131.
(19) Hawkins, I., and Meyer, D.M. 1989, Ap. J., 303, 401.
(20) Januzzi, B.T., Black, J.H., Lada, C.J., and van Dishoeck, E.F. 1988, Ap. J.,
     332, 995.
(21) Magnani, L., Blitz, L., and Mundy, L. 1985, Ap. J., 295, 405.
(22) Magnani, L., and Salzer, J.J. 1989, A. J., 98, 926.
(23) Mattila, K. 1986, Astr. Ap., 160, 157.
(24) Meyer, D.M., and Roth, K.C. 1990, Ap. J., 349, 91.
(25) Meyer, D.M., Roth, K.C., and Hawkins, I. 1989, Ap. J. (Letters), 343, L1.
(26) Millar, T.J., and Hobbs, L.M. 1988, M.N.R.A.S., 231, 953.
(27) Pineau des Forets, G., Flower, D.R., Hartquist, T.W., and Dalgarno, A. 1986,
     M.N.R.A.S., 220, 801.
(28) Snow, T.P., Seab, C.G., and Joseph, C.L. 1988, Ap. J., 335, 185.
(29) Stahl, O., Wilson, T.L., Henkel, C., and Appenzeller, I. 1989, Astr. Ap., 221,
     321.
(30) Tarafdar, S.P., Prasad, S.S., Huntress, W.T., Villere, K.R., and Black, D.C.
     1985, Ap. J., 289, 220.
(31) Vanden Bout, P.A. 1972, Ap. J. (Letters), 176, L127.
(32) Vanden Bout, P.A., and Snell, R.L. 1980, Ap. J., 236, 460; erratum, 1981,
     Ap. J., 246, 1045.
(33) van Dishoeck, E.F., and Black, J.H. 1986, Ap. J. Suppl., 62, 109.
(34) ___ 1989, Ap. J., 340, 273.
(35) Watson, W.D., Anicich, V.G., and Huntress, W.T. 1976, Ap. J. (Letters), 205,
     L165.
(36) Welty, D.E., Hobbs, L.M., Blitz, L., and Penprase, B.E. 1989, Ap. J., 346,
     232.

Chemistry and Physics in Warm Molecular Cloud Cores: Are Grain Surface Reactions Necessary to Explain Our Observations?

Åke HJALMARSON* and Per BERGMAN

Onsala Space Observatory, S-439 00 ONSALA, Sweden

This paper is dedicated to Hiroko Suzuki, with whom ÅH shared some serious thoughts on astrochemistry and life itself ...

Abstract

We discuss recent CH_3CN mapping, paired with radiative transfer and statistical equilibrium excitation modeling, of the molecular cloud cores in Orion A, Sgr B2, and G327.3-0.6. These warm, dense cores are very small or clumpy. The CH_3CN concentration in the clumps is approximately 100 times larger than "normal" and the velocity gradients (as yet of unknown nature) are as large as 100 km s^{-1} pc^{-1}. We also highlight the first results from an ongoing spectral line survey around 230 GHz of three nearby positions in Sgr B2, using SEST.

1. Introduction

The discovery of the first polyatomic interstellar molecules (NH_3, H_2O, and H_2CO) very soon lead to the suggestion that grain surface reactions were important (cf. Watson and Salpeter 1972 [1,2], Duley [3]). However, already the tentative (!) identification of the strong, accidentally detected, line U89.2 (GHz) with the molecular ion HCO^+ — widespread among dense interstellar clouds — triggered the development of the (more efficient) ion-molecule reaction schemes, which still are our common belief (cf. Herbst and Klemperer 1973 [4], Rydbeck and Hjalmarson [5], Herbst [6]). It was then concluded that grain surface reactions may not be so important after all, except for H_2 which is unambiguously known to be formed on grain surfaces.
More recently, elevated concentrations of many complex species have been observed in warm molecular cloud cores, which seem hard to explain by ion-molecule reactions alone (cf. Irvine *et al.* [7]). An NH_3 study of Orion A lead Schweizer [8] to propose that the large fractional abundance deduced ($\approx 10^{-6}$ - 10^{-5}, compared to 10^{-7} in cold, dark clouds) was a result of evaporation of NH_3 *ice* from grain surfaces (cf. also Pauls *et al.* [9]). Similar abundance enhancements of HDO (and hence H_2O) and CH_3OH were observed in Orion A and also in other "hot" core sources (cf. Johansson *et al.* [10], Olofsson [11], Menten *et al.* [12], Jacq *et al.* [13]). In the W51M hot core Millar *et al.* [14] have found concentrations of $(CH_3)_2O$ and CH_3CH_2OH, which seem 2-3 orders of magnitude larger than expected from "todays" gas phase chemistry. The recent chemical models by Brown *et al.* [15, 16] — involving grain surface reactions and liberation of molecular grain mantles by heating due to the formation of nearby stars — may provide an explanation.

2. Mapping of physical and chemical conditions in warm molecular cloud cores using CH_3CN

Orion A $CH_3CN(J=6-5)$ mapping of relatively high quality, paired with statistical equilibrium excitation (SE) calculations, have lead to fairly accurate estimates of temperatures and column densities for the

---

* also Astrophysics Group, Institute of Theoretical Physics, University of Göteborg

clearly separable cloud components (Andersson [17]); the extended 8 and 10 km s^{-1} ridge clouds have $T_{kin} \approx 40$ - 50 K (similar to estimates using CH_3C_2H, cf. Askne *et al.* [18]), while the so called compact ridge clump and the hot core have temperatures around 150 K (cf. Johansson *et al.* [10] and Blake *et al.* [19]). As estimated by Irvine *et al.* [7] and Blake *et al.* the abundance of CH_3CN is enhanced by an order of magnitude, or so, in the hot core, but not clearly so in the warm compact ridge clump. These results were based upon the assumption of low optical depth. However, recent Onsala CH_3CN and $CH_3^{13}CN$ data (Bergman *et al.* [20]) clearly reveal that $\tau > 1$ and hence that photon trapping has to be accounted for in the radiative transfer as well as in the SE analysis. The estimated CH_3CN abundance in the hot core increases to 10^{-8} - 10^{-7} in a 10" source (observed with a 35" antenna beam). In the hot core also the abundances of NH_3, HDO (H_2O), HC_3N, CH_2CHCN, and CH_3CH_2CN are high [7, 19]. In the warm compact ridge clump, on the other hand, the concentrations of HDO (H_2O), CH_3OH, $HCOOCH_3$, and $(CH_3)_2O$ are higher than "normal", but this is still uncertain for CH_3CN.

An even more pronounced CH_3CN abundance enhancement is apparent in recent SEST (Swedish ESO Submillimeter Telescope, in Chile) mapping of the molecular cloud core in G327.3-0.6 (Figure 1). In a small (≈ 0.07 pc), dense ($\approx 10^6$ cm^{-3}), warm (≈ 115 K) clump (or clumps) the estimated CH_3CN concentration is 10^{-7} - 10^{-6} (!). This result is obtained separately from the $CH_3^{13}CN$, $^{13}CH_3CN$, and $CH_3CN(J=6-5)$ spectra (cf. Figure 2). The opacities (from LVG-SE analysis) of the lower energy CH_3CN lines are 5-10, which (almost surprisingly!) matches the fact that the CH_3CN lines are twice as broad as those of the ^{13}C species. The CH_3CN concentration in the surrounding cloud ($T_{kin} \approx 30$ - 40 K) appears to be much lower ($\leq 10^{-9}$). Here the line widths are only 3 km s^{-1}, compared to 4.5 and 8 km s^{-1} in the warm core for $CH_3^{13}CN$ and CH_3CN, respectively.

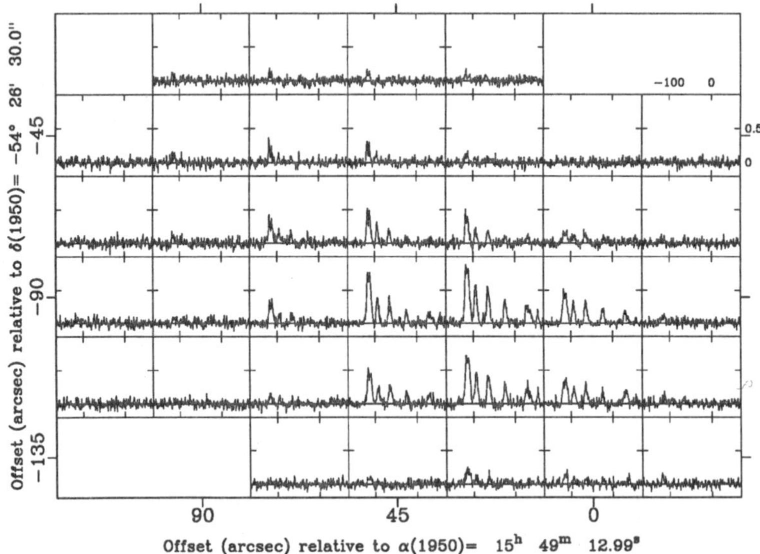

FIG. 1.—$CH_3CN(J=6-5)$ spectra toward G327.3-0.6.

Mapping of the $CH_3CN(J=6-5$ and $J=13-12)$ transitions of the Sgr B2 core region also has been performed at SEST. The $CH_3C_2H(J=14-13)$ lines were mapped simultaneously. These data are presently being analysed. A preliminary (but rather convincing) result is that the temperature and the CH_3CN abundance is higher in Sgr B2(N) — $T_{kin} \geq 130$ K, $X[CH_3CN] \approx 10^{-8}$ - 10^{-7} — than in the nearby Sgr B2(M) source. However, this distant and massive molecular cloud core is complicated and the physical and chemical conditions are not so easily separable ([7], Goldsmith *et al.* [21], Vogel *et al.* [22]). It has e.g. been proposed that Sgr B2(N) is a hot core source, while Sgr B2(M) shows signs of outflow [22]. The latter source then would be expected to exhibit enhanced abundances of SiO, SO, SO_2, similar to the conditions in the warm Orion outflow. For SO such a trend may be present in the high resolution Nobeyama data of Goldsmith *et al.* ([21], cf. discussion in [7]) and SO_2 lines also are peaking here (see § 3).

To improve our "statistics" on increasing CH_3CN concentrations in warm, dense cloud cores we refer to Bergman and Hjalmarson [23], who estimate fractional abundances in the range 10^{-8} - 10^{-6} in the warm (75 - 150 K) sources W51M, W51N, and G34.3+0.2.

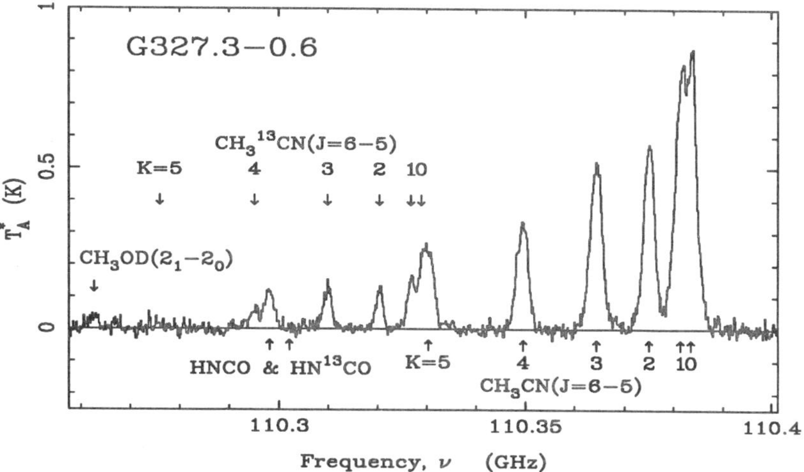

FIG. 2.—$CH_3^{13}CN$ and $CH_3CN(J=6-5)$ spectrum in peak position of G327.3-0.6.

3. The SEST spectral scan of Sgr B2(M), Sgr B2(N), and Sgr B2(2'N)

We here report the first results from a collaboration between Bergman, Friberg, and Hjalmarson (Onsala), Bill Irvine (Univ. of Massachusetts, Amherst), Tom Millar (UMIST, Manchester), and Ohishi-san (Nobeyama & Toyama University). To gain an improved understanding of the chemical and physical conditions in the Sgr B2 core region we are performing a spectral line survey around 230 GHz of three nearby positions, see Figure 3. While the positions Sgr B2(M) and (N) were rather obvious choices, based upon the high resolution data of Goldsmith *et al.* ([21], Nobeyama) and Vogel *et al.* ([22], VLA), our third position 2'N of (M) was chosen because the $HOCO^+$ brightness is peaking here (Minh *et al.* [24]).

Previous lower frequency spectral scans of Sgr B2 (Cummins *et al.* [25], Turner [26]) had been performed toward Sgr B2(OH), about 30" south of (M). However, due to the larger antenna beam widths they also partly sample the conditions in (M). In fact, Cummins *et al.* in their lower frequency range also would observe lines emanating in (N). SEST spectra towards the positions (M), (N), and (2'N) for the frequency band 225.65 - 226.15 GHz are shown in Figure 4. The trend visible here is typical for the whole frequency range observed until now (235 - 245 GHz + some additional bands around 226 GHz). In the (N) position the line density is on average 23 lines per GHz, in (M) $\approx$ 12 and in (2'N) $\approx$ 2. In the (N) source we are already *at the confusion limit* and severe blends of transitions from different molecules are apparent. E.g. the $HDO(3_{12}-2_{11})$ line is "too strong" because of a blend with CH_2CHCN and $HCOOCH_3$, see Figure 5. In fact, *a considerable part of the continuum level* observed towards Sgr B2(N), and also (M), *could be due to a line emission forest* — similar to the apparent conditions in the Owens Valley 1.3 mm line survey of Orion (Sutton *et al.* [27]). Although firm conclusions in terms of abundances would be premature here, it is clear that most lines observed are several times stronger in Sgr B2(N), compared to (M). However, this is definitely not the case for the SO_2 lines which peak in (M). As yet only lines from H_2CO, CH_3OH, CS, OCS, H_2CS, CN, HC_3N, CH_3CN, and CH_3C_2H have been identified at the $HOCO^+$ maximum (2'N). For "calibration" against the earlier lower frequency spectral scans we plan to observe a few important bands towards Sgr B2(OH).

Based upon experience from earlier spectral surveys (cf. Hjalmarson [28]) we may expect i) rather accurate information on the relative abundances of various (optically thin) molecules in the physically different cloud regions, ii) better knowledge of the physical conditions in the Sgr B2 core region, especially when we combine the scan data with our CH_3CN & CH_3C_2H mapping information (see § 2), and iii) a number of U-lines — hopefully leading to the identification of new interstellar molecules. Here our "access" to unpublished Nobeyama spectral surveys and detailed maps will be crucial.

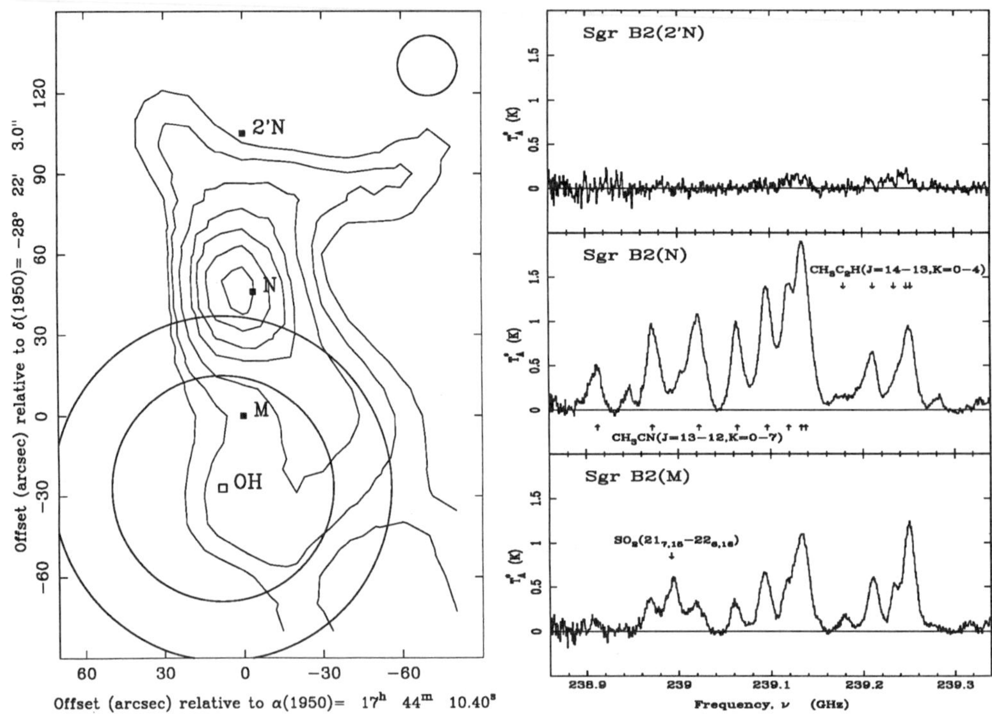

FIG. 3.—(*Left*) MEM-deconvolved SEST map of the $CH_3CN(J=6-5,K=4)$ transition towards the Sgr B2 cloud core region. The high temperature of Sgr B2(N) is clearly evident from this line (the K=0 line is equally strong in M and N), since the upper state excitation energy is 130 K. Spectral scans around 230 GHz are performed towards the positions Main, North, and 2' North using SEST (telescope beam size of 22" is indicated in upper right corner). Beam sizes at median frequency for the line surveys towards Sgr B2(OH) by Cummins *et al.* ([25], Bell Labs., 128") and Turner ([26], NRAO, 83") are also drawn. (*Right*) $CH_3CN(J=13-12)$ and $CH_3C_2H(J=14-13)$ spectra toward the 2'N, N, and M positions. Note that the SO_2 line is clearly stronger in M compared to N.

4. A note on the physical conditions in warm, dense cloud cores

The most important results from our CH_3CN analysis probably are not the high temperatures, densities and abundances. It has become very clear that the core regions are *clumpy*, with clump sizes < 0.05 - 0.20 pc and *large velocity gradients* in the range 50 - 200 km s^{-1} pc^{-1}, the nature of which are as yet unclear to us (cf. Pauls *et al.* [9], Wilson [29], Mundy and Bååth [30]).

Acknowledgements. Onsala Space Observatory (OSO) is operated by Chalmers University of Technology with financial support from NFR — the Swedish Natural Science Research Council. SEST is a joint venture between NFR and ESO, operated by OSO. Dedicated NFR grants for research in astrochemistry/astrophysics have been crucial for the research reported here.

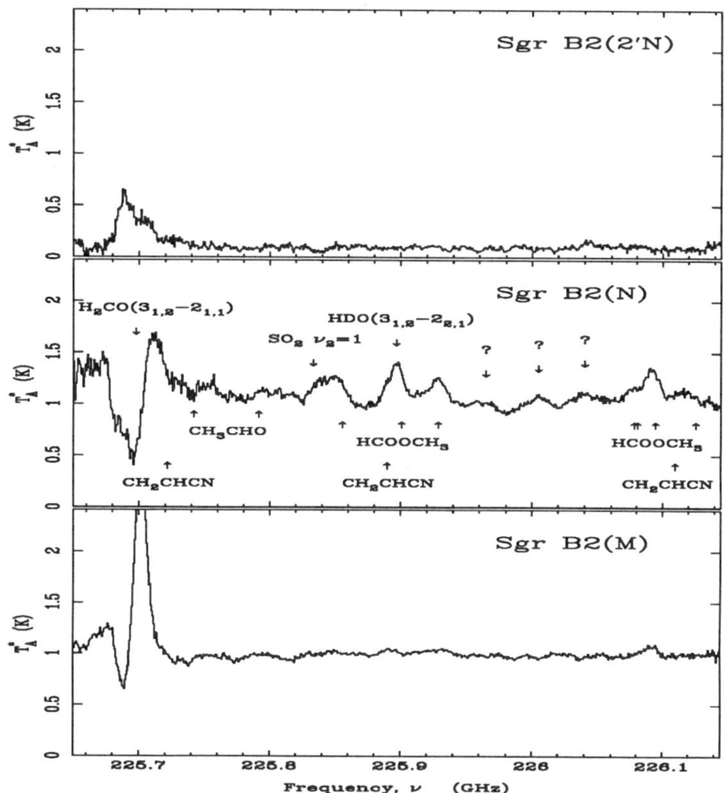

FIG. 4.—SEST spectral scan sample spectra at 226 GHz toward the three positions 2'N, N, and M. Spectral resolution is 1 MHz. Tentative identifications are included.

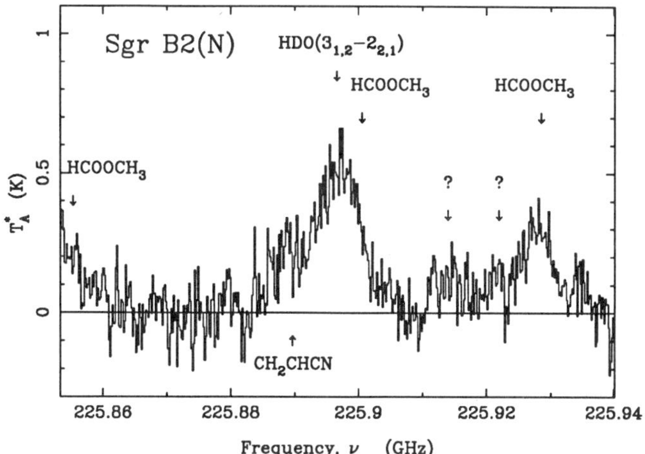

FIG. 5.—SEST spectra of the HDO(3_{12}-2_{11}) line at 225.9 GHz. Note the blend with the CH$_2$CHCN(10_{38}-10_{29}) and HCOOCH$_3$(6_{60}-5_{50}) lines. Spectral resolution is 0.17 MHz.

References

1. Watson, W. D., and Salpeter, E. E. 1972, *Ap. J.*, **174**, 321.
2. Watson, W. D., and Salpeter, E. E. 1972, *Ap. J.*, **175**, 659.
3. Duley, W. W. 1989, in *The Physics and Chemistry of Interstellar Molecular Clouds*, ed. G. Winnewisser and J. T. Armstrong (Berlin: Springer-Verlag), p. 353.
4. Herbst, E. and Klemperer, W. 1973, *Ap. J.*, **185**, 505.
5. Rydbeck, O. E. H., and Hjalmarson, Å. 1985, in *Molecular Astrophysics*, ed. G. H. F. Diercksen, W. F. Huebner, and P. W. Langhoff (Dordrecht: Reidel), p. 45.
6. Herbst, E. 1989, in *The Physics and Chemistry of Interstellar Molecular Clouds*, ed. G. Winnewisser and J. T. Armstrong (Berlin: Springer-Verlag), p. 344.
7. Irvine, W. M., Goldsmith, P. F., and Hjalmarson, Å. 1987, in *Interstellar Processes*, ed. D. J. Hollenbach and H. A. Thronson (Dordrecht: Reidel), p. 561.
8. Schweizer, J. S. 1978, *Ap. J.*, **225**, 116.
9. Pauls, T. A., Wilson, T. L., Bieging, J. H., and Martin, R. N. 1983, *Astr. Ap.*, **124**, 123.
10. Johansson, L. E. B, Andersson, C., Elldér, J., Friberg, P., Hjalmarson, Å., Irvine, W. M., Olofsson, H., and Rydbeck, G. 1984, *Astr. Ap.*, **130**, 227 ("the Onsala spectral scan").
11. Olofsson, H. 1984, *Astr. Ap.*, **134**, 36.
12. Menten, K. M., Walmsley, C. M., Henkel, C., and Wilson, T. L. 1986, *Astr. Ap.*, **183**, 109.
13. Jacq, T., Walmsley, C. M., Henkel, C., Baudry, A., Mauersberger, R., and Jewel, P. R. 1989, IRAM preprint No 164, submitted to *Astr. Ap.*
14. Millar, T. J., Olofsson, H., Hjalmarson, Å., and Brown, P. D. 1988, *Astr. Ap.*, **205**, L5.
15. Brown, P. D., Charnley, S. B., and Millar, T. J. 1988, *M.N.R.A.S.*, **231**, 409.
16. Brown, P. D., and Millar, T. J. 1989, *M.N.R.A.S.*, **237**, 661.
17. Andersson, M. 1985, in *Submillimeter Astronomy*, ed. P. A. Shaver and K. Kjär (ESO conference and Workshop Proc. No 22), p. 355.
18. Askne, J., Höglund, B., Hjalmarson, Å, and Irvine, W. M. 1984, *Astr. Ap.*, **130**, 311.
19. Blake, G. A., Sutton, E. C., Masson, C. R., and Phillips, T. G. 1987, *Ap. J.*, **315**, 621.
20. Bergman, P., Hjalmarson, Å., and Andersson, M. 1990, in preparation.
21. Goldsmith, P. F., Snell, R. L., Hasegawa, T., and Ukita, N. 1987, *Ap. J.*, **314**, 525.
22. Vogel, S. N., Genzel, R., and Palmer, P. 1987, *Ap. J.*, **316**, 243.
23. Bergman, P., and Hjalmarson, Å. 1989, in *The Physics and Chemistry of Interstellar Molecular Clouds*, ed. G. Winnewisser and J. T. Armstrong (Berlin: Springer-Verlag), p. 124.
24. Minh, Y. C., Irvine, W. M., and Ziurys, L. M. 1988, *Ap. J.*, **334**, 175.
25. Cummins, S. E., Linke, R. A., and Thaddeus, P. 1986, *Ap. J. Suppl.*, **60**, 819.
26. Turner, B. E. 1989, *Ap. J. Suppl.*, **70**, 539.
27. Sutton, E. C., Blake, G. A., Masson, C. R., and Phillips, T. G. 1984, *Ap. J. (Letters)*, **283**, L41.
28. Hjalmarson, Å. 1989, in *The Physics and Chemistry of Interstellar Molecular Clouds*, ed. G. Winnewisser and J. T. Armstrong (Berlin: Springer-Verlag), p. 73.
29. Wilson, T. L. 1985, *Comments Ap.*, **11**, 83.
30. Mundy, L. G., and Bååth, L. B. (1989), in *Molecular Clouds in the Milky Way and External Galaxies*, ed. R. L. Dickman, R. L Snell, and J. S. Young (Berlin: Springer-Verlag), p. 235.

Cold, Dark Interstellar Clouds: Can Gas-Phase Reactions Explain the Observations?

Welliam M. Irvine

Five College Radio Astronomy Observatory, University of Massachusetts, Amherst, MA 01003,
U. S. A. and Nobeyama Radio Observatory, Minamisaku, Nagano 384-13, Japan

The chemical composition of cold, dark interstellar clouds is reviewed. Recent observations of NO, H_2S, HCOOH, H_2CS, C_3H_2, CH_3CH_2CN, and CH_2CHCN are discussed in relation to abundances predicted by current gas-phase ion-molecule chemical models. In general, the observed abundances are in agreement with the predictions, although the situation is unclear for H_2S and for the ratio of ortho-to-para species of H_2CS and C_3H_2.

INTRODUCTION

One of Hiroko Suzuki's active research areas was the chemistry of dark interstellar clouds (1). These objects are indeed fascinating chemical laboratories, with physical conditions that are extreme by terrestrial standards. Kinetic temperatures are approximately 10 K, densities in core regions are typically between 10^4 and 10^5 molecules per cubic centimeter, and extinction by the interstellar dust particles is thought to exclude the ambient stellar ultraviolet radiation field from interior regions (2). Since some of these sources are the nearest dense interstellar clouds to the solar system, they may be studied with high angular resolution. In addition, some are formation sites for solar-type stars. The dark clouds lack, however, the high-luminosity embedded stars and protostars which form in giant molecular clouds and are the source for a variety of energetic processes which may alter the chemistry in the gas phase from that in more quiescent regions. The gas-phase chemistry in dark cloud cores is thus limited to binary reactions which are primarily exothermic and which lack activation barriers, with the ultimate energy source being cosmic rays. These are just the conditions postulated in theoretical models of "ion-molecule chemistry", so that the dark clouds probably provide the closest physical match to the assumptions of this paradigm of interstellar chemistry.

In spite of this apparent simplicity in physical conditions, the inventory of known chemical species in these cold clouds is a rich one. More than 50 of the 80 to 90 definitively identified interstellar molecules are present in dark clouds, including species with up to 13 atoms and weights up to 147 AMU. Molecular hydrogen is by far the dominant gas-phase constituent, as might be expected on grounds of cosmic abundance. The next most abundant molecular constituent is CO, about 10^4 times less abundant by number than H_2. In spite of the highly reducing conditions, many of the observed molecular species are highly unsaturated, including the cyanopolyynes $H(C{\equiv}C)_{2n}CN$ [n=0,...5], the hydrocarbon radicals C_nH [n=0,...6], the small ring C_3H_2, and others. Many other species which are "unstable" by standards of the terrestrial laboratory are also present, including other radicals and positive ions. The chemistry is characterized by very large isotopic fractionation for hydrogen, with some molecular species having ratios of the deuterated-to-normal-hydrogen form up to 10^4 times the cosmic D/H abundance ratio. This is reminiscent of the large (up to of order 100) hydrogen isotopic fractionation observed in certain organic fractions of carbonaceous chondrites, which it has been suggested is the result of preservation of interstellar molecular material in these primitive solar system objects (3, 4, 5; see also Table I).

Although there are many practical problems in determining molecular abundances in dark clouds (6), it appears that these abundances are matched in general quite well by the theoretical gas-phase models (e.g., 7). The models themselves, however, exhibit a number of problems. There are many poorly defined parameters, as a result of the lack of laboratory data on reaction rates and products, including particularly rates at low temperatures. Moreover, they are clearly oversimplified from the physical point of view, since models with the largest number of included reactions and molecular species do not include physical evolution of the cloud, as well as processes and constituents which may at least in some cases be significant (e.g., an internal ultraviolet radiation field generated by cosmic ray excitation of H_2 or the presence of "large molecules" which may carry the bulk of the negative charge; 8, 9). Perhaps the most significant omission from the largest (in terms of number of reactions and species) current models is any role for the interstellar grains other than that as formation sites for molecular hydrogen. The grains are clearly present, as manifested by the extinction of background starlight, and simple calculations show that molecular constituents other than H_2 ought to collide with and "freeze out" onto the grains in times short compared with expected cloud lifetimes (10). Although there must in consequence be mechanisms for recycling grain mantles back to the gas phase, observable effects of gas-grain interaction have been notoriously difficult to identify. In other words, it is perhaps surprising that the gas-phase models agree with observations as well as they do (see also 11, 12, 13).

It is thus important to obtain additional constraints on the models through further observations. One method for doing this is to identify and determine the abundance of new molecular constituents of dark clouds. Table I provides a list of such new molecular species identified in the past five or six years in these regions; the contribution of Hiroko Suzuki to this research is manifestly evident. It is also significant at an international conference that scientists from many countries have contributed to the identifications in Table I, including Australia, Canada, the Federal Republic of Germany, France, Japan, Spain, Sweden, and the U.S.A.

A second way to test the models is to investigate <u>differences</u> in chemical composition both among and within the dark clouds. The viable models must explain observed variations as a result of plausible differences in the model parameters. Two of the most thoroughly investigated dark clouds are TMC-1 and L134N, and some of the qualitative differences in chemical composition between these regions are illustrated in Table I. TMC-1 is more abundant in highly unsaturated organic species, sometimes by very large factors. L134N, on the other hand, seems to be more abundant in some oxygen-containing molecules and in such fully saturated species as H_2S and NH_3. Some organic molecules containing oxygen are roughly equally abundant in the two sources, including CH_3CHO, OCS, and CH_3OH.

A third way of testing the models is to investigate the relative abundance of closely related species such as isomers, isotopomers, ortho- and para-species of the same molecule, and related species exhibiting different amounts of saturation.

The present paper describes some examples of recent observations carried out with a view to testing current models along the lines discussed above. Three molecular species recently identified for the first time in dark clouds are discussed, including one case where published models provide good agreement with the observations (NO), one case where observed abundances seem to exceed predictions (H_2S), and one case where the observations initially appeared to yield too little relative to published models (HCOOH). We then consider the abundance ratio of ortho-to-para H_2CS (thioformaldehyde), which at first sight appears to offer evidence of significant interchange between the gas phase and grain mantles. A similar character in the ortho-to-para ratio has been found recently for C_3H_2 and was reported earlier for formaldehyde. We then consider the relative abundance of the related species HC_3N, CH_2CHCN, and CH_3CH_2CN in the cold cloud TMC-1 and the much

Table I

First Detections in Dark Clouds

H. Suzuki	Species	Year	TMC-1	L134N	Reference
	SO_2	1983		+	(37)
	CH_2CHCN	1983	•		(38, 27)
	CH_3CN	1983	•		(39)
	CH_3C_4H	1984	•		(40, 41, 42)
	CH_3C_3N	1984	•		(43)
√	C_3O	1984	•		(44, 45)
	$HC_{11}N$	1985	•		(46)
	CH_3CHO	1985	≈	≈	(47)
	ℓ-C_3H	1985	•		(48)
	C_3H_2	1985	+	•	(28, 49, 50)
	H_2C_2O	1986	•		(51, 52)
	C_5H	1986	•		(53)
√	C_6H	1986	+		(54, 55, 56)
	OCS	1987	≈	≈	(57)
√	c-C_3H	1987	•		(58)
√	C_2S	1987	+	•	(59, 56)
√	C_3S	1987	•		(60)
	CH_3OH	1988	≈	≈	(61)
√	HC_2CHO	1988	•		(62)
√	CH_2CN	1988	+		(63)
	H_2S	1989	•	+	(21)
	H_2CS	1989	+	•	(52)
√	C_2O	1989	•		(64)
	$HCOOH$	1990		•	(24)
	NO	1990		•	(17)

Symbols signify the following:
√ -- participation by H. Suzuki in detection; + -- of order 5 or more times more abundant in this cloud; ≈ -- approximately equal abundance in the two clouds; • -- detected in this cloud; blank -- only upper limit (see References) or no data.

hotter, active region known as the Orion hot core, using new observations of TMC-1. The Orion observations have been interpreted in terms of grain surface formation for the more saturated members of this series.

DETECTION OF THREE NEW DARK CLOUD MOLECULAR SPECIES

Nitric Oxide (NO)

As a simple diatomic molecule containing two abundant elements, it might be expected that the chemistry of NO in interstellar clouds would be well-understood. Recent models predict that NO will be the most abundant nitrogen-containing molecule after N_2 if, in accord with cosmic abundances, the elemental gas-phase abundance of oxygen is greater than that of carbon (14). Presumably because its lowest rotational transition lies in the 2mm wavelength band, where most radio observatories lack state-of-the-art receivers, there have been very few published studies of NO in the interstellar medium. Detection is also hindered by the small dipole moment of NO (μ = 0.15 D). Nitric oxide was first observed in the galactic center source Sgr B2 in 1978 (15), and there have subsequently been tentative detections in the Orion molecular cloud. It is interesting that models suggest that both the principal production route and two of the main destruction routes for NO are neutral-neutral reactions, rather than ion-molecule reactions. Specifically, production is thought to occur primarily through $N + OH \rightarrow NO + H$, while important destruction mechanisms include $NO + C$ and $NO + N$. Given these reactions, it is not surprising that the predicted abundance of NO is a strong function of the gas-phase availability of carbon versus oxygen (16).

Using a new 2mm wavelength receiver on the FCRAO 14m telescope at the University of Massachusetts, McGonagle et al. (17) have recently made the first dark cloud detection of NO. The strongest hyperfine component of each lambda doublet member of the lowest frequency transition at about 150 GHz was detected in the cloud L134N. The corresponding fractional abundance relative to molecular hydrogen, $6 (10)^{-8}$, agrees very well with recent model calculations for dark cloud conditions (7, 18, 19). It is interesting that NO was not detected in the archetypical cloud TMC-1, although the limit was not very stringent (about one half the detected abundance in L134N). From comparison with the published abundance of protonated molecular nitrogen, it was deduced that NO is ~ 100 times less abundant than N_2, so that the latter is the principal gas-phase reservoir for nitrogen in cold clouds. The failure to detect NO in TMC-1 can be understood either in terms of this cloud being in an earlier evolutionary state than L134N, or in terms of its being relatively richer in carbon compared with oxygen. Overall, the results are in good agreement with gas-phase models.

Hydrogen Sulfide (H_2S)

There have been suggestions in the past that sulfur chemistry may be dominated in interstellar clouds by grain surface reactions, at least for chemically reduced species (20). Grain processes were proposed because the reactions of both S^+ and S with molecular hydrogen are endothermic. H_2S was observed toward several giant molecular clouds in 1972, but there have been very few subsequent studies. As in the case of NO, this is at least partly the result of the lowest frequency rotational transition falling in the poorly observed 2mm wavelength region.

The first dark cloud detection of H_2S was recently made at FCRAO by Minh, Irvine, and Ziurys (21). Hydrogen sulfide was detected at two positions in L134N, and one in TMC-1. In particular, it was not observed at the so-called cyanopolyyne peak in TMC-1, where the upper limit was between 4 and 5 times less than the peak abundance observed in L134N. In fact, the L134N abundance is a factor of 4 or 5 times larger than model results, even when the model has been "pushed" quite hard by increasing the (unmeasured) rate-limiting step in the production by an order of magnitude and taking a relatively small depletion of sulfur (22). Given the lack of laboratory data and the possibility of variable sulfur abundances in the gas phase, however, this result can perhaps only be taken as weak evidence for the formation of H_2S on grain surfaces. We do note, however, that H_2S has been tentatively identified in grain mantles from infrared observations toward W33A (23).

Formic Acid (HCOOH)

At first sight an apparent discrepancy between theory and observation in the opposite sense of that for H_2S appears to arise in the case of formic acid. This species had previously been observed in the massive star formation regions Sgr B2 and Orion-KL, but only weakly in the latter case. This is somewhat curious, since the methylated relative of formic acid, methyl formate, is very prominent in the millimeter wave spectra of Sgr B2, Orion-KL, and W51. Recent gas-phase chemical models of dark clouds predict a relatively high abundance of HCOOH (7), in the range $2 (10)^{-9}$ to $2 (10)^{-8}$.

Using the Nobeyama Radio Observatory 45m telescope so familiar to Hiroko Suzuki, as well as observations at NRAO and FCRAO, Irvine et al. (24) have recently confirmed the detection of HCOOH in L134N. The observed abundance, however, is between one tenth and one one-hundredth of the published model predictions. The observed upper limit in TMC-1 was only slightly less than the detected abundance in L134N, so that it is impossible to say whether formic acid falls into the category of molecules which are approximately equally abundant in these two clouds, or into the oxygen-rich category which is more abundant in L134N.

The apparent discrepancy between theory and observation is removed, however, by the recent realization that the principal production route in the model (the radiative association of HCO^+ and H_2O followed by dissociative electron recombination) will not occur. If, as now seems likely, the principal production pathway is through the ion-molecule reaction $CH_4 + O_2^+$, again followed by dissociative electron recombination, the resultant abundance agrees very well with the observed value in L134N and the upper limit toward TMC-1 (E. Herbst, private communication).

THE ORTHO-TO-PARA RATIO FOR MOLECULAR SPECIES IN DARK CLOUDS

It is generally believed that the interconversion between ortho- and para-symmetry states for molecules with identical hydrogen nuclei such as H_2, H_2O, NH_3, H_2CO, etc. is very slow in the gas phase, because both radiative and collisional conversion are strictly forbidden and gas-phase interconversion mechanisms such as proton exchange are thought to be slow (25). It has thus been suggested that the ortho-to-para abundance ratio would preserve information on molecular formation processes, unless heterogeneous processes were important subsequent to molecule formation (26). For example, Kahane et al. (26) measured an ortho-to-para ratio for H_2CO in TMC-1 and L134N and concluded that it corresponded to equilibrium at a temperature of $\sim$ 10 K, which is the approximate kinetic temperature and grain temperature in dark clouds. These authors offered reasons for believing that purely gas-phase processes would have led to an ortho-to-para ratio much larger than that observed, and they therefore concluded that thermalization was probably taking place on grain mantles and that interchange between the grains and the gas phase must be efficient. Because the observed lines appeared not to be optically thin, however, the analysis relied on a specific and perhaps non-unique radiative transfer model. Note that if equilibrium prevailed at molecular formation, the small energy difference between the lowest ortho- and the lowest para-states (of order 15 K) would lead to an ortho/para abundance ratio of 3:1, the ratio of statistical weights.

Recently data has become available for two other molecular species in cold, dark clouds which exhibit ortho- and para-states. Two ortho- and two para-transitions for H_2CS have been measured in TMC-1 by Minh (27), and measurements have been made towards four dark clouds for several C_3H_2 transitions by Madden (28). The results are collected in Table II, which lists both the ortho-to-para ratio expected for the particular molecular species in thermal equilibrium at 10 K and the observed ratio. It is seen that the astronomical results are generally consistent with the thermal equilibrium values, not with the ratio of statistical weights (3:1) which corresponds to equilibrium at the high temperature of molecular formation.

In analogy with the ortho- and para- symmetry species for molecules with two symmetric hydrogen nuclei, the A and E species for molecules with torsional symmetry are also distinguished by the lack of allowed radiative and collisional interconversion in the gas phase. Such molecules detected in dark clouds include NH_3, CH_3C_2H, CH_3CN, CH_3C_4H, CH_3C_3N, CH_3CHO, and CH_3OH. To our knowledge the ratios of A to E species for CH_3CN, NH_3, and CH_3C_3N have not been measured in dark clouds. For CH_3OH the ratio is consistent with equilibrium at 10 K (61), while for CH_3CHO the energy difference between the lowest A and E states is so small that the ratio expected at 10K is already the high temperature limit of 1:1 (47). In contrast, the ratio for CH_3C_2H appears to correspond to the ratio of statistical weights and not to equilibrium at the local kinetic temperatures (65); a similar result may hold for CH_3C_4H, although the uncertainties are necessarily greater (42). In other words, the situation for CH_3C_2H and perhaps for CH_3C_4H appears to contrast with that for the molecules listed in Table II, and perhaps with that for CH_3OH (although the energy level structure for CH_3OH leads to complex problems in interpretation).

Table II

Ortho/Para Ratios for Three Molecules in Dark Clouds

Species	Source	TE O/P	Obs. O/P	Reference
H_2CO	TMC-1/L134N	1.5-2	1-2	(26)
H_2CS	TMC-1	1.5-2	1.8±0.3	(27)
C_3H_2	4 clouds	2.6-2.7	2.4±0.2	(28)

Note: TE O/P is ortho/para ratio for thermal equilibrium at 10 K; Obs. O/P is observed value.

How should these results be interpreted? Following Kahane et al. (26) and limiting the discussion to the species in Table II, the ratios may provide evidence for efficient interchange between the gas phase and grain mantles. Even if the assumption of no interchange between ortho- and para-species in the gas phase subsequent to molecular formation is correct, however, it may be that the initial ratio is constrained by details of the formation process to a non-equilibrium value. Detailed quantum mechanical studies of all significant processes for the relevant species are presumably necessary to determine if this is the case. Clearly this is a very difficult problem, since it involves identifying the principal production paths and the spin states of the precursor molecules, and analyzing the energy surface for these reactions. Another possibility is that there are, in fact, gas-phase processes which can equilibrate the ortho/para ratio during the lifetime of a dark cloud. We understand that studies of possible processes for H_2CO are underway by J. Black and E. Van Dishoeck [private communication; cf. (29)].

We conclude that the results in Table II must stand at present as interesting observational facts, with the interpretation to be still determined by future research. Major uncertainties include the rates of ortho-para conversion for reactions of the species of interest with H^+, and the proton abundance in dark cloud cores.

RELATIVE CHEMICAL SATURATION

Hydrogenation of small hydrocarbons and hydrocarbon radicals is difficult in the gas phase in cold clouds because reactions of such species with H_2 frequently involve activation energies (30). The best example is probably the series $HC{\equiv}CCN$, $H_2C{=}CHCN$, and CH_3CH_2CN. HC_3N is rather abundant in dark clouds, particularly in TMC-1 and other regions in the Taurus cloud complex. CH_2CHCN has also been detected, but only at about 5% the abundance of HC_3N. A rather different picture emerges in the so-called Orion "hot core", which apparently consists of clumps of dense, hot gas in the immediate vicinity of the luminous infrared source (protostar) IRC2 in the core of the Orion molecular cloud. There CH_2CHCN and HC_3N have roughly equal abundances, which are themselves a factor of 5 or 10 less than the abundance of CH_3CH_2CN (31, 32). It has been suggested that the predominance of more saturated species in the hot Orion gas is the result of formation on grain surfaces, where hydrogenation is relatively easy, followed by evaporation into the gas phase (32, 33).

In order to investigate this situation further, Minh (27) has recently made a deep search for CH_3CH_2CN in TMC-1, as well as re-observing CH_2CHCN in this source. He finds that the CH_3CH_2CN abundance is less than 2% that of HC_3N and less than half that of CH_2CHCN in this dark cloud. These results are quite consistent with the abundances predicted by gas-phase models for cold clouds.

<u>CONCLUSION</u>

The recent results reinforce earlier conclusions that there are significant differences in the trace constituent composition of the clouds TMC-1 and L134N, with the former richer in highly unsaturated organic compounds and the latter in some oxygen-containing species and in fully saturated molecules like NH_3 and H_2S (see Table I). It has been suggested that these differences might be an effect of evolutionary state, with TMC-1 a "younger" region than L134N in some models, and an "older" region in others (17, 34). Although there are chemical gradients <u>within</u> each of these clouds which to some extent mimic the differences between them, it nonetheless would seem possible that certain portions of the cloud (for example, denser ones) might evolve faster than other portions, even if the cloud as a whole was "formed" at a certain time. Alternatively, it has been suggested (35, 36) that these differences might reflect differing abundances of carbon relative to oxygen in the gas phase. This might be brought about by differential depletion of these elements onto grains, for example in the form of H_2O and CH_4. Some such effect would be expected as a function of temperature because of the different volatility of these two species (32). In fact, both these explanations produce similar effects, because as clouds evolve, the available carbon tends to become locked up in CO, which is highly stable. As a result, calculations predict that the abundance of many organic compounds will reach a maximum and then decrease very strongly with time, to some extent mimicking the result of calculations which begin with less available gas-phase carbon (14). Nonetheless, we may hope that a detailed comparison of observations and models might be able to distinguish between these, and perhaps other, hypotheses. Nitric oxide may be a good probe in this regard, if a deeper search can be conducted in TMC-1 (17).

We conclude that in general gas-phase models do remarkably well in producing molecular abundances in agreement with observations. There is little hard-evidence for molecules in the gas having been produced on or in the interstellar grains in cold clouds, apart from the theoretical conclusions that H_2 must be primarily formed on grain surfaces and the timescale for "freeze out" of gas constituents onto the grains [see, however, (10, 66, 67)]. If more complete laboratory data becomes available on the reactions relevant to H_2S production and destruction, this situation may change; it does seem that gas-phase models are stretched to the limit in order to match the abundances which have been reported in dark clouds, particularly in L134N. Likewise, it is possible that a clearer theoretical understanding of the processes involved in determining the ortho-to-para ratio for molecules exhibiting these two symmetry species may confirm that gas-phase interconversion is too slow to be important and that initial formation ratios should be larger than those observed (Table II). Such a result would be strong evidence for efficient interchange between the gas phase and grain mantles.

I am grateful to Y. Minh for the use of unpublished data and for discussion of many of the results reported here. This research was supported in part by NASA grant NAGW-436 and NSF grant 88-15406. I am very grateful to the National Astronomical Observatory of Japan for a Foreign Research Fellowship during the period when this paper was written, and to Jan Whitaker for preparing the manuscript.

<u>REFERENCES</u>

(1) Suzuki, H. 1989, <u>Toward Interstellar Chemistry</u>, ed. N. Kaifu (Tokyo: Univ. Tokyo).
(2) Goldsmith, P.F. 1987, in <u>Interstellar Processes</u>, ed. D. Hollenbach and H. Thronson (Dordrecht: D. Reidel), p. 51.
(3) Irvine, W.M. and Knacke, R.F. 1989, in <u>Origin and Evolution of Planetary and Satellite Atmospheres</u>, ed. S.K. Atreya, J.B. Pollack, and M.S. Matthews (Tucson: U. Arizona), p. 3.

(4) Zinner, E. 1988, in <u>Meteorites and the Early Solar System</u>, ed. J.F. Kerridge and M.S. Matthews (Tucson: U. Arizona), p. 956.

(5) Irvine, W.M., Avery, L.W., Friberg, P., Matthews, H.E., and Ziurys, L.M. 1988, <u>Ap. Letters Comm.</u>, <u>26</u>, 167.

(6) Irvine, W.M., Schloerb, F.P., Hjalmarson, Å., and Herbst, E. 1985, in <u>Protostars and Planets II</u>, ed. D.C. Black and M.S. Matthews (Tucson: U. Arizona), p. 579.

(7) Herbst, E. and Leung, C.M. 1989, <u>Ap. J. Suppl.</u>, <u>69</u>, 271.

(8) Gredel, R., Lepp, S., and Dalgarno, A. 1987, <u>Ap. J. (Letters)</u>, <u>323</u>, L137.

(9) Lepp, S. and Dalgarno, A. 1988, <u>Ap. J.</u>, <u>324</u>, 553.

(10) Millar, T.J. 1989, <u>Symp. Infrared Spectroscopy in Astronomy</u> (ESA SP-290), p. 109.

(11) Allamandola, L.J. 1990, this volume.

(12) Herbst, E. 1990, this volume.

(13) Millar, T.J. 1990, this volume.

(14) Herbst, E. and Leung, C.M. 1986, <u>M.N.R.A.S.</u>, <u>222</u>, 689.

(15) Liszt, H.S. and Turner, B.E. 1978, <u>Ap. J. (Letters)</u>, <u>224</u>, L73.

(16) Leung, C.M. 1989, Private Communication.

(17) McGonagle, D., Ziurys, L.M., Irvine, W.M., and Minh, Y.C. 1990, <u>Ap. J.</u>, in press.

(18) Langer, W.D. and Graedel, T.E. 1989, <u>Ap. J. Suppl.</u>, <u>69</u>, 241.

(19) Millar, T.J., Leung, C.M., and Herbst, E. 1987, <u>Ast. Ap.</u>, <u>183</u>, 109.

(20) Millar, T.J. 1982, <u>M.N.R.A.S.</u>, <u>199</u>, 309.

(21) Minh, Y.C., Irvine, W.M., and Ziurys, L.M. 1989, <u>Ap. J. (Letters)</u>, <u>345</u>, L63.

(22) Millar, T.J. and Herbst, E. 1990, <u>M.N.R.A.S.</u>, in press.

(23) Geballe, T.R., Baas, F., Greenberg, J.M., and Schutte, W. 1985, <u>Astr. Ap.</u>, <u>146</u>, L6.

(24) Irvine, W.M., Friberg, P., Kaifu, N., Matthews, H.E., Minh, Y.C., Ohishi, M., and Ishikawa, S. 1990, <u>Astr. Ap.</u>, in press.

(25) Flower, D.R. and Watt, G.D. 1984, <u>M.N.R.A.S.</u>, <u>209</u>, 25; erratum <u>213</u>, 991.

(26) Kahane, C., Frerking, M.S., Langer, W.D., Encrenaz, P., and Lucas, R. 1984, <u>Astr. Ap.</u>, <u>137</u>, 211.

(27) Minh, Y.C. 1990, Ph.D. Thesis, University of Massachusetts.

(28) Madden, S.C. 1990, Ph.D. Thesis, University of Massachusetts.

(29) Dalgarno, A., Black, J.H., and Weisheit, J.C. 1973, <u>Ap. Letters,</u> <u>14</u>, 77.

(30) Herbst, E., Adams, N.G., and Smith, D. 1984, <u>Ap. J.</u>, <u>285</u>, 618.

(31) Johansson, L.E.B. et al. 1984, <u>Astr. Ap.</u>, <u>130</u>, 227.

(32) Blake, G.A., Sutton, E.C., Masson, C.R., and Phillips, T.G. 1987, <u>Ap. J.</u>, <u>315</u>, 621.

(33) Irvine, W.M. and Hjalmarson, Å. 1984, <u>Origins Life</u>, <u>14</u>, 15.

(34) Stahler, S.W. 1984, <u>Ap. J.</u>, <u>281</u>, 209.

(35) Millar, T.J. and Freeman, A. 1984, <u>M.N.R.A.S.</u>, <u>207</u>, 425.

(36) Irvine, W.M., Goldsmith, P.F., and Hjalmarson, Å. 1987, in <u>Interstellar Processes</u>, ed. D.J. Hollenbach and H.A. Thronson (Dordrecht: D. Reidel), p. 561.

(37) Irvine, W.M., Good, J.C., and Schloerb, F.P. 1983, <u>Astr. Ap.</u>, <u>127</u>, L10.

(38) Matthews, H.E. and Sears, T.J. 1983, <u>Ap. J.</u>, <u>272</u>, 149.

(39) Matthews, H.E. and Sears, T.J. 1983, <u>Ap. J. (Letters)</u>, <u>267</u>, L53.

(40) MacLeod, J.M., Avery, L.W., and Broten, N.W. 1984, <u>Ap. J. (Letters)</u>, <u>282</u>, L89.

(41) Loren, R.B., Wootten, A., and Mundy, L.G. 1984, <u>Ap. J. (Letters)</u>, <u>286</u>, L23.

(42) Walmsley, C.M., Jewell, P.R., Snyder, L.E., and Winnewisser, G. 1984, <u>Astr. Ap.</u>, <u>134</u>, L11.

(43) Broten, N.W., MacLeod, J.M., Avery, L.W., Irvine, W.M., Höglund, B., Friberg, P., and Hjalmarson, Å. 1984, <u>Ap. J. (Letters)</u>, <u>276</u>, L25.

(44) Matthews, H.E., Irvine, W.M., Friberg, P., Brown, R.D., and Godfrey, P.D. 1984, <u>Nature</u>, <u>310</u>, 125.

(45) Suzuki, H., Ohishi, M., Morimoto, M., Kaifu, N., Friberg, P., Irvine, W.M., Matthews, H.E., and Saito, S. 1985, in <u>The Search for Extraterrestrial Life, Recent Developments</u>, ed. M.D. Papagiannis (Dordrecht: D. Reidel), p. 139.

(46) Bell, M.B. and Matthews, H.E. 1985, Ap. J. (Letters), 291, L63.
(47) Matthews, H.E., Friberg, P., and Irvine, W.M. 1985, Ap. J., 290, 609.
(48) Thaddeus, P., Gottlieb, C.A., Hjalmarson, Å., Johansson, L.E.B., Irvine, W.M., Friberg, P., and Linke, R.A. 1985, Ap. J. (Letters), 294, L49.
(49) Thaddeus, P., Vrtilek, J.M., and Gottlieb, C.A. 1985, Ap. J. (Letters), 299, L63.
(50) Matthews, H.E. and Irvine, W.M. 1985, Ap. J. (Letters), 298, L61.
(51) Matthews, H.E. and Sears, T.J. 1986, Ap. J., 300, 766.
(52) Irvine, W.M., et al. 1989, Ap. J., 342, 871.
(53) Cernicharo, J., Guélin, M., Menten, K.M., and Walmsley, C.M. 1986, Astr. Ap., 172, L5.
(54) Suzuki, H., Ohishi, M., Kaifu, N., Ishikawa, S., and Kasuga, T. 1986, Publ. Ast. Soc. Japan, 38, 911.
(55) Guélin, M., Cernicharo, J., Kahane, C., Gomez-Gonzalez, J., and Walmsley, C.M. 1987, Astr. Ap., 175, L5.
(56) Suzuki, H., Yamamoto, S., Ohishi, M., Kaifu, N., Ishikawa, S., Hirahara, Y., and Takano, S. 1990, in preparation.
(57) Matthews, H.E., MacLeod, J.M., Broten, N.W., Madden, S.C., and Friberg, P. 1987, Ap. J., 315, 646.
(58) Yamamoto, S., Saito, S., Ohishi, M., Suzuki, H., Ishikawa, S.I., Kaifu, N., and Murakami, A. 1987, Ap. J. (Letters), 322, L55.
(59) Saito, S., Kawaguchi, K., Yamamoto, S., Ohishi, M., Suzuki, H., and Kaifu, N. 1987, Ap. J. (Letters), 317, L115.
(60) Yamamoto, S., Saito, S., Kawaguchi, K., Kaifu, N., Suzuki, H., and Ohishi, M. 1987, Ap. J. (Letters), 317, L119.
(61) Friberg, P., Madden, S.C., Irvine, W.M., and Hjalmarson, Å. 1988, Astr. Ap., 195, 281.
(62) Irvine, W.M., et al. 1988, Ap. J. (Letters), 335, L89.
(63) Irvine, W.M., et al. 1988, Ap. J. (Letters), 334, L107.
(64) Ohishi, M. et al., in preparation.
(65) Askne, J., Höglund, B., Hjalmarson, Å., and Irvine, W.M. 1984, Astr. Ap., 130, 311.
(66) Greenberg, J.M. 1990, in Chemistry in Space, ed. J.M. Greenberg and V. Pironello (Dordrecht: Kluwer), in press.
(67) Walmsley, C.M. 1989, in Interstellar Dust, ed. L.J. Allamandola and A.G.G.M. Tielens (Dordrecht: Kluwer), p. 263.

The Orion Nebula

Geoffrey A. BLAKE

Div. of Geological and Planetary Sciences, California Institute of Technology 170-25, Pasadena, CA 91125, U. S. A.

ABSTRACT.   As one of the nearest and brightest massive star forming molecular clouds, the core of the Orion nebula presents a unique opportunity to examine the complexity and evolution of interstellar chemistry. We present here an overview of the results gleaned from high resolution millimeter and submillimeter spectral line surveys of the region along with aperture synthesis maps of molecular emission obtained with interferometers such as the Owens Valley (OVRO) and Hat Creek millimeter-wave instruments and the VLA. Emphasis will be placed on the alteration of the chemical composition of the parent molecular cloud by the star formation process.

1. THE SOURCE

The Orion Nebula is but a small part of the vast Orion-Monoceros giant molecular cloud complex, lying roughly 500 parsecs from the Sun and covering some 30 square degrees of the sky (1). This subregion (called Orion/KL or OMC-1), centered on the Trapezium stars in the sword of the Orion region and including the luminous infrared sources BN and IRc2, is perhaps the most widely known and best studied example of a young stellar nebula. Excellent reviews covering the wealth of observational studies of this region and the relationship of the Orion Nebula to the rest of the giant molecular cloud complex may be found in Genzel and Stutzki (2) and in the proceedings of the Henry Draper Symposium (3). Indeed, this short contribution can in no way do justice to the tremendous amount of research which has contributed to our understanding of the origin and fate of this unique collection of stars, gas, and dust. Rather, we shall mainly be concerned with how high resolution, in both an angular and spectral sense, molecular emission line observations can be used to explore the chemistry which occurs during and after massive star formation.

Because of its proximity, large intrinsic luminosity ($2\text{-}10 \times 10^4$ $L_\odot$), and enormous quantity of surrounding material, the IRc2 region is one of the brightest compact objects in the sky from mid-infrared to millimeter wavelengths. Its unique strength in molecular line emission is obvious from its central role in molecular line searches since the early days of molecular radio astronomy (4-10). In more recent years, these strong lines have made IRc2 an irresistible source for millimeter interferometer studies of molecular spatial distributions (11-14). As a result of these efforts, a great deal has been learned about IRc2 and its environs. The next two sections outline the results from high angular resolution interferometric observations and high spectral resolution molecular emission line surveys of the Orion cloud core.

2. INTERFEROMETRIC OBSERVATIONS

The connection of the Orion/KL, or OMC-1, core with the Orion-Monoceros giant molecular cloud complex is visible in maps of many molecular species as a strip, or "ridge," of material running roughly N-S near the Trapezium. The linewidths in the ridge material are at most 4-5 km s^{-1}, and decrease with increasing spatial resolution. Condensations in the Orion ridge are nicely sampled by high density tracers such as CS, as is demonstrated in Figure 1. This mosaic, obtained by combining OVRO interferometric data with single dish observations from the Onsala Space Observatory (15), contains *all* of the flux on size scales larger than 7.5″ in the 3.5′ field mapped. Three condensations are clearly visible, with dimensions from 0.03 to 0.11 pc and virial masses from 30 to 80 $M_\odot$. The largest condensation is centered on the main source in the region, IRc2, a roughly 25 $M_\odot$ star deeply embedded in the local material (2,3). The 1.5′S source is the only other clump in Figure 1 which also shows obvious signposts of star formation activity (16).

In addition to the "spike" or ridge gas, a number of other distinct kinematic components are detected near IRc2. For example, centered approximately 2″ south of IRc2 is the "hot core" component which is a $\sim$10 M$_\odot$ clumping of gas and dust. This material is associated with and strongly heated by IRc2 to at least 100-200 K, but does not appear to be a circumstellar disk (2,14). A strong bipolar outflow from IRc2 has interacted with the structure in the local molecular cloud to create a doughnut of slowly expanding shocked cloud material (the "low velocity plateau" component, Δv$\leq$25 km s^{-1}) roughly normal to the flow axis. Along the flow axis the velocities are correspondingly higher due to the faster density decrease (the "high velocity plateau"), as is evident in the emission from CO, HCO$^+$, and H$_2$ (12,13). Finally, a compressed clump of cloud material called the "compact ridge" is visible some 10-15″ to the southwest of IRc2.

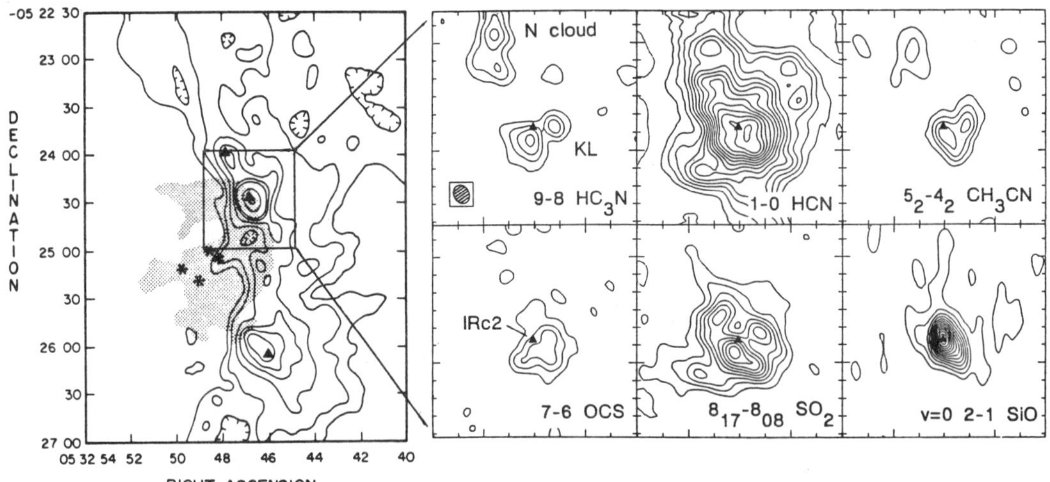

Figures 1(left) and 2(right). *(1) OVRO/Onsala image of the integrated CS J = 2 → 1 emission from the Orion ridge. The asterisks denote the positions of the Trapezium stars, while the triangles mark the positions of the millimeter continuum sources (15). (2) Hat Creek aperture synthesis maps molecular emission towards Orion/KL. The contour levels for the first four 1′ square maps step by 5 K, while for the last two the interval is 10 K (17).*

Chemistry adds considerably to the complexity of the region as variations in temperature, atomic abundances, and chemical history drive variations in the molecular composition of each component. For example, interferometer maps of species are often wildly different in appearance, as Figure 2 demonstrates. This Figure, obtained by the Hat Creek millimeter-wave interferometer (now called the Berkeley-Illinois-Maryland Array, or BIMA), outlines dramatic differences in molecular emission on the 3-4″ scale. When combined with the temperature and overall abundance constraints provided by the spectral line surveys outlined below, these maps demand that true chemical heterogeneity exists on at most the 2000 AU scale.

3. SINGLE DISH SPECTRAL LINE SURVEYS

Due to the opaque nature of the atmosphere in many wavelength regions and the relatively insensitive nature of early receiver/telescope combinations, much of the initial work on the chemistry of dense molecular clouds proceeded through a biased selection of molecules on a line-by-line basis. However, recent developments in receiver and telescope technology have allowed unbiased surveys of small regions in the millimeter-wave spectrum to be recorded for the brightest objects in the sky such as the Orion and Sagittarius cloud cores (4-10). Advantages of line surveys versus selected observations include completeness, better calibration, certainty of identification, and measurement of the total integrated line flux.

For example, the construction of higher frequency SIS receivers, and the improved performance possible with the OVRO 10.4m telescopes, forerunners of the Caltech Submillimeter Observatory (CSO), permitted a high sensitivity survey of the Orion molecular cloud (OMC-1) between 208-263 GHz (7,8). This survey was conducted, in part, as an attempt to preview the broad band millimeter and submillimeter wave spectra of dense molecular clouds. Nearly 900 resolved spectral features, arising from some thirty species in sixteen varying isotopic substitutions, were detected. Of these, less than 15 remain currently unidentified! Analyses of the intensities of the emission features after a deconvolution of their line profiles have resulted in accurate estimates of molecular abundances in different regions of the cloud (see below).

The total observing time required for the OVRO Orion spectrum was approximately 28 nights. Data reduction took several hundred hours of computer time and months of human analysis. Indeed, identifications of the many lines left after comparison with publicly available spectral line catalogs would not have been possible without a vigorous, complementary laboratory program at Duke University and JPL (18-19). Catalog incompleteness will continue to be a problem, especially as the line searches move into the submillimeter region where laboratory observations are sparse, and an essential part of future programs of this type must be continued close ties and access to state-of-the-art laboratory spectrometers.

Several factors combine to make spectral line searches with the CSO much more efficient than the 1.3 mm Orion/KL survey conducted at OVRO. The major "hardware" improvements are, of course, the telescope surface and the atmospheric transmission (for the latest results from the JCMT please see the contribution by Matthews in this volume). We have therefore begun to undertake spectral line surveys in the 1.3 and 0.87 mm windows of a number of representative sources, and are utilizing the sensitivity improvements in two distinct ways. For brighter sources such as OMC-1 and Sgr-B2, confusion limited scans, in the double sideband sense, require sensitivities of only 0.1-0.5 K. Essentially complete spectral surveys of these objects *in a number of positions* are being acquired to examine the nature of chemical heterogeneity in giant molecular clouds and their star-forming cores. To reach similar abundance sensitivities in cold clouds, circumstellar shells, and protostellar nebulae requires surveys with much lower noise and better spectral resolution. More limited surveys with deep integrations are therefore being performed in the 1.3 and 0.87 mm windows in sources such as IRC+10216, T Tau, and IRAS16293-2422.

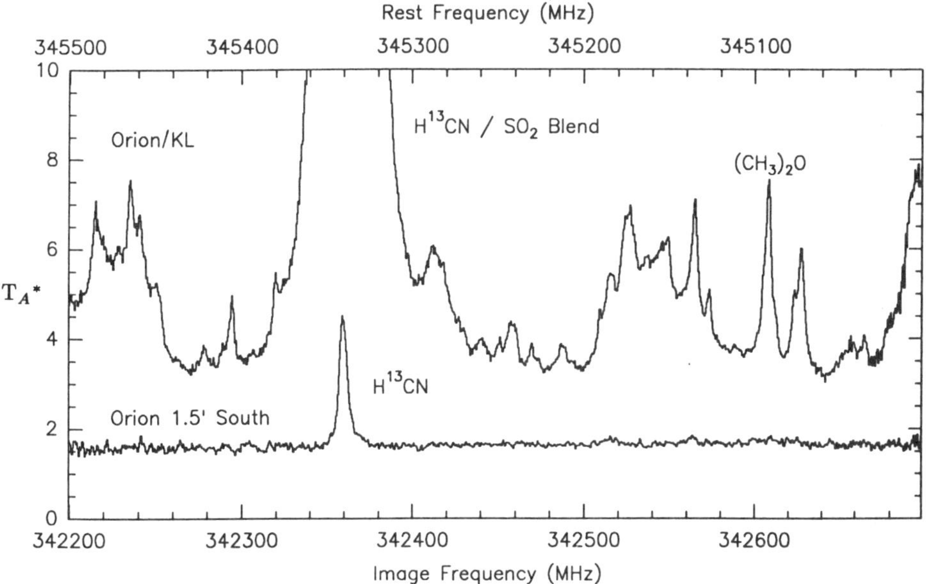

Figure 3. *CSO 500 MHz double sideband spectrum of the Orion/KL and Orion 1.5'S sources near the $H^{13}CN$ $J = 4 \rightarrow 3$ transition at 345.34 GHz. Both spectra have the same vertical scale, and the total integration time was less than 10 minutes.*

The CSO 345 GHz survey of the Orion/KL region yields dramatically more intense emission than is prevalent at longer wavelengths due to the high temperature and large size/mass of the central source. Two positions have been observed to examine the nature of chemical abundance variations on small size scales, in this case the classical Orion/KL "plateau" source and the warm core 1.5'S (Figure 1). The survey currently covers the Orion/KL region completely between 330-358 GHz, with approximately 40% coverage of the 1.5'S source. The differences in the spectra between the two sources are striking, as is illustrated by Figure 3.

Indeed, the access to high frequency atmospheric windows and the boost in system sensitivity have revealed an incredible complexity for the millimeter-wave spectrum of this brightest of molecular cloud cores. The density of molecular lines in the spectra obtained greatly impedes the proper interpretation of the physics and chemistry of the region in that the crowding of myriads of lines, mostly arising from high energy levels (>300 K) of asymmetric top molecules such as methyl formate, sulfur dioxide, or methanol, is such that only in a few selected bands can the true continuum level be reached. We typically observe an approximately 1.7 degree offset in good weather at all LO settings, which we interpret as the true continuum level in our 20″ beam. The offset (and therefore the total H_2 column density) is much the same in the southern source, but the line density is at least an order of magnitude lower. Thus, when observed with "continuum" instruments such as bolometers, any allusions to variations in the gas-to-dust ratio in hot molecular cloud cores must be viewed with caution if they are similar to the Orion Hot Core (the line/continuum ratio differs by nearly a factor of ten in the CSO survey for the two positions observed).

The CSO 345 GHz survey of Orion/KL is also quite different from the NRAO 12m survey of the same region, reflecting the large difference in efficiencies between the two telescopes. We typically observe a factor of 5-10 increase in $T_A{}^*$ for lines from compact sources such as the hot core and the compact ridge. Thus, whereas the observed Kitt Peak line density is *lower* than the OVRO 1.3mm survey, for the CSO survey the line density and strength increase, as expected. The difference in the total line count is also substantial. A preliminary analysis of the chemistry yields no major surprises, although the lack of catalog completeness is a severe problem indeed. All of the species detected in the OVRO 1.3 mm survey are still detected, most of them with *stronger* emission. This is due partly to the small beamsize of the 345 GHz survey as compared to the 230 GHz survey, but mostly it is due to the dense, warm nature of the Orion/KL region.

The most striking difference between the south source and the Orion/KL region is the complete absence of SO_2 in the former, despite easily observable SO and SiO spectral lines. In addition, several of the species more traditionally associated with cooler molecular clouds, such as CCH, are observed to have stronger emission lines and greater abundances in the southern source. Much of the Orion spectral line survey data have been collected only very recently, and are still in the process of final reduction. Once the complete surveys have been acquired, excitation/abundance analyses will be performed on all of the identifiable peaks to yield quantitative measures of the chemical differentiation in the two sources.

4. CHEMICAL HETEROGENEITY

The observed steep abundance gradients noted above and inferred time dependence stress more and more the close interdependence between source dynamics, energy balance, and chemical composition. No longer can the ultimate stages of star formation be studied ignoring the chemical changes and vice versa, for example. All this demonstrates that observational astrochemistry will continue to mature only if high angular and spectral resolution measurements can be obtained on a statistically significant selection of objects at the highest achievable sensitivity. Nevertheless, Orion/KL will likely remain a detailed "testing ground" for models of interstellar chemistry, and it is heartening to note that the tremendously heterogeneous composition of the Orion cloud core as revealed by the single dish abundances and high resolution millimeter-wave interferometric images may be consistently interpreted in the framework of the interaction between a quiescent molecular cloud chemistry and that induced by massive star formation (20).

The chemistry of the extended ridge material is similar to that observed in a number of other cool, quiescent clouds. Contributions by Kaifu, Ohishi, Irvine, Federman, Millar, and Herbst in these proceedings deal directly with the observational and theoretical aspects of chemistry in such regions; and as such we concentrate on the chemistry of the Orion core. We note, however, that the abundances of cyanopolyynes and other reactive carbonaceous species so beautifully observed and modeled by Hiroko Suzuki, to whose memory this book is dedicated, peak in the ridge gas.

Whereas gas in the exterior regions appears to be controlled primarily by ion-molecule chemistry and depletion effects, the chemical composition of the Orion/KL is dominated by high temperature chemistry, including shocks and grain mantle evaporation. A graphical demonstration of the differences between the quiescent ridge chemistry and that of the plateau and hot core sources is depicted in Figure 4. The most striking result for the plateau source is the tremendous drop in the abundances of carbon rich species like C, CCH, and CN and the concomitant rise in the abundance of species such as H_2O, HCN, SiO, H_2S, and SO_2. While the extended ridge clouds appear observationally to have a deficit of oxygen in the gas phase, the plateau gas seems to have a gas phase C/O ratio that is solar or lower. Indeed, the observed value for molecules other that CO is $C/O \leq 0.5$ not including water, and $<<1$ for all reasonable estimates of the residual deuterium fractionation in HDO (as constrained by the recent IRAM 183 GHz H_2O data (21)). In contrast, the same ratio is greater than unity in the ridge material. The higher oxygen fugacity and the very high temperatures behind shock fronts should drive most of the carbon into stable forms such as CO, but will also enhance the abundances of more refractory species such as sulfur and silicon via grain disruption or evaporation (as is more thoroughly discussed by Ziurys in these proceedings). Finally, the considerable abundance of "fragile" species such as H_2CO and the large residual deuterium enhancement in water imply an extremely clumpy density structure in the outflowing wind.

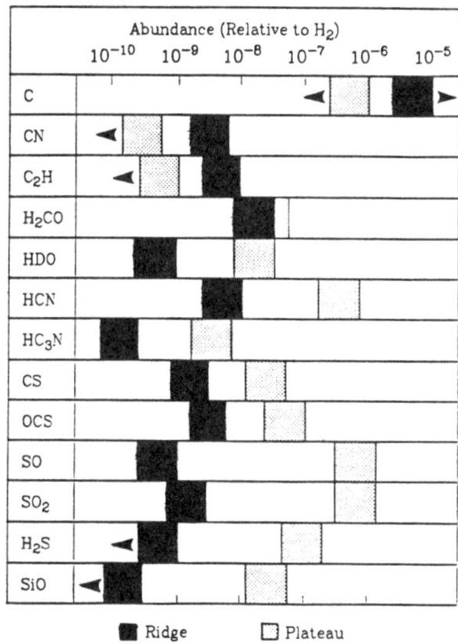
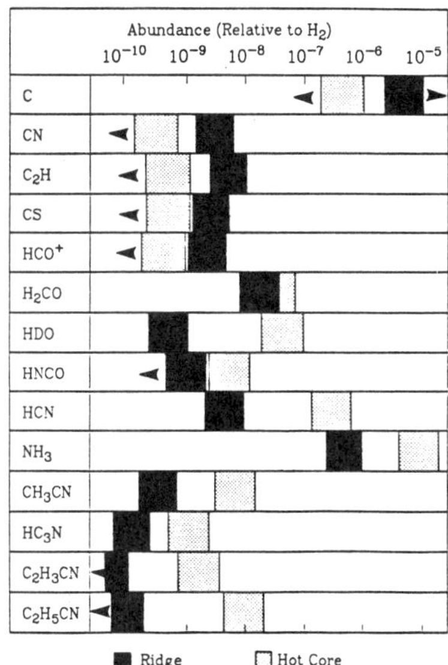

Figure 4. *Graphical presentation of the Orion ridge, plateau, and hot core abundances as derived from the OVRO 1.3 mm spectral line survey (20).*

High temperature chemistry is also evident in observations of the hot core, but with a considerably different signature (see Figure 4). Again, the abundances of reactive, carbon rich species are significantly reduced, with the observed C/O ratio in molecular form less than unity (as in the plateau gas the actual ratio depends sensitively on the assumed deuterium fractionation in HDO). Because the heating is passive, however, grain vaporization does not occur and the abundances of the refractory sulfur and silicon species remain low. The high densities and temperatures in the hot core reduce the effectiveness of ion-molecule networks, and are sufficient to release much of the grain mantle material stored over the cloud lifetime. Indeed, the composition of the hot core is most consistent with chemistry that is dominated by kinetic equilibrium and evaporation from grains as recent detailed numerical modeling by Brown and Millar has confirmed (22). Water and ammonia in particular are observed to be tremendously enhanced in the hot core to levels near that inferred from infrared solid state studies of grain mantles. The HDO and NH_2D (23) observations are especially noteworthy in that significant fractionation is retained even though the kinetic temperature is well above 100 K. The most likely explanation is that the observed HDO/NH_2D is "fossil" water/ammonia trapped onto grains throughout the cloud lifetime and released when star formation has raised the local temperatures sufficiently to evaporate their icy grain mantles (14,22). Further, the conversion of polyacetylenes into more hydrogenated species like ethyl cyanide must occur on grains because the relevant gas phase routes are highly endothermic. For more extensive discussions of "hot core chemistry," please see the contributions by Hjalmarson, Brown, Walmsley, and Breuker herein.

Finally, the importance of mixing in active cores like Orion/KL is demonstrated by the composition of the compact ridge source. It is here that the abundances of complex oxygen containing organics such as CH_3OH and $HCOOCH_3$ peak. These species could either be produced by grain mantle hydrogenation of CO, etc., or by gas phase routes. An earlier gas phase suggestion involving mixing of water from the plateau source with ion-rich gas from the ridge (19) does not appear to explain the observed abundances via an ion-molecule route when subjected to detailed numerical tests, although the direct injection of methanol appears to be more promising (see Herbst, this volume). Grain mantle models could most likely account for some of the abundances, but so little is known about chemistry on cryogenic, disordered surfaces that it will be very difficult with such models to quantitatively account for the observed selectivity (namely, that dimethyl ether and methyl formate are found to be much more abundant than their isomers ethanol and acetic acid). Another point in favor of a different mechanism than that proposed for the hot core is provided by recent high angular resolution observations of methanol in Orion/KL from Hat Creek (14). If methanol and the other related species were produced by grain mantle catalysis, then they should show observational correlation with other grain mantle products. For example, ammonia, ethyl cyanide, and water show striking similarities in Orion. Methanol, however, appears to be *anti*correlated with these species; and thus any model which accounts for its chemistry cannot be similar in detail to the hot core.

5. CONCLUDING REMARKS

Observational studies of the chemistry of interstellar, circumstellar, and protostellar matter have reached a pivotal stage. By combining high angular resolution interferometric observations to image molecular distributions with broad band spectroscopic surveys to accurately constrain excitation and abundances, it has become possible to examine in detail the nature of chemical evolution in the brightest sources in the sky. It is now clear that a number of processes act to alter the composition from that predicted by the simplest theories of interstellar chemistry, and as such it is imperative that experiment, observation, and theory continue their close ties. Advances in detector and telescope fabrication now make it possible to examine ever fainter and more distant objects. For example, in recent CSO surveys of the 27 $L_\odot$ source IRAS16293-2422, the same suite of molecules used to infer high temperature chemistry in Orion (SiO, SO, SO_2, NH_2D, etc.) are easily visible (24). The first great leap forward in astrochemistry occurred with the radio detections of the polyatomic molecules NH_3, H_2O, and H_2CO in dense galactic clouds, which shattered the theoretical dogma of the day that only diatomic molecules could survive the harsh environment of the interstellar medium. Modern instrumentation has now set the stage for the next quantum leap in observational astrochemistry: the ability to spectroscopically probe the very cores of a statistically significant number of circum/protostellar, interstellar, and extragalactic cloud regions and observe the driving chemical reactions that steer processes ranging from the formation of galaxies through the formation of stars and planets. It is the challenge of the coming century to both the observationalist and theoretician alike to unravel the chemical threads that constitute the tapestry of our universe, and to thereby honor the memory and life's work of Hiroko Suzuki.

ACKNOWLEDGMENT.   It is a pleasure to acknowledge the tireless efforts of a number of coworkers, longstanding members of "the Orion community" (as R. Genzel has described it), who have contributed much to this work. In particular, T. Groesbeck played a critical role in the CSO 345 GHz surveys, while E. Sutton, C. Masson, and T. Phillips were essential to the OVRO 230 GHz spectral line survey. The author also acknowledges many helpful discussions and successful OVRO sessions with L. Mundy and A. Sargent. Figure 2 from the Hat Creek Array was kindly provided by R. Plambeck. Astronomy at the OVRO and at the CSO is support by the NSF, grants #AST 84-12473 and #AST 83-11849

REFERENCES

(1) Kutner, M.L., Tucker, K.D., Chin, G., and Thaddeus, P. 1977, *Ap.J.*, **215**, 521.
(2) Genzel, R., and Stutzki, J. 1989, *A.R.A.A.*, **27**, 41.
(3) Glassgold, A.E., Huggins, P.J., and Shucking, E.L., eds. 1982, *Ann. N.Y. Acad. Sci.*, **395**, .
(4) Lovas, F. J., Johnson, D. R., Buhl, D., and Snyder, L. E. 1976, *Ap.J.*, **209**, 770.
(5) Lovas, F. J., Snyder, L. E., and Johnson, D. R. 1979, *Ap.J.Suppl.*, **41**, 451, and references therein.
(6) Johansson, L. E. B. *et al.* 1984, *Astr.Ap.*, **130**, 227.
(7) Sutton, E. C., Blake, G. A., Masson, C. R., Phillips, T. G. 1985, *Ap.J.Suppl.*, **58**, 341.
(8) Blake, G. A., Sutton, E.C., Masson, C.R., and Phillips, T.G. 1986, *Ap.J.Suppl.*, **60**, 357.
(9) Turner, B. E. 1989, *Ap.J.Suppl.*, **70**, 539.
(10) Jewell, P. R., Hollis, J. M., Lovas, F. J., and Snyder, L. E. 1989, *Ap.J.Suppl.*, **70**, 833.
(11) Plambeck, R. L. *et al.* 1985, in *URSI Symposium on Millimeter and Submillimeter Astronomy*, ed. (Granada: URSI), p.235.
(12) Vogel, S. N., Bieging, J. H., Plambeck, R. L., Welch, W. J., Wright, M. C. H. 1985, *Ap.J.*, **296**, 600.
(13) Masson, C. R., and Mundy, L. G. 1988, *Ap.J.*, **324**, 538.
(14) Plambeck, R.L. and Wright, M.C.H. 1988, *Ap.J.(Letters)*, **317**, L101.
(15) Mundy, L.G. *et al.* 1988, *Ap.J.*, **325**, 382.
(16) Ziurys, L., Friberg, P., and Irvine, W. M. 1989, *Ap.J.*, **343**, 201.
(17) Plambeck, R.L. and Wright, M.C.H. 1988, in *Molecular Clouds in the Milky Way and External Galaxies*, ed. Dickman, R.L., Snell, R.L., and Young, J.S. (Munich: Springer-Verlag), p.182.
(18) Plummer, G.M., Herbst, E., and De Lucia, and Blake, G.A. 1984, *Ap.J.Suppl.*, **55**, 633.
(19) Sutton, E.C. and Herbst, E. 1989, *Ap.J.*, **387**, 654.
(20) Blake, G. A., Sutton, E. C., Masson, C. R., and Phillips, T.G. 1987, *Ap.J.*, **315**, 621.
(21) Jacq, T., Jewell, P.R., Henkel, C., Walmsley, C.M., and Baudry, A. 1988, *Astr.Ap.*, **199**, L5.
(22) Brown, P.D., Charnley, S.B., and Millar, T.J. 1988, *Mon.Not.R.Astr.Soc.*, **231**, 409.
(23) Walmsley, C.M. *et al.* 1987, *Astr.Ap.*, **172**, 311.
(24) Blake, G.A., Groesbeck, T., and Mundy, L.G. 1990, *Ap.J.*, in preparation.

Comet Halley and Interstellar Chemistry

Lewis E. Snyder[1], Patrick Palmer[2], and Imke de Pater[3]

[1] *Department of Astronomy, University of Illinois, Urbana, IL 61801, U. S. A.*
[2] *Department of Astronomy and Astrophysics, University of Chicago, Chicago, IL 60637, U. S. A.*
[3] *Astronomy Department, University of California, Berkeley CA 94720, U. S. A.*

How complex is the chemistry of the interstellar medium? How far does it
evolve and how has it interacted with the chemistry of the solar system? Are the
galactic chemical processes terminated, continued, or even enhanced in comets? Are
biogenic molecules formed in space and have these molecules or their formation
mechanisms interacted in any way with prebiotic organic chemical processes on the
early earth? Radio molecular studies of comets are important for probing deep into
the coma and nuclear region and thus may help answer these questions. Comets are
believed to be pristine samples of the debris left from the formation of the solar
system and may have been the carrier between interstellar and terrestrial prebiotic
chemistries. Recent observations of Comet Halley and subsequent comets have given us
an excellent opportunity to study the relationship between interstellar molecular
chemistry and cometary chemistry.

INTRODUCTION

At least 86 molecular species have been found to date in the interstellar
clouds of gas and dust grains which pervade our galaxy. The chemical processes which
generate interstellar molecules are efficient over large regions of the clouds and
appear to be intimately related to the presence of both the grains and stars
enveloped in the interstellar clouds. In very dense clouds, the grains may provide
catalytic surfaces which enhance the production of polyatomic molecules whereas in
more tenuous regions there is evidence that radiative association could be the
predominant mechanism for molecular formation. The usual restrictions on molecular
stability placed on the laboratory spectroscopist by wall·collisions are negligible
in the interstellar clouds. The total gas content of the galaxy is quite large,
perhaps 3 per cent of the total mass, but cloud dimensions are of the order of light
years; the resulting densities are very low; a cloud with only 10^6 hydrogen
molecules cm^{-3} is considered very dense. Low densities and average kinetic
temperatures between 10 and 100 K are indications that mean free paths between 10^9
and 10^{15} cm are not unusual. The resulting times between collisions (10^4-10^{10} s) are
of the order of relaxation times for rotationally-excited molecules; thus electronic
and vibrational relaxation times (typically of the order of 10^{-8} and 10^{-2} s,
respectively) usually are too short to influence collisional formation processes. In
addition, the central regions of dense clouds may provide sufficient shielding from
photodestructive processes to allow fairly complex molecular species, once formed,
to survive for years. For example, some dark dust clouds have a remarkably high
abundance of long carbon chain molecules - the largest and heaviest known
interstellar species. How complex is the chemistry of the interstellar medium? How
far does it evolve and how has it interacted with the chemistry of the solar system?
Most significantly, if biologically relevant molecules are forming in space, can we
detect and identify them and learn whether their formation mechanisms have
interacted in any way with prebiotic organic chemical processes on the early earth?
Radio molecular studies of comets are important for probing deep into the coma and
nuclear region and thus may help answer these questions. Comets are believed to be

pristine samples of the debris left from the formation of the solar system and may have been the carrier between interstellar and terrestrial prebiotic chemistries. Recent observations of Comet Halley and subsequent comets have given us an excellent opportunity to study the relationship between interstellar molecular chemistry and cometary chemistry.

As a guide to begin to discuss this topic, Table 1 gives a list of the 86 interstellar and circumstellar molecular species which have been reported as of December, 1989, but it does not include C_5O which was reported by B. E. Turner at this meeting. The molecules are ordered in rows by number of atoms and are arranged in alphabetical order by empirical formulae according to the modified Hill system which is employed in the <u>Chemical Abstracts Formula Index</u>. Individual isotopes and a few unconfirmed species have not been listed. Square brackets mean that further confirmation of the detection is needed. Ring molecules are preceded by c- but where the structure is also linear an l- is used. Current lists of the transition frequencies have been prepared by Lovas [1,2]. Many of the interstellar molecular species have also been found in circumstellar shells. Table 1 does not include the polycyclic aromatic hydrocarbons (PAHs), which are inferred to exist in interstellar grains [3].

Table I. The 86 Reported Interstellar and Circumstellar Molecules Listed in Order by Number of Atoms as of December, 1989.

<u>2</u> CH CH^+ C_2 CN CO CP CS CSi H_2 HCl AlCl AlF KCl NO NS NaCl OH PN SO SO^+ SiO SiS

<u>3</u> HCN HNC HCO HCO^+ HCS^+ C_2H OCS C_2S c-SiC_2 H_3^+ H_2O H_2S N_2H^+ HNO SO_2

<u>4</u> $HCNH^+$ HNCO HNCS $HOCO^+$ H_2CO H_2CS HC_2H c-C_3H 1-C_3H C_3N C_3O C_3S NH_3 [H_3O^+]

<u>5</u> HCOOH H_2CHN H_2NCN CH_4 CH_2CN H_2C_2O HC_3N c-C_3H_2 C_4H C_4Si SiH_4 <u>6</u> $HCONH_2$ CH_3OH

CH_3SH CH_3CN CH_3NC C_2H_4 C_5H <u>7</u> NH_2CH_3 $HCOCH_3$ CH_2CHCN CH_3C_2H HC_5N C_6H

<u>8</u> $HCOOCH_3$ CH_3C_3N <u>9</u> CH_3CH_2OH $(CH_3)_2O$ CH_3CH_2CN CH_3C_4H HC_7N <u>10</u> CH_3C_5N $(CH_3)_2CO$

<u>11</u> HC_9N <u>13</u> $HC_{11}N$

Almost any modern scenario links the composition of comets to the composition of the interstellar medium, albeit with many uncertainties. As Whipple [4] pointed out, the presence of chemically ill-mated carbon molecules and the apparent low abundance of methane show that comets are a mixture of low temperature compounds which are not formed by slow cooling from a hot gas. Since the same carbon compounds are present in the interstellar gas, it is believed that there is a similarity between the interstellar medium and cometary gas. While comets may have formed in the Kuiper belt or may even be ejecta from the primordial planetary region, neither direct contributions from the primordial interstellar cloud that formed the solar system nor contributions of accreted comets from fragmented interstellar clouds connected gravitationally with the Sun have been ruled out.

While it is clear that cometary spectroscopy has made important contributions toward understanding the composition of the volatile species in the nucleus, the growing body of evidence from interstellar and circumstellar molecular spectroscopy, the Comet Halley flyby mass spectra, and from meteoritic chemistry all suggest that cometary nuclei are much more chemically evolved than the current spectroscopy suggests. The first clue for this comes from the fact that almost any modern scenario links the composition of comets to the composition of the interstellar medium but the currently known interstellar and circumstellar molecules (Table I) are much more chemically complicated than the known cometary molecules (Table II).

Table II. Ground based cometary spectroscopy in the ultraviolet, optical, infrared, and radio has led to the identification of at least 24 cometary molecules, listed in order by number of atoms.

<u>2</u> CH CH^+ C_2 CN CN^+ CO CO^+ CS NH N_2+ OH OH^+ S_2 <u>3</u> HCN NH_2 C_3 CO_2 CO_2+ H_2O H_2O^+

<u>4</u> H_2CO NH_3 <u>5</u> CH_4 <u>6</u> CH_3CN

A second clue that cometary nuclei are relatively complex is found in the data acquired by the Giotto spacecraft. As one example, the heavy ion analyzer, RPA2-PICCA, on board the Giotto spacecraft recorded mass spectra intensity peaks from 14 amu up to 110 amu in the gas flowing from the nucleus of Comet Halley [5]. While most of these mass spectra peaks remain unidentified, the proposed identification of short polymer chains of polyoxymethylene (POM), or polymerized formaldehyde, by Huebner [6] which is based on mass spectra data is perhaps the most exotic chemical identification to be suggested to date for Comet Halley molecular data. Perhaps a bit less exotic is the suggestion [7] that the nucleus of Comet Halley is covered with a surface layer of very low albedo hydrogen cyanide polymers and related compounds. This covering would make Comet Halley one of the darkest bodies in the solar system.

The third clue comes from the chemical analysis of meteorites. Hahn _et al._ [8] find that at the ionization wavelength of 266 nanometers, parent peaks of polycyclic aromatic hydrocarbons (PAHs) such as naphthalene (mass 128) dominate the mass spectra of the Murchison meteorite. Cronin, Pizzarello, and Cruikshank [9] have discussed the reasons for believing that the meteoritic amino acids and other organic matter (and by inference that in comets, asteroids and satellites) were synthesized abiotically early in the solar system. They have argued that the surfaces of comets, asteroids, and some planetary satellites are probably covered in part by materials representative of the low albedo substances found in primitive carbonaceous chondrites. Indeed, Kerridge and Anders [10] have discussed the current evidence for the survival of both circumstellar and interstellar cloud material in meteorites and Anders [11] has estimated that meteorites may have contributed as much organic carbon to the early earth as did the Urey-Miller reactions in the Earth's atmosphere. Because many paths have been suggested that may link meteoritic and cometary chemistry, it is clear that the cometary analysis lags far behind.

Finally, it should be noted that Hoyle and Wickramasinghe [12] have argued that the dust grains in interstellar clouds contain bacteria such as E coli, which survive in comets when the clouds condense to form stars and planets, and they have suggested that cometary impact could have led to the start of terrestrial life. Much of the experimental evidence for bacterial interstellar grains has centered on the comparison of the shapes of interstellar extinction curves with laboratory opacity curves for various preparations of E coli. These comparisons have been controversial (for example, see Moore and Donn [13]), mostly because solid state spectra generally do not have any sharply localized spectral features like the well defined rotational or vibrational-rotational lines found in gas phase spectra. Therefore, it appears that the bacterial grain controversy is at a standstill unless it can be formulated in terms of gas phase spectroscopy, where well- defined patterns of spectral lines can uniquely define interstellar compounds. However, the general proposition that terrestrial prebiotic organic chemistry is a highly specialized outgrowth of interstellar chemistry can not be dismissed easily. It is recognized that many of the small molecules in Table I which are important for regions of star formation (e.g. HCN, H_2O, NH_3, H_2CO) are also important for primitive biological systems on earth.

RADIO SPECTROSCOPY OF COMETS

a) The Pre-Halley Observations. Prior to the last apparition of Comet Halley, the situation in cometary radio spectroscopy had been discouraging. Only OH, NH_3, H_2O, HCN and possibly CH, CH_3CN, and HC_3N had been reported in the radio spectra of comets. Of these molecules only OH was well-confirmed because it had been observed in 16 comets, starting with Kohoutek (1973 XII). In marked contrast, HCN had been detected only in Kohoutek and possibly IRAS-Araki-Alcock (1983 VII); NH_3 in only IRAS-Araki-Alcock; H_2O in Bradfield (1974 III) and in IRAS-Araki-Alcock; CH and CH_3CN possibly in Kohoutek; and HC_3N possibly in Sugano-Saigusa-Fujikawa (1983 V) [14,15,16]. The relative ease of detecting interstellar molecules and the failure to detect a comparable number of cometary molecules had caused some to doubt that comets actually have a very complex chemistry. But even prior to the spacecraft observations of Comet Halley, the radio observations of cometary OH suggested that this pessimistic viewpoint was incorrect and that part of the problem with cometary

molecular detections was in understanding the relevant excitation mechanisms. For
example, the actual radio observations of cometary OH are tricky because OH is
pumped by solar ultraviolet radiation. Time variations in the OH pumping mechanism
(the Swings effect) happen when cometary OH molecules absorb Doppler-shifted solar
ultraviolet Fraunhofer bands. This causes steady-state fluorescent pumping of OH
molecules from the $X^2\Pi$ ground state Λ doublet levels to the excited $A^2\Sigma^+$ state. The
OH molecules cascade back to the ground state, thereby establishing the relative
populations of the ground state Λ doublets. This mechanism determines whether
cometary radio OH signals will be observed in absorption, emission, or not at all.
It is quite likely that other cometary molecules with greater chemical complexity
may also undergo radiative or collisional pumping which ultimately determines
whether they are detectable with current observational approaches. It is also
possible that the previous searches at radio wavelengths failed because single
element telescopes have excessive beam dilution. For a moderately bright comet at a
distance of 1 AU, the molecular coma may vary in size from ~100 km (for species such
as S_2 and NH_3) to ~100,000 km (HCN), corresponding to angular scales of ~0.1" to
~100". Hence, even prior to the current Halley observations, there were suggestions
that low abundance may not have been the only reason for the past failures to detect
cometary molecules.

b) The Halley Observations and Beyond. With the acquisition of data obtained with
new technologies at all wavelengths from the ultraviolet through the radio, with the
coordinated collection of data by International Halley Watch, and with the
absolutely unique in situ measurements by space probes to Comet Giacobini-Zinner
(preliminary summaries in the 18 April 1986 issue of Science) and to Comet Halley
(preliminary summaries in the 15 May 1986 issue of Nature) we are on the threshold
of a major step in our understanding of comets. There were several successful
molecular observations of Comet Halley which are very encouraging for future
cometary radio spectroscopy. Cometary HCN was observed with three different radio
telescopes [17,18,19] and H_2O was detected with the high resolution infrared
spectrometer on board the NASA KAO [20].

The first successful VLA observations of OH emission from any comet were our
observations of the 1667.359 MHz transition of OH in Comet Halley [21]. These were
the first cometary radio maps made with resolution better than several arcminutes
and, close to the nucleus, they show small scale emission structure that was
previously unknown. Inspection of the high resolution images showed two new results:
the emission was "clumpy" rather than smoothly distributed; and there was a clear
absence of emission from the direction of the nucleus, which was interpreted as the
result of quenching of the OH maser emission by collisions in the high density part
of the coma immediately surrounding the nucleus. The radio maps show that the ring
of OH emission clumps is $\approx$ 60,000 km from the nucleus [22]. Whether or not this is
significant will remain an open question until a data base can be built from
observations of other comets. However, this clump radius is contained by the heavy
ion mantle and cometopause region at $\approx 10^5$ km, as measured for Comet Halley in 1986
March with the plasma instrument packages (PLASMAG-1) aboard the Vega 1 and Vega 2
spacecraft by Gringauz et al. [23]. It is also within the magnetic field pile-up
region at $\approx 10^5$ km measured with the Giotto magnetometer by Neubauer et al. [24].

Recently, the VLA was used for OH imaging observations of a new comet, Comet
Wilson (19861), and these data show some remarkable differences with those for Comet
Halley, an old comet [25]. In comparison with Comet Halley, the radio OH emission
from Comet Wilson behaved very erratically, changing rapidly in position as well as
in velocity, while the emission and brightness distribution from Comet Halley
displayed apparent stability. A few months later, nearer perihelion, just the
opposite behavior was observed at UV wavelengths. Another difference between the two
comets is that the OH emission from Comet Halley seems confined to a region a few
times 10^5 km in size, while the emission from Comet Wilson shows up in sporadic
blobs, with variable intensities and velocities, at distances as far as 10^6 km from
the nucleus. This behavior in Comet Wilson may be associated with the disintegration
of the outer frosting associated with new comets and possibly with the fragmentation
and ejection of cometesmals from the nucleus.

Shortly after the detection of interstellar formaldehyde in 1969, several radio searches for cometary formaldehyde were initiated with single element radio telescopes but all were unsuccessful [14]. Despite this, modern theoretical models of the cometary nucleus often include a large component of formaldehyde because they are based on the assumption that comets form either in the outer parts of the presolar nebula or in a companion fragment of the parent interstellar cloud from which the solar system evolved. In 1986 January, the VLA was used to detect the 1_{11} - 1_{10} transition of formaldehyde (H_2CO) at 4829.659 MHz (6 cm. wavelength) in emission from Comet Halley [26]. In order to compute the H_2CO production rate it was necessary to extend the H_2CO scalelength, r_{es}, in the equation of transfer from 1.44×10^3 to 1.5×10^4 km; thus initially r_{es} was identical to the CO scalelength found by Eberhardt et al. [27], and later to the H_2CO scalelength found by Krankowsky [28] from the NMS data. This confirmed that the H_2CO emission in Comet Halley was produced from an extended source in the coma as well as directly from the nucleus. The VLA was used again in September, 1988, to detect this same 6 cm transition of H_2CO in emission from Comet Machholz, a new comet [29]. The derived H_2CO production rate for Comet Machholz is 2×10^{29} mol s^{-1}, and for Comet Halley it is about an order of magnitude smaller, 1.5×10^{28} mol s^{-1}. Unfortunately, Comet Machholz was not recovered after its perihelion passage (September 17.57) and it faded considerably only days after the H_2CO observations, which suggests that it fell apart. Indeed, the Comet Machholz H_2CO line is $\approx$ 2.5 times wider than the Comet Halley H_2CO line, which may indicate that the comet actually had started to come apart during these VLA observations. While the body of data available for this comet is far less than for Comet Halley, it offered the rare chance to observe a bright new comet, possibly as it was disintegrating.

SUMMARY

We have summarized the growing body of experimental evidence from interstellar and circumstellar molecular spectroscopy, from the Comet Halley flyby mass spectra, and from meteoritic chemistry that suggests cometary nuclei are much more chemically evolved than the current gas-phase cometary spectroscopy suggests. Radio observations of comets are important for probing deep into the coma and nuclear region and thus are crucial for understanding the relationship between interstellar molecular chemistry and cometary chemistry, but prior to the 1985-86 apparition of Comet Halley, the situation in cometary radio spectroscopy had been discouraging. However, the recent observations of Comet Halley and subsequent comets have given us an excellent opportunity to study the relationship between interstellar molecular chemistry and cometary chemistry. With the data obtained through new technologies at all wavelengths from the ultraviolet through the radio, and with the absolutely unique in situ measurements by space probes, we are on the threshold of a major step in our understanding of cometary chemistry. In particular, we are intrigued by the various scalesizes which were measured at various times for Comet Halley: an H_2CO and CO scalelength of 15,000 km (VLA, 1986 Jan.; Giotto, 1986 March); an OH clump radius of $\approx$ 60,000 km (VLA, 1985 Nov.,1986 Jan.); the heavy ion mantle and the cometopause region at $\approx$ 100,000 km (Vega 1 and Vega 2, 1986 March); and the magnetic field pile-up region at $\approx$ 100,000 km (Giotto, 1986 March). Whether or not it is significant that the radio molecular activity appears to be contained well within the boundaries of the heavy ion mantle, the cometopause region, and the magnetic field pile-up region is presently an open question. If it is significant, this apparent hierarchical structure may contain important clues about the role of galactic chemical processes in comets and whether comets effectively connect interstellar and terrestrial prebiotic chemistries.

L.E.S. and P.P acknowledge support from NASA grant NAGW-1131 to the University of Illinois. I.d.P. acknowledges support from NSF grant AST-8900156 to the University of California, Berkeley.

REFERENCES
1. Lovas, F. J. 1986, <u>J. Phys. Chem. Ref. Data</u> 15, 251.
2. Lovas, F. J. 1987, <u>J. Phys. Chem. Ref. Data</u> 16, 153.

3.Allamandola, L. J., Sandford, S. A., and Wopenka, B. 1987, <u>Science</u> 237, 56.
4.Whipple, F. L. 1976, <u>Nature</u> 263, 15.
5.Korth, A., Richter, A. K., Mendis, D. A., Anderson, K. A., Carlson, C. W.,Curtis, D. W., Lin, R. P., Mitchell, D. L., Rème, H., Sauvaud, J. A., and d'Uston, C. 1987, <u>Astr. Ap.</u> 187, 149.
6.Huebner, W. F. 1987, <u>Science</u> 237, 628.
7.Matthews, C. N., and Ludicky, R. 1986, in <u>20th ESLAB Symposium on the Exploration of Halley's Comet</u> (ESA SP-250), ed. B. Battrick, E. J. Rolfe, and R. Reinhard (Noordwijk:ESA), p. 273.
8.Hahn, J. H., Zenobi, R., Bada, J. L., and Zare, R. N. 1988, <u>Science</u> 239, 1523.
9.Cronin,J. R., Pizzarello, S., and Cruikshank, D. P. 1988, in <u>Meteorites and the Early Solar System</u>, eds. J. F. Kerridge and M. S. Matthews (Tucson: University of Arizona Press), p. 819.
10.Kerridge, J. F., and Anders, E. 1988, in <u>Meteorites and the Early Solar System</u>, eds. J. F. Kerridge and M. S. Matthews (Tucson: University of Arizona Press), p. 1149.
11.Anders, E., 1989, Nature, 342, 255.
12.Hoyle, F.W., and Wickramasinghe, C. 1984, <u>From Grains to Bacteria</u> (Cardiff: University College Cardiff Press).
13.Moore, M.H., and Donn, B. 1982, <u>Ap. J. (Letters)</u> 257, L47.
14.Snyder, L.E. 1982, <u>Icarus</u>, 51, 1.
15.Snyder, L.E. 1986, <u>A. J.</u>, 91, 163.
16.Crovisier, J. 1985, <u>A. J.</u>, 90, 670.
17.Despois, D., Crovisier, J., Bockelée-Morvan, D., Schraml, J., Forveille, T., and Gérard, E. 1986, <u>Astr. Ap.</u>, 160, L11.
18.Schloerb, F. P., Kinzel, W. M., Swade, D. A., and Irvine, W. M. 1986, <u>Ap. J. Letters)</u>, 310, L55.
19.Winnberg, A., Ekelund, L., and Ekelund A. 1987, <u>Astr. Ap.</u>, 172, 335.
20.Mumma, M. J., Weaver, H. A., Larson, H. P., Davis, D. S., and Williams, M. 1986, <u>Science</u>, 232, 1523.
21.de Pater, I., Palmer, P., and Snyder, L. E. 1986, <u>Ap. J. (Letters)</u>, 304, L33.
22.de Pater, I., Palmer, P., and Snyder, L. E. 1990, in <u>Comets in the Post-Halley Era</u>, eds. R. L. Newburn, M. Neugebauer, and J. Rahe (Dordrecht: Kluwer), in press.
23.Gringauz, K. I., Gombosi, T. I., Remizov, A. P., Apáthy, I., Szemerey, I., Verigin, M. I., Denchikova, L. I., Dyachkov, A. V., Keppler, E., Klimenko, I. N., Richter, A. K., Somogyi, A. J., Szego, K., Szendro, S., Tátrallyay, M., Varga, A., Vladimirova, G. A. 1986, <u>Nature</u>, 321, 282.
24.Neubauer, F. M., Glassmeier, K. H., Pohl, M., Raeder, J., Acuna, M. H., Burlaga, L. F., Ness, N. F., Musmann, G., Mariani, F., Wallis, M. K., Ungstrup, E., and Schmidt, H. U. 1986, <u>Nature</u>, 321, 352.
25.Palmer, P., de Pater, I., and Snyder, L. E. 1989, <u>A. J.</u>, 97, 1791.
26.Snyder, L. E., Palmer, P., and de Pater, I. 1989, <u>A. J.</u>, 97, 246.
27.Eberhardt, P., Krankowsky, D., Schulte, W., Dolder, U., Lämmerzahl,P., Berthelier, J. J., Woweries, J., Stubbemann, U., Hodges, R. R., Hoffman, J. H., and Illiano, J. M. 1987, <u>Astr. Ap.</u> 187, 481.
28.Krankowsky, D., 1990, in <u>Comets in the Post-Halley Era</u>, eds. R. L. Newburn, M. Neugebauer, and J. Rahe (Dordrecht: Kluwer), in press.
29.Snyder, L. E., Palmer, P., and de Pater, I. 1990, <u>Icarus</u>, in press.

The Chemistry of Translucent and High-Latitude Molecular Clouds

Ewine F. van DISHOECK

Dev. of Geological and Planetary Sciences, California Institute of Technology 170-25, Pasadena, CA 91125, U. S. A.

ABSTRACT. The chemical and physical structure of translucent and high–latitude interstellar clouds is reviewed with reference to recent observational results. Current steady–state gas–phase models can explain the measured abundances of simple molecules such as CH, C_2 and OH, but fail to reproduce the observed column densities of more complex molecules such as H_2CO and C_3H_2. The importance of grain–surface chemistry and shock–heated chemistry in these clouds is discussed.

1. INTRODUCTION

Translucent clouds are interstellar clouds with total visual extinctions $A_V^{tot} \approx 2$–5 mag. Not only are these clouds interesting as a class by themselves, but they also provide the important link between the classical diffuse clouds, such as ζ Oph ($A_V^{tot} \approx 1$ mag), and the classical dense clouds, such as TMC–1 ($A_V^{tot} \gtrsim 10$ mag). They can be either isolated small clouds by themselves, or represent the outer parts of dense, dark clouds. The clouds are called "translucent", because they are thin enough that optical absorption line observations can be performed, provided that there is a bright background star located fortuitously behind the cloud. This also implies that radiation can penetrate the clouds, so that photoprocesses play an important role in the chemistry. In these aspects, translucent clouds are similar to the classical diffuse clouds. On the other hand, they are thick enough to permit millimeter emission line observations. Also, the importance of the photoprocesses diminishes toward the center, so that in these aspects the clouds resemble more the dense, dark clouds. An additional motivation for studying translucent clouds is the growing evidence that dense molecular clouds have a "clumpy" structure through which the ultraviolet radiation can penetrate (1). The individual clumps are estimated to be similar in size to the translucent clouds, so that a detailed study of their chemistry may be of paramount importance to understanding the chemistry of larger cloud complexes.

A similar class of clouds is formed by the high–latitude molecular clouds, which were discovered through their millimeter CO emission (2,3). They are also associated with the more extended IRAS 100 μm cirrus emission (4,5). Most of them are very nearby, with distances of 100 pc or less (6,7). Their averaged extinctions from star counts are found to be similar to those of the classical diffuse clouds, $A_V^{tot} \approx 0.5$ mag (8), but it is likely that the extinctions are somewhat higher on small scales, especially in places where CO is detected.

The aim of translucent and high–latitude cloud studies is to: (a) determine the physical structure of these clouds; are the densities and temperatures similar to those of diffuse clouds? (b) investigate the validity of the steady–state chemistry for the abundances of the simplest molecules in these clouds; is there evidence for shock–heated or grain–surface chemistry? (c) bridge the gap between the diffuse and dense cloud chemistries.

Many reviews of the chemistry in classical diffuse clouds have recently been presented (9–14), some of which also briefly discuss the translucent and high–latitude clouds (12–14). The reader is referred to these reviews for more background information, which cannot be presented here due to space limitations.

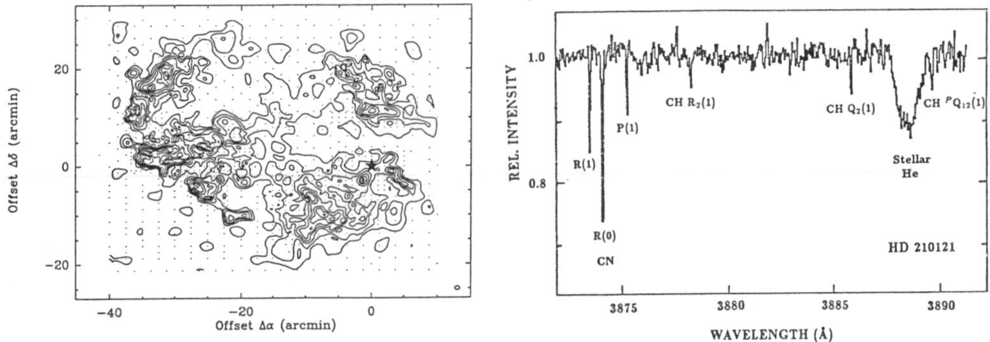

Figure 1. *Millimeter and optical observations toward HD 210121 (from ref. 16) (a) CO 1–0 map with the position of the star indicated; (b) CN and CH absorption lines toward the star.*

2. OBSERVATIONS

The combination of the optical and millimeter observational techniques for the study of translucent and high–latitude clouds has several advantages over studies based on either technique alone (15). Briefly, the big advantages of the millimeter observations are: (i) the lines are in emission, so that mapping of the surroundings is possible; (ii) the high spectral resolution ($\nu/\Delta\nu \geq 10^6$) ensures that the lines are resolved, in contrast with the optical lines; and (iii) a larger variety of molecules, especially the more complex ones, have been observed at millimeter wavelengths. On the other hand, the advantages of the optical observations are: (i) very high angular resolution ($\leq 0.01''$), which ensures that the same material is sampled for all species; (ii) straight–forward derivation of column densities, if the lines are not too saturated; (iii) observations of molecules with great diagnostic value, such as H_2 and C_2, which cannot be done at radio wavelengths; (iv) measurements of lines of atoms in various stages of ionization, which yield directly the amount of depletion of the elements and the electron abundances; and (v) observations of the extinction and polarization, which provide information on the properties of the dust. The big disadvantage of the optical observations is, of course, the requirement of a bright star located fortuitously behind the cloud.

The observational work on translucent and high–latitude clouds has been approached from either side. For translucent clouds, most studies started with the identification of highly–reddened, bright stars ($V \lesssim 8$ mag) for use in optical absorption line studies (15, 17–21 and references cited). About 25 stars, many of which are distant supergiants, have so far been identified which show strong molecular absorption lines. Some of these shine through the outer edges of dense molecular clouds (such as the well–studied star HD 29647, which lies behind most of the Taurus clouds (15)), whereas others appear to lie behind isolated translucent clouds (such as the star HD 169454 (22)). More recently, these optical observations are being complemented by millimeter emission line observations, mostly of CO. Some interesting studies of translucent clouds have also been made by absorption line studies at millimeter wavelengths against bright (extra)galactic continuum background sources (23). An excellent example is provided by the line of sight toward 3C 123, which lies behind a region with $A_V^{tot} \approx 2$–3 mag in the Taurus clouds. This line of sight should therefore be very similar in properties to that toward HD 29647.

The high–latitude clouds have so far been studied mostly by their millimeter CO emission (2,3), but absorption line observations are starting to become available for a few lines of sight (6,7,16,24). Figure 1 shows an example of observations for the high–latitude cloud toward HD 210121.

Table I. Diagnostics of Physical Conditions

Species	Phys. Condition Probed[a]	Diffuse Clouds	High–Latitude Clouds	Translucent Clouds
H_2 low J	T	$+$	$-$	$-$
H_2 high J	I_{UV}	$+$	$-$	$-$
H_2/H	$n_H y_f / I_{UV}$	$+$	$-$	$-$
C_2 low J	T	$+$	$+$	$+$
C_2 high J	I_R / n_H	$+$	$+$	$+$
C, C^+, O J	n_H, T	$+$	$-$	$-$
CO low J	n_H, T	$+^b$	$+^c$	$+^c$

[a] I_{UV}=scaling factor for radiation field at ultraviolet wavelengths; I_R=scaling factor for radiation field at far–red wavelengths; y_f=grain formation efficiency (12).
[b] From ultraviolet absorption line observations.
[c] From millimeter emission line observations.

3. PHYSICAL CONDITITIONS

The physical conditions in translucent and high–latitude clouds can be analyzed along the same lines as for the classical diffuse clouds. Table I compares the diagnostic species available for the various clouds. The biggest difference with the diffuse clouds studied with the *Copernicus* satellite is that ultraviolet observations of the H_2 rotational excitation are currently lacking. This implies that there are no constraints on the scaling factor for the ultraviolet radiation field I_{UV}, which is an important parameter in the models. The most useful probe is probably the C_2 molecule, for which the absorption lines lie around 8750 Å. At these long wavelengths, the star suffers less extinction and the C_2 lines can be conveniently observed from Earth. The C_2 rotational populations result from the competition between collisional processes and radiative pumping at far–red wavelengths. The central temperatures derived from the lower J lines are typically found to be somewhat lower, $T \approx 15$–25 K, and the densities somewhat higher, $n_H \approx 700$–1500 cm^{-3}, in the translucent clouds than in the classical diffuse clouds (21). C_2 has very recently been detected in the high–latitude cloud toward HD 210121 (16). The analysis suggests that the temperature is quite low, $T \lesssim 20$ K, and the density similar to that in the translucent clouds.

Another possible probe of the density is provided by the CO molecule, for which the excitation is dominated by collisions with H and H_2. In diffuse clouds, the CO rotational excitation can be obtained from ultraviolet absorption lines in the A–X band around 1300–1500 Å. For translucent and high–latitude clouds, such data are not yet available, but the CO excitation can also be inferred from millimeter observations of the molecule. In particular, the CO J=1–0/J=3–2 antenna temperature ratio is a sensitive function of the density in the cloud, and the observed ratios result in densities that are similar to, or slightly higher than, those found from the C_2 excitation (25).

In contrast with CO, the excitation of high dipole moment molecules such as CN and CS is governed by electron collisions in these clouds (29). Although their excitation temperatures are found to be very close to that of the cosmic background, the recent very accurate determination of the cosmic background radiation temperature T_{bg}=2.735±0.06 K by COBE (26) can be used to place limits on the electron densities in the clouds (27).

4. MOLECULAR ABUNDANCES

Table II compares the observed column densities for two classical diffuse clouds, the high–latitude cloud toward HD 210121, and the translucent cloud toward HD 169454. Also included are the results of millimeter absorption line observations of the translucent cloud toward 3C 123, and of emission line studies toward the dark cloud TMC–1 (28). Note that the latter are particularly uncertain. The main conclusion to be drawn from this table is that for many species, such as CH, C_2, CN, CO and probably CS, there is a gradual increase in molecular column densities from the diffuse clouds to the translucent clouds. The column densities in

Table II. Observed column densities (in cm^{-2})

Species	Diffuse		High–Latitude	Translucent		Dark
	ζ Per	ζ Oph	HD 210121	HD 169454	3C 123	TMC–1
H_2	$(4.8\pm1.0)(20)$	$(4.2\pm0.3)(20)$	$(6–10)(20)$	$(1.5–2.2)(21)$	$(1–2)(21)$:	$1(22)$:
H	$(6.5\pm0.7)(20)$	$(5.2\pm0.2)(20)$	$\geq1(20)$	$\geq2(20)$	$\leq1.7(21)$	...
CH	$(2.0\pm0.3)(13)$	$(2.5\pm0.3)(13)$	$(3.5\pm0.5)(13)$	$(5.8\pm0.8)(13)$	$7(13)$	$2(14)$
CH^+	$(3.5\pm0.4)(12)$	$(2.9\pm0.3)(13)$	$(6\pm2)(12)$	$(3.0\pm1.0)(13)$	...	...
C_2	$(1.9\pm0.3)(13)$	$(2.4\pm0.3)(13)$	$(8\pm4)(13)$	$(7.0\pm1.4)(13)$	...	...
CN	$(3.0\pm0.3)(12)$	$(2.9\pm0.3)(12)$	$(1.4\pm0.6)(13)$	$(5.5\pm0.6)(13)$	...	$3(14)$
OH	$(4.2\pm0.5)(13)$	$(4.8\pm0.5)(13)$	...	...	$2(14)$	$1(15)$
CO	$(6.1\pm3.0)(14)$	$(2.0\pm0.3)(15)$	$(2–8)(15)$	$(1–5)(16)$	$2(17)$	$8(17)$
CS	...	$(0.7–5)(12)$	$(0.4–8)(13)$	$(0.4–7)(13)$	...	$1(14)$
H_2CO ...	...	...	...	...	$1(13)$	$1(14)$
C_3H_2	...	...	...	...	$4(12)$	$5(13)$
A_V (mag)	1.0	0.8	1.1	3.3	2–3	15:
Ref.	12	12	16,24	22	23	28

the high–latitude cloud fall in the middle, and do not appear anomalous. The CO column density increases more steeply than that of other species, which is well understood from the theoretical models (31). The CH^+ column density is surprisingly small toward HD 210121 (24), and suggests that if shocks are responsible for the formation of this ion, they affect the chemistry to a smaller extent in this high–latitude cloud than in the classical diffuse clouds.

Table II illustrates that care has to be taken in the comparison of molecular abundances in diffuse and translucent clouds, especially with respect to claims of "normal" or "abnormal" abundances. Because the concentrations of many species change drastically with depth (e.g. the fraction of carbon in CO can change from less than 1% to 100%), the only correct method is to compare the total observed *column* densities. In contrast with dense clouds, column density ratios cannot in general be equated with local density ratios in the center of the clouds. Also, the abundances of some species like CH are large only in the outer zone of the cloud, where photoprocesses play a role. Thus, the column densities of these species do not continue to grow with cloud thickness, but will level off. Once this limit has been reached, the column density ratio of such a molecule with respect to that of molecules like H_2 or CO will start to decrease. This could lead to incorrect conclusions about the molecule being underabundant. Tabulations of *average* abundances as functions of *average* extinctions determined from e.g. star counts are of little use for comparison with models, since these clouds show structure on very small scales (cf. Figure 1).

5. MODELS

Some of the first chemical models of clouds with $A_V^{tot} \approx 1–4$ mag were developed by Hiroko Suzuki (30), in which she tried to explain the formation of long carbon–chain molecules. Since the fraction of ionized carbon is high in translucent clouds, the insertion of C^+ into chains can efficiently lead to larger chains in these regions, if the photodissociation rates of the chains are low.

Because the relative importance of the chemical processes changes rapidly with depth, proper models of translucent and high–latitude clouds must take the depth dependence of the atomic and molecular abundances into account. In particular, the $C^+ \rightarrow C \rightarrow CO$ transition is governed by the rate at which CO is photodissociated, which in itself is a quite complicated process (31,32). Models which treat these processes in detail can reproduce the abundances of molecules such as CH and C_2 quite well, provided that reasonable parameters are chosen for unknown reaction rates (21). In fact, the observed CH abundance can be used as a measure of the total H_2 column density along the line of sight. The CO abundances, and their rapid increase with total cloud thickness, can also be reproduced in the models provided that the ultraviolet radiation incident on the translucent and high–latitude clouds is not enhanced over the standard background radiation field (31,14). As mentioned in §3, we currently have no observational constraints on this scaling factor I_{UV}.

The observed CN column densities in translucent clouds are somewhat puzzling, since large variations are observed between clouds which otherwise appear chemically very similar. Although the average column densities are well reproduced in the models, the large variations are not yet understood. One possible solution may lie in differences in the amount of ultraviolet extinction: CN photodissociates mostly at $\lambda < 1000$ Å, so that variations in the shape of the extinction curve at the shortest wavelengths can result in orders of magnitude different photorates at the center of a translucent cloud. However, this scenario only works if atomic oxygen is substantially depleted inside the clouds, since it otherwise removes CN more rapidly than photodissociation.

The OH radical has been seen in classical diffuse clouds by its ultraviolet absorption lines, and the measured column densities can be reproduced well with a cosmic ray ionization rate $\zeta_o \approx 6 \times 10^{-17}$ s^{-1}, if the H$_3^+$ dissociative recombination is slow. If a similar value of ζ_o is adopted in the models for the translucent and high–latitude clouds, the OH column densities are found to be quite large. This is due mostly to the reduced radiation field in these clouds compared with the diffuse clouds, which lowers not only the OH photodissociation rate, but also the abundance of C$^+$, the other principal destroyer of OH. OH column densities of a few times 10^{14} cm^{-2} have been found in high–latitude clouds and in the translucent cloud toward 3C 123, which are well reproduced in the models.

The CS molecule has been detected in translucent and high–latitude clouds by its millimeter emission (29), and its abundance can be well understood in models in which the molecule is formed through the reaction of S$^+$ with CH and C$_2$ at low temperatures.

H$_2$CO has been seen by its millimeter absorption lines against the 2.7 K background radiation (33,34), and in absorption toward galactic and extragalactic radio continuum sources (35). The inferred column densities are typically a few times 10^{12} cm^{-2}, although these numbers may need to be revised slightly if excitation by electrons is taken into account. H$_2$CO has not yet been seen in classical diffuse clouds, so it is not known whether such column densities are "abnormal" or not. However, the current models fail to reproduce the observed column densities in the thicker clouds by 2 to 3 orders of magnitude.

C$_3$H$_2$ has been detected in emission in a few high–latitude cores (34) and in absorption toward radio continuum sources (23). The inferred C$_3$H$_2$ column densities are about $(3–5)\times10^{12}$ cm^{-2} for the radio continuum lines of sight through Taurus, and somewhat lower for the high–latitude cores. The model predictions are typically 10^{10} cm^{-2} for a translucent cloud like HD 169454, about 2 orders of magnitude below observations. Both the H$_2$CO and C$_3$H$_2$ column densities would be increased substantially if an additional source of CH$_4$ (or some other hydrocarbon) were available in the clouds, e.g. through grain surface formation or reactions with large molecules. The current models do not produce large amounts of carbon–chain molecules either, mostly because the photodissociation of the smaller chains like C$_3$ is thought to be rapid (36).

The observations of CH$^+$ in high–latitude and translucent clouds add to the mystery of the origin of this molecule. The measured CH$^+$ column densities in translucent clouds vary considerably from cloud to cloud (19), but still continue to increase with total thickness of the cloud. Such an increase would not be expected on the basis of shock models, unless the number of shocked layers also increases with total thickness. On the other hand, some translucent and high–latitude clouds show very little CH$^+$ absorption. The inferred densities for these clouds are on the high side of the observed range ($n_H \gtrsim 1000$ cm^{-3}). Could the CH$^+$ formation be "quenched" at higher densities (19)? Of particular interest are also new high–resolution observations of CH and CH$^+$ which indicate velocity differences of less than 1 km s^{-1} (37, 38), a feature which is difficult to explain in shock models unless a very special geometry is adopted. Finally, there is growing evidence that diffuse, translucent and high–latitude clouds show significant variations in density, column density and/or velocity on scales as small as 0.01 pc. It will be interesting to investigate whether collisions between these "clumps" can be responsible for the CH$^+$ formation.

6. CONCLUDING REMARKS

In summary, it appears that the abundances of simple molecules such as CH, C$_2$ and OH in translucent and high–latitude clouds can be understood with steady–state gas–phase chemical networks, although lingering uncertainties in crucial reaction rates prevent firm conclusions. The abundances of more complicated species such as H$_2$CO and C$_3$H$_2$ cannot be reproduced in the current models, in which the photodissociation rates of these species are high. It may be that grain surface processes are indirectly responsible for the large abundances of these molecules, by enhancing the abundances of hydrocarbons such as CH$_4$. Grains also have an indirect effect on the abundances of other molecules like CN, since they are responsible for the shape of the ultraviolet extinction curve. The only indicator of shock–heated chemistry is the CH$^+$ ion, although current shock models have difficulties in accounting for some of the recent observational data. To what extent these processes can contribute to the abundances of other simple molecules is still an open question.

New information on the chemistry in translucent and high–latitude clouds will hopefully come from high–resolution ultraviolet observations with the Hubble Space Telescope. This should permit direct observation of C, C^+, O, CO, and possibly H_2 and other molecules in these clouds. Additional input will come from ground–based observations of weak millimeter emission lines toward the stars, and from detailed observations of the small–scale structure of the clouds.

REFERENCES

1. Stutzki, J., Stacey, G.J., Genzel, R., Harris, A.I., Jaffe, D.T., and Lugten, J.B. 1988, *Ap. J.*, **332**, 379.
2. Magnani, L., Blitz, L., and Mundy, L. 1985, *Ap. J.*, **295**, 402.
3. Keto, E.R. and Myers, P.C. 1986, *Ap. J.*, **304**, 466.
4. Weiland, J.L., Blitz, L., Dwek, E., Hauser, M.G., Magnani, L., and Rickard, L.J. 1986, *Ap. J. (Letters)*, **306**, L101.
5. Désert, F.X., Bazell, D., and Boulanger, F. 1988, *Ap. J.*, **334**, 815.
6. Hobbs, L.M., Blitz, L., and Magnani, L. 1986, *Ap. J. (Letters)*, **306**, L109.
7. Hobbs, L.M., Blitz, L., Penprase, B.E., Magnani, L., and Welty, D.E. 1988, *Ap. J.*, **327**, 356.
8. Magnani, L. and de Vries, C.P. 1986, *Astr. Ap.*, **168**, 271.
9. Crutcher, R.M. and Watson, W.D. 1985, in *Molecular Astrophysics*, eds. G.H.F. Diercksen, W.F. Huebner and P.W. Langhoff, NATO ASI Series **157** (Reidel, Dordrecht), p. 255.
10. Black, J.H. 1987, in IAU Symposium **120**, *Astrochemistry*, eds. M.S. Vardya and S.P. Tarafdar (Reidel, Dordrecht), p. 217.
11. Dalgarno, A. 1988, *Astro. Lett. and Communications*, **26**, 153.
12. van Dishoeck, E.F. and Black, J.H. 1988, in *Rate Coefficients in Astrochemistry*, eds. T.J. Millar and D.A. Williams (Kluwer, Dordrecht), p. 209.
13. van Dishoeck, E.F. 1990, to appear in *Molecular Astrophysics —A volume honoring Alexander Dalgarno*, ed. T.W. Hartquist (Cambridge University).
14. van Dishoeck, E.F. 1990, to appear in *The Evolution of the Interstellar Medium*, ed. L. Blitz (A.S.P. Centennial Meeting).
15. Crutcher, R.M. 1985, *Ap. J.*, **288**, 604.
16. Gredel, R., van Dishoeck, E.F., Black, J.H., and de Vries, C.P. 1990, in preparation.
17. Cohen, J.G. 1973, *Ap. J.*, **186**, 149.
18. Cardelli, J.A. and Wallerstein, G. 1986, *Ap. J.*, **302**, 492.
19. Cardelli, J.A. and Wallerstein, G. 1990, *Ap. J.*, in press.
20. Federman, S.R. and Lambert, D.L. 1988, *Ap. J.*, **328**, 777.
21. van Dishoeck, E.F. and Black, J.H. 1989, *Ap. J.*, **340**, 273.
22. Jannuzi, B.T., Black, J.H., Lada, C.J., and van Dishoeck, E.F. 1988, *Ap. J.*, **332**, 995.
23. Cox, P., Güsten, R., and Henkel, C. 1988, *Astr. Ap.*, **206**, 108.
24. de Vries, C.P. and van Dishoeck, E.F. 1988, *Astr. Ap.*, **203**, L23.
25. van Dishoeck, E.F., Black, J.H., Phillips, T.G., and Gredel, R. 1990, *Ap. J.*, submitted.
26. Mather, J. *et al.* 1990, *Ap. J. (Letters)*, in press.
27. Black, J.H. and van Dishoeck, E.F. 1990, in preparation.
28. Irvine, W.M., Goldsmith, P.F., and Hjalmarson, Å 1987, in *Interstellar Processes*, eds. D. Hollenbach and H.A. Thronson (Reidel, Dordrecht), p. 561.
29. Drdla, K., Knapp, G.R., and van Dishoeck, E.F. 1989, *Ap. J.*, **345**, 815.
30. Suzuki, H. 1983, *Ap. J.*, **272**, 579.
31. van Dishoeck, E.F. and Black, J.H. 1988, *Ap. J.*, **334**, 771.
32. Viala, Y.P., Letzelter, C., Eidelsberg, M., and Rostas, F. 1988, *Astr. Ap.*, **193**, 265.
33. Magnani, L., Blitz, L., and Wouterloot, J.G.A. 1988, *Ap. J.*, **326**, 909.
34. Turner, B.E., Rickard, L.J., and Lan–ping, X. 1989, *Ap. J.*, **344**, 292.
35. Colgan, S.W.J., Salpeter, E.E., and Terzian, Y. 1986, *Astron. J.*, **91**, 107.
36. van Dishoeck, E.F. 1988, in *Rate Coefficients in Astrochemistry*, eds. T.J. Millar and D.A. Williams (Kluwer, Dordrecht), p. 49.
37. Crawford, I.A. 1989, *M. N. R. A. S.*, **241**, 575.
38. Hawkins, I. and Craig, N. 1989, private communication.

What Molecules Remain to be Seen?

B. E. Turner

National Radio Astronomy Observatory, Charlottesville, Virginia, U. S. A.

We review what species remain to be seen for several types of astrochemistry–thermochemical equilibrium (TE) in Circumstellar Envelopes (CSEs), photochemistry in CSEs, ion-molecule chemistry in cold interstellar clouds, evaporation of grain mantles, and disruption of grains (producing refractory species) near star-forming regions. Since astrochemistry is as yet not very predictive, we regard the identified species more as tests of current models, than as predictions. Several of these test species have recently been detected. We discuss: 1) the detection of CP in IRC10216 and its implications for both TE and photochemistry in CSEs; 2) progress in searches for H_3O^+; 3) the probable detection of C_5O in TMC-1 as predicted by ion-molecule models; 4) the probable detections of D_2CO and NHD_2 as indicators of the importance of grain mantle evaporation and of surface chemistry; and 5) predictions of refractory element species as products of grain disruption, based on observations of several P and Si species and an analysis of their chemistry.

I. CIRCUMSTELLAR ENVELOPES

CSEs offer several advantages over interstellar clouds as test objects for astrochemical models. In their innermost regions, the temperature and density are high enough to assure TE, under which individual reaction rates need not be known. The products of TE flow outward in the expanding shell rapidly enough to "freeze out", i.e. to minimize further reaction by the time they reach the outer envelope. In the outer envelope, photochemistry occurs under the action of a reasonably well understood interstellar UV field. The geometry and dynamics of the CSEs are much better defined than for interstellar clouds. Nevertheless, several aspects of CSE chemistry remain uncertain and must be tested: i) condensation occurs whenever the partial pressure of a species exceeds its vapor pressure; this is much less understood than gas phase TE chemistry; ii) the omission of *any* important species in the TE model may result in incorrect equilibrium compositions; iii) although TE models require only temperature, elemental abundances, and dissociation energies of all relevant species, these are not all known with high precision.

TE Chemistry

Observation and models compare quite well, but with several qualifications, the most important being that predictions are generally available only for O-rich objects, while most observations are of the C-rich envelope IRC10216. We consider the second and third row elements, whose cosmic abundances may be grouped as (Fe,Si,Mg,S), (Al,Ca,Na),(P,K,Ti), each group being $\sim$ 10 times more abundant than the following one. In O-rich objects the most abundant molecular species are predicted (Tsuji 1973) to be FeS, FeO, $Fe(OH)_2$, SiO∗,SiS∗, SiH_4, MgOH†, AlOH∗†, AlF, AlCl, CaOH†, NaCl, PS, KCl, TiO. Here, ∗ denotes a dominant species for the given element, † denotes a very high predicted abundance which is detectable with current technology. In C-rich envelopes the only existing predictions (Tsuji 1987; Turner et al 1989) are SiO∗, SiS∗, Si_2C, HCP, and TiO. Predicted condensates (Tarafdar 1987) are FeS, Fe_2SiO_4, SiO, SiS, $MgSiO_3$, Mg_2SiO_4, Al_2O_3, AlCl for O-rich envelopes, and Fe_3C, SiC, AlN, TiC, TiN for C-rich objects.

Observationally, SiO and SiS dominate the Si species, as predicted. However, SiH_4 is also seen in IRC10216, although it is predicted to be abundant only in O-rich objects. The metallic compounds AlF, AlCl, NaCl, KCl are also observed in IRC10216, at 10 times or so smaller abundance than predicted in O-rich envelopes. CaOH is not detected in either C- or O- rich objects, at levels roughly equal to the predicted abundance. TiO and FeO are not seen either, but the current sensitivities are inadequate to test the models. Species which currently have no measured frequencies are FeS, $Fe(OH)_2$, Si_2C, MgOH, AlOH. These species should eventually provide definitive tests.

The TE models for Si, Al, K, Na thus appear quite satisfactory, when (uncertain) allowance is made for loss of these elements to condensation or subsequent adsorption of their gas phase molecules onto grains. The predictions for Fe, Mg, Ca, Ti are not yet tested.

Phosphorus raises interesting questions. Tsuji (see Turner et al 1990) calculates that all P should be in the form HCP in IRC10216, while searches for HCP establish limits at close to the predicted values. However, CP has recently been detected in IRC10216 (Guelin et al 1990) at an abundance level $CP/P \sim$ 0.04 and with a source size < 12" diameter as deduced from line profile shapes. It is possible that most of the HCP is adsorbed onto grains, with $\sim 4\%$ photodissociated to CP at radius $r = 6$". By contrast, however, HCN is not believed to be heavily adsorbed. The main question is, what happens to the CP at r > 6"? Note that CN, a photodissociation product of HCN, is highly abundant throughout the envelope of IRC10216. Based on interferometric observations and on line- profile information, one deduces that:
- HCN dissociates to CN at r = 6"; CN persists in the outer envelope.
- HCP dissociates to CP at r $\leq$ 6"; CP is somehow lost almost immediately.
- SiS dissociates at r = 12".
- AlF, AlCl do not dissociate at r = 12" (source diameter > 24").

Polyatomic species (HCN, HCP) are expected to be most easily photodissociated, covalent-bonded diatomics (SiS) to be less easily photodissociated, and ionic-bonded diatomics (AlF, AlCl) to be most resistant to photodissociation. The observed photodissociation radii fit these expectations. The fate of CP remains a mystery.

Photochemistry of the Outer Envelope.

The volatile first-row products of TE chemistry are predicted to be predominantly CO, C_2H_2, HCN, N_2, CH_4 for C-rich envelopes, and H_2O, CO, N_2 for O- rich envelopes. These species are the progenitors of the photochemistry that dominates the outer envelope. In IRC10216, photo- and ion-chemistry explain satisfactorily such species as CN, HNC, but produce much too little HC_3N, C_3N, and other cyanopolyynes. The observed similar distribution of C_3N and HC_3N suggests that they have a common precursor, HC_3NH^+, formed by the reaction $C_2H_2^+ + HCN$. If so, HC_3NH^+ may be detectable in IRC10216, though only with a large telescope. The recent consensus that HC_3N does not occur in the inner envelope of IRC10216 (Bieging, priv. comm.) obviates the need for any neutral-neutral reactions such as $CN + C_2H_2$ or $C_2H + HCN$, the first of which has recently been invoked in the interstellar context to help explain the HC_3N abundance. Presence of such reactions in the inner CSE context would imply a central source of UV such as a warm chromosphere. The presence of CP and not HCP might similarly be explained. Absence of a central HC_3N distribution also precludes grain surface reactions such as $C_2H_2 + HCN \rightarrow HC_3N$ in the inner envelope. It is noteworthy that the highly disparate interstellar and CSE pathways to HC_3N, ($C_nH_m^+ + N$ and $C_2H_2 + HCN$ respectively) both fail by large measure to produce the observed HC_3N abundances. The cyanopolyyne problem remains one of the most vexing in all of astrochemistry.

The photochemistry of C_2H_2 explains the C_nH family quite well, except for C_2H, which is observed with 10 times smaller abundance than predicted from the photodissociation of C_2H_2. Ions such as $C_2H_2^+$, $C_2H_3^+$ are expected to produce C_3O by reaction with CO, but recent searches for C_3O establish limits below expected levels. These results suggest a depletion of C_2H_2 itself in the outer envelope, by unknown mechanisms (adsorption onto grains is unlikely). The photochemistry of CH_4 should lead predominantly to CH_2 and should also produce detectable amounts of H_2CCO via reaction of CH_3^+ with CO. H_2CCO is not seen to levels below those predicted, while searches for CH_2 are hampered by a small dipole moment. Searches with large telescopes should, however, detect CH_2, even if the CH_4 abundance reflects only TE formation (cf. Nejad and Millar 1987). Additional CH_4 may form outside the TE zone by catalytic reactions on grains, as might C_2H_4, which should be searched in the IR to ascertain the possible role of grain reactions.

Si chemistry in the outer envelope is of current interest because of recent detections of SiC_2, SiC, SiC_4. Silane (SiH_4) is observed in the inner envelope in quantities greater than expected under TE, and some catalytic formation on grains may be indicated. SiH_4 is the only Si species volatile enough to reach the outer envelope in significant quantities in gas phase. It photodissociates primarily to SiH_2, which should be detectable. Searches are planned. SiH would be only a minor constituent. However, Si^+ is apparently needed in reaction with C_2H_2 to form the SiC_n family (of which SiC is a photoproduct). It is believed that SiS photoionizes to produce the Si^+. This scheme is consistent with the observed ratios $SiS/Si \sim 0.03$, $SiC_2/Si \sim 0.01$ in the outer envelope. Reactions such as $SiC_nH_m^+ + C_2H_2 \rightarrow SiC_{n+2}H_{m+1}^+$ may form the higher-order members of the SiC_n family, though with decreasing abundance. Linear versions of SiC_2, SiC_3, SiC_5 may all occur (Ohishi et al 1990).

The recent detections of C_3, C_5 likewise suggest many other members of the C_n family. The observed abundance of C_3 agrees well with the ion-molecule prediction of Nejad and Millar (1987) who also predict comparable amounts of C_4. C_5 and higher order C_n are not expected in significant quantities unless photolysis of large C-clusters is involved.

II. INTERSTELLAR ION-MOLECULE CHEMISTRY IN COLD CLOUDS

It is a fundamental prediction that nearly all complex interstellar molecules ($n \geq 3$ atoms) have protonated precursors. So far, the observed molecular ions (all protonated species) are HCO^+, N_2H^+, HCS^+, $HOCO^+$, $HCNH^+$, H_3O^+ (?), and H_2D^+ (??). Not seen despite sensitive searches are H_3S^+, HC_3NH^+, CH_3CNH^+, $HOCS^+$, $HSiO^+$, H_2Cl^+, H_2COH^+. No other protonated species have available microwave spectra at this time.

H_3O^+ is central to the entire oxy-hydride family in ion-molecule chemistry. Early searches in 1986 detected a line in Ori(KL) and SgrB2 coincident with the $J,K = 2,1 \rightarrow 1,1$ transition (at 90 K excitation), but this line could also be identified with a high-lying torsionally excited CH_3OH transition. Recent searches by Wootten et al (1989) for the $3,2 \rightarrow 2,2$ transition of H_3O^+ at 364.8 GHz have detected a line in SgrB2 at about the expected intensity, but not in Ori(KL). It is not surprising that SgrB2 shows this higher-energy line, as it has the largest known column densities, is known to possess regions of high excitation, and is the only known source of $HOCO^+$. These results may also be consistent with H_3O^+ in Ori(KL), because the 307 GHz line occurs at $v = 8$ km/s, and thus arises in the "compact ridge" component where the kinetic temperature is $\leq$ 90K, insufficient to excite the 365 GHz transition at an energy of 150 K. The abundance of H_3O^+ deduced from the 307 GHz line alone is consistent with model predictions. A final confirmation of H_3O^+ awaits detection of the $3,0 \rightarrow 2,0$ transition at 396 GHz (177 K excitation). Plans are underway.

Complex carbon-chain species such as the C_nH, C_nN, C_nO, C_nS, HC_nN families, are a hallmark of ion-molecule chemistry in cold clouds. Until recently C_3O was the only known member of the C_nO group, but recent laboratory studies of the reactions of hydrocarbon ions with CO (Adams et al 1989) have predicted that the families C_nO, HC_nO, H_2C_nO will have detectable abundances in a cold cloud such as TMC-1 for n up to at least 6. Based on ab initio calculations, Turner et al (1990c) have recently made a probable detection of C_5O in TMC-1. Three lines have been seen, identified with the $J = 7-6, 8-7,$ and $9-8$ transitions, and yielding constants $B_0 = 1354.870$, $D_0 = 2.3544 \times 10^{-3}$, $H_0 = 5.188 \times 10^{-6}$ MHz. The need for a higher-order centrifugal distortion constant (H_0) for a satisfactory fit may cast doubt on the identification (although the well-known terrestrial species $O=C=C=C=O$ is also fit with an H_0 term). Confirmation will require detection of the $J = 4-3$ transition at 13.5 GHz, and a search is planned. Assuming C_5O, the observed abundance is $C_5O/C_3O \sim 0.2$, consistent with prediction given the uncertainty of the number of H atoms lost in the dissociative recombination (DR) of $C_3H_3O^+$ and $C_5H_2O^+$, the dominant precursor ions. Assuming that C_5O is confirmed, it is predicted that C_4O, HC_4O, H_2C_4O, C_6O, HC_6O, and possibly C_7O will also be detectable in TMC-1, once frequencies are known. Ratios such as $C_4O/HC_4O/H_2C_4O$ will tell us much about DR processes.

Other recent tests of ion-molecule chemistry in cold clouds involve deuteration of complex hydrocarbons. The recent detection of C_4D (Turner 1989) leads to a dilemma, given the reactions believed relevant in the chemistry of C_3H_2, HC_3N, and C_4H (see Fig. 2 of Turner 1989). These reactions predict that $R(C_4H)/R(HC_3N)/R(C_3H_2) = 1/3/6$, while the observed ratios in TMC-1 are $1/3.5/18.6$ ($R = XD/XH$). Both HC_3N and C_3H_2 are formed from $C_3H_3^+$, while C_4H is now believed formed from C_3H_2 via $C^+ + C_3H_2 \rightarrow C_4H^+$ followed by reaction with H_2. Thus a large value of $R(C_3H_2)$ should produce a large value of $R(C_4H)$, contrary to observation, unless the reactions leading from C_3H_2 to C_4H favor H over D. This would be contrary to the Gellene-Porter effect, invoked to explain the large value of $R(C_3H_2)$, in which the reaction $C_3H_2D^+ + e$ preferentially detaches H atoms. On the other hand, it is now believed that DC_3N cannot arise from the usually invoked reaction $H_2D^+ + HC_3N \rightarrow DC_3N$ because the HC_3NH^+ ion is now known to be linear, so that $H_2D^+ + HC_3N$ produces HC_3ND^+ with D at the wrong end (Millar 1989, priv. comm). Thus DC_3N must arise from $C_3H_2D^+ + N \rightarrow DC_3NH^+$, from which one deduces $R(HC_3N) = 1/6$ $R(C_3H_3^+)$. This demands a *large* value of $R(C_3H_3^+)$ in order to produce the observed value of $R(HC_3N)$, and would explain the large observed value of $R(C_3H_2)$, but not the smaller observed $R(C_4H)$. The dilemma remains unresolved in terms of currently adopted reactions. Observations of deuterated species can also help to decide which of several possible routes lead from C_4H to higher C-number hydrocarbons. For example, the sequence $C_4H + C^+ \rightarrow C_5^+$ will suppress the deuteration of C_5 species, and the reaction $C_5H^+ + C \rightarrow C_6^+$ suppresses the deuteration of C_6 and higher-order species, while the process $C_4H_2^+ + C_2H_2 \rightarrow C_6H_3^+$ retains a significant deuteration of C_6 and higher species. Deuteration reactions thus remain important diagnostics in our understanding of ion-molecule reactions.

HC_3N and the cyanopolyynes continue to pose additional problems for ion-molecule chemistry in addition to the difficulty of explaining their large abundances. Standard ion-molecule models predict detectable abundances of HC_3NH^+ and possibly of CH_3CNH^+, but recent searches for these and the $HOCS^+$ ion have been negative (Turner et al 1990a). The observed limit for HC_3NH^+/HC_3N is about 4 times lower than predicted, while an earlier detection of $HCNH^+$ gives a ratio $HCNH^+/HCN$ several times higher than predicted. The disparity in these ratios, which are predicted to be about equal, appears to argue against reactions of the type $HC_nNH^+ + C_2H_2$ or $HC_nN + C_2H_2^+$ for forming higher-order cyanopolyynes from lower order ones, and to favor reactions of the type $C_nH_m^+ + N$ even though laboratory work suggests that these are inefficient for n > 3 or 4. There is an additional problem. The so-called "low reaction rates" for certain ion-polar reactions in the current models can just explain the observed low HC_3NH^+/HC_3N ratio (but not the high $HCNH^+/HCN$ ratio, which suggests an unknown but efficient destruction for HCN). However, the low rates then cannot produce the observed high HC_3N abundance. Put another way, no matter how HC_3N is formed, it is difficult to see how it fails to be protonated by reaction with HCO^+ and other ions, yielding a ratio HC_3NH^+/HC_3N larger than the current observed upper limit. A large destruction rate for HC_3NH^+ (which does not lead to HC_3N) seems most likely. The reaction $HC_3NH^+ + C_2H_2 \rightarrow$? is probably fast enough at low temperatures, but must lead to products other than HC_5N in order not to produce too much HC_5N. Problems with the cyanopolyyne family thus remain numerous and unresolved.

The larger species are in general much less well understood than smaller molecules. Radiative associative reactions between large fragments are generally believed responsible, now that it is recognized that such reactions can proceed efficiently at low temperatures if there are no activation barriers or too many competing exit channels. Thus, abundant ions such as CH_3^+, CH_5^+, H_3O^+, $C_2H_2^+$, $C_2H_3^+$, $C_4H_2^+$, $C_4H_3^+$, HCO^+, H_3CO^+, NH_3^+ can react with abundant neutral species such as CO, H_2O, H_2CO, C_2H_2, CH_4, C_2H_4, CH_3OH, C_3H_2, C_4H_2 to produce a large variety of complex products. We may compare expected products with species observed in the cold TMC-1 cloud as follows:
- observed species explainable by radiative association reactions: CH_3OH, CH_3CHO, H_2CCO, C_3O, C_5O, HC_2CHO, $HCOOH$, CH_3C_2H, CH_3CN, $HNCO$, C_2S, C_3S, H_2S.
- observed species not explainable by radiative association reactions: CH_2CN, HC_3N, VyCN, HC_5N, HC_7N
- species not observed: VyOH, CH_3OHCO, NH_2CHO, EtCN
- species not yet searched which should be seen: EtOH, $(CH_3)_2O$, $(CH_3)_2CO$, NH_2CN.

Note that all species in the second catagory contain N, perhaps suggesting the importance of the $C_nH_m^+ + N$ reactions in forming complex nitrocarbon species. Of the species in the third catagory , the absense of VyOH suggests the reaction $H_3O^+ + C_2H_2$ has a competing exit channel, since both H_3O^+ and C_2H_2 should be abundant. Absence of CH_3OCHO and NH_2CHO is less surprising, since the reactants (H_3CO^+, NH_3^+) have low abundances. EtCN is a special case. It apparently cannot be formed by any radiative associative or other ion-molecule reaction, and therefore its large abundance in *warm* molecular clouds suggests formation on grain surfaces.

III. GRAIN PROCESSES (EVAPORATION)

Other, more direct evidence of grain processes is the observation in several warm star-forming regions of copious amounts of HDO and NH_2D. Such large deuteration cannot have occurred in the gas phase at the current high temperatures, but may have arisen at an earlier, cooler phase and subsequently been frozen onto grains which are now evaporating. Large amounts of other species (HCN, CH_3OH, EtCN) in the Orion hot core also support this conclusion.

A prediction arising from the role of grain mantle processing is that *doubly deuterated* species will be efficiently formed. In fact, assuming that surface reactions actually occur in addition to the role of grains as storage units, one expects $(XD_2/XH_2)_{grain} > (XD_2/XH_2)_{gas}$ for several reasons: 1) the sticking coefficient for D is larger than for H; 2) single-D species on grains do not lose D in subsequent reactions; 3) D abstraction by H or D atoms from D-species has higher activation barrier than H abstraction from H-species. Thus the highly frequent abstraction reactions that occur on surfaces insure the continued buildup of deuterium. Tielens and Allamandola (1987) actually predicted $NHD_2/NH_3 \sim 1$ under some conditions. Other models, however, obtain a much smaller ratio, $\sim 3 \times 10^{-5}$ (Brown and Millar 1989).

We have made recent detections of both D_2CO and NHD_2 in the Orion hot core (Turner 1990) which appear to certify the importance of grain mantle processing in addition to simple freeze-out onto grains. The observed abundances, $NHD_2/NH_3 \sim 3 \times 10^{-3}$, $D_2CO/H_2CO \sim 8 \times 10^{-2}$, lie between the two extremes of little or no surface processing, and highly efficient surface abstraction reactions.

Given these results, it seems clear that D_2O will also be produced at detectable levels, although abstraction reactions are believed to be less efficient for H_2O than for H_2CO or NH_3. Singly-deuterated complex species such as D-EtCN, D-EtOH should also occur, as follows. If the abundance of gas phase atomic hydrogen, X(H), is large, then H_3C- is abundant on surfaces and species such as H_3C-O-CH_3 form readily. Large X(H) corresponds to small $(D/H)_{gas}$, so insignificant deuteration of $(CH_3)_2O$ is expected. But if X(H) is smaller, then C-, HC-, H_2C- are more prevalent on surfaces, and will participate in *linear* growth in two directions, e.g. C-C-O (Millar et al 1988), or C-C-N, leading to EtCN. Smaller X(H) means larger $(D/H)_{gas}$ so significant deuteration of EtOH, EtCN may occur. The presence of EtCN but not of $(CH_3)_2O$ in the Orion hot core suggests the latter, small X(H) case at the time of grain mantle processing. A detection of D-EtOH, D-EtCN, and perhaps D-VyCN would strongly suggest these complex species are formed mainly on grains rather than by gas- phase radiative association reactions. Efforts to measure laboratory frequencies for these species are needed.

IV. REFRACTORY-ELEMENT SPECIES: DISRUPTION OF GRAINS IN STAR- FORMING REGIONS

Our current understanding of refractory-element astrochemistry comes from laboratory studies and observations of only two elements: Si and P. Nevertheless, an interesting picture emerges.

The interstellar ion-molecule chemistry of P as developed by Thorne et al (1984) and Millar et al (1987) predicts that $P/PN/HPO = 0.77/0.20/0.023$ with all other species negligible. These predictions are borne out by observations of HPO, HCP, PH_3 and PN by Turner et al (1990b). The observed PN and HPO abundances indicate a depletion factor $\delta(P)$ for P of 1000 in dense, star-forming molecular clouds, and $>$ 10^5 in cold clouds. Similarly, the gas-phase chemistry of Si (Herbst et al 1989) predicts that $SiO/Si \sim 0.98$, with no other Si species significant. Recent searches by Turner et al (1989) for $HSiO^+$ are consistent with this picture. The observed abundances of SiO then give $\delta(Si) = 1750$ in warm clouds, $\geq 10^6$ in TMC-1. Thus $\delta(P) = \delta(Si)$ within the uncertainties.

The observed depletions of the refractory elements are given in table 1. Those for the diffuse ISM are taken from the literature, those for dense clouds from the above-mentioned study of the chemistry.

Table 1. Depletions of Refractories ($\log X/X_{solar}$)					
Cloud Type	S	Mg	P	Si	Ca
diffuse (warm)	0	-0.2	-0.09	-0.6	-1.2
diffuse (cold)	?	-0.7	-0.7	-1.6	-3.6
dense (warm)	-1.7	?	-3.	-3.2	?
dense (cold)	-1.7?	?	<-5.	<-6.	?

To explain these depletions, we first note that interstellar dust grains cannot *condense*, they can only accrete. Thus the grain "seeds" must originate from the material condensed in CSEs. This condensed material has the form $MgSiO_3$, Mg_2SiO_4 for Si and Mg; Al_2O_3 for Al; HCP or CP (gas phase), or Fe_3P (solid) for P; $CaAl_2Si_2O_8$ for Ca; and $NaAlSi_2O_8$ for Na (Fegley and Lewis 1980). The silicates of Al, Ca, Ti as well as Al_2O_3 condense at the highest temperatures, nearest the star, and subsequently grow by becoming clad by more abundant Mg/Fe silicates at lower temperatures, further from the star. After ejection into the ISM, photoerosion first releases the outer layers of the grains, explaining why Si and Mg are less depleted than Na, Ca, and Al. The small depletion of P may suggest that it resides mostly as HCP on grain surfaces in CSEs, and is easily photodesorbed (and dissociated) in the diffuse ISM.

The undepleted fraction of the refractory elements combine with plentiful H atoms on grain surfaces to produce saturated hydrides. Those of Si, Mg, P, Al are covalently bonded and easily photodesorbed and photodissociated in the diffuse ISM, but when they enter dense clouds the hydrides will remain on grain surfaces, where some portion will be converted to hydroxides assuming that O is the most mobile surface atom other than H. The remaining free atoms will form PN, SiO and unknown species of Mg, Al. These heavier species will be strongly adsorbed on grains. Near energetic star-forming regions at least the hydrides and possibly the hydroxides will be desorbed, then ionized by cosmic rays (or UV). Reactive gas-phase products are produced, which, according to our understanding of the ion-molecule processes, will form PN (20% of the total P) and SiO (98% of the Si). The failure to observe PN or SiO in cold clouds (TMC-1) then implies that not only PN and SiO, but also the hydrides PH_3, SiH_4 are more strongly adsorbed on grains than are volatile species such as H_2O and NH_3, which explains why O, N are little depleted even in dense cold clouds. On this picture a general prediction may be made, viz., that $\delta(Si) \sim \delta(P) \sim \delta(Mg) \sim \delta(Al)$. Near star-forming regions, therefore, *compounds of Mg and Al are expected to occur in detectable quantities* so long as only one or two dominant compounds of each element are formed. Current observations rule out MgO, MgS, and MgH, while searches for AlO are pending. Laboratory studies of the gas-phase chemistry of Mg and Al are urgently needed to establish their dominant species.

Ca differs from the above-mentioned elements in that CaH_2 has ionic bonding and is unlikely to photodesorb from grains even in diffuse clouds. As a permanent resident on grains, it is eventually converted entirely to $Ca(OH)_2$, which is likely difficult to desorb even near star-forming regions, unless the grains are totally disrupted, contrary to the observations. Thus *Ca compounds are much less likely than those of*

Mg, Al to be detected in a star-forming region like Ori(KL). So far, CaH and CaOH have been searched unsuccessfully.

Na occurs as Na^+ in diffuse clouds and does not react with H on grains, owing to its inert-gas electronic structure. Tielens and Allamandola (1987) argue that Na will probably desorb readily from grains under the action of recombining with electrons on the grain surfaces. Gas-phase Na^+ can form NaH by radiative association with H_2, but NaH reacts rapidly with O to form NaO. In dense clouds NaO is probably strongly adsorbed onto grains, where it may convert partly to NaOH. Searches for both NaO and NaOH as well as atomic Na have been unsuccessful. If both NaO and NaOH are strongly bound, Na (like Ca) may not be capable of a gas-phase existence unless the grains are totally disrupted.

Transition metals such as Fe, Ti are likely to be strongly adsorbed to grains because their unfilled d orbitals can form strong complexes with electron-rich material (Tielens and Allamandola 1987). They may form mono- or dihydrides on grain surfaces but these are expected to be strongly bound also. Thus Fe, Ti compounds, like those of Ca, Na, should be hard to detect, in contrast to those of Si, P, Mg, Al.

REFERENCES

Adams, N.G., Smith, D., Giles, D., and Herbst, E. 1989, A.&A. 220,269

Brown, P.D., and Millar, T.J. 1989, M.N.R.A.S. 240, 25P

Fegley, B., and Lewis, J.S. 1980, Icarus 41, 439

Guelin, M., Cernicharo, J., Paubert, G., and Turner, B.E. 1990, Nature, in press

Herbst, E., Millar, T.J., Wlodek, S., and Bohme, D.K. 1989, A.&A. 222, 205

Millar, T.J., Bennett, A., and Herbst, E. 1987, M.N.R.A.S. 229, 41P

Millar, T.J., Olofsson, H., Hjalmarson, A., and Brown, P.D. 1988, A.&A. 205, L5

Nejad, L.A.M., and Millar, T.J. 1987, A.&A. 183, 279

Ohishi, M., Kaifu, N., Kawaguchi, K., Murakami, A., Saito, S., Yamamoto, S.,
  Ishikawa, S., Fujita, Y., Shiratori, Y., and Irvine, W.M. 1989, Ap.J. (Letters) 345, L83

Tarafdar, S.P. 1987, in ASTROCHEMISTRY, Eds. M.S. Vardya and S.P. Tarafdar,
  D. Reidel

Thorne, L.R., Anicich, V.G., Prasad, S.S., and Huntress, W.T. 1984, Ap.J. 280, 139

Tielens, A.G.G.M., and Allamandola, L.J. 1987, in INTERSTELLAR PROCESSES,
  Eds. H.A. Thronson and D.J. Hollenbach, Astrophysics and Space Science Library

Tsuju, T. 1973, A.&A. 23, 411

Turner, B.E. 1989, Ap.J. (Letters) 347, L39

Turner, B.E. 1990, in preparation

Turner, B.E., Amano, T., and Feldman, P.A. 1990a, Ap.J. (Jan. 20 issue)

Turner, B.E., Tsuji, T., Bally, J., Guelin, M., and Cernicharo, J. 1990b, Ap.J.submitted

Turner, B.E., DeFrees, D.F., and McLean, A.D. 1990c, in preparation

Wootten, H.A., Turner, B.E., and Watt, G. 1990, in preparation

New Developments in Radio Astronomy in the Submillimeter-Wave Region

Henry E. Matthews

Joint Astronomy Centre, 665 Komohana Street, Hilo, Hawaii 96720, U. S. A. and Herzberg Institute of Astrophysics, National Research Council of Canada, Ottawa, Ontario K1A 0R6, Canada

Submillimeter-wave astronomy is a technically demanding discipline which has seen the recent construction of several large ground-based facilities and major advances in receiver technology. The status of the field is reviewed with attention being paid to the contributions to astrophysics and to the major difficulties facing the observer at these wavelengths. The results of surveys for molecular species, and observations of dense, warm gas and of cold dust are discussed. Both continuum and spectral lines have great potential at these wavelengths particularly for the investigation of protostellar objects, and of the late stages of stellar evolution. Prospects for the future development of submillimeter astronomy are bright; the next years should see a wealth of new information obtained in this wavelength range.

Introduction

The submillimeter-wave region occupies an interesting niche in the spectra of astronomical objects. Cool (<30 K) dust is particularly visible at wavelengths less than 1 mm, and at the shortest wavelengths accessible to ground-based telescopes (about 300μm) reaches the peak of its emission curve. High-excitation rotational lines of common molecules such as CO and HCN allow the cores of dense clouds to be studied with great facility. Although heavier molecules are of diminishing importance as the frequency increases, important diatomic species such as the metal hydrides have their lowest transitions in this band, and vibrationally excited transitions of species commonly seen at lower frequencies should be observed. At the high end of the ground-based window vibration-rotation lines of polyaromatic hydrocarbons and other large molecules should be observable. Phillips[19] has succinctly summarized this information in graphical form in a recent article. In addition, all classes of solar system bodies including comets and asteroids become observable at sub-millimeter wavelengths, and provide a potential wealth of information about their surface layers.

In this review, I will confine myself to frequencies greater than 300 GHz, corresponding to wavelengths of 1 mm and less. Since I work with the James Clerk Maxwell Telescope this will also introduce a certain natural bias to the material presented; however, what is to be said applies largely to other similar facilities as well. Furthermore, since the greatest advances in recent years have been in large ground-based facilities, most of my subject matter will be in reference to them, and related developments.

There are already in print a number of excellent discussions of various aspects of submillimeter astronomy. Three recent symposia have a high content in this regard: the Stirling Summer School[33], the Zermatt meeting[31], and the Kona conference[28]. There are a number of other articles; for instance, the one by Payne[18] is particularly good on technical matters.

Technical Developments

Overcoming the Atmosphere

For the most part, sub-mm signals from extraterrestrial sources are weak in comparison to other, local, sources of noise. Astronomical signals must compete with much stronger ones which are (a) internally generated by the amplifier/receiver system, (b) resulting from losses due to the antenna structure and its imperfections, and (c) due to the atmosphere. The first and second noise sources provide a powerful driving force behind the rapid pace of the development in both receiver and antenna technology, as does the need to detect ever weaker signals. As necessary as it is to provide improvements in technology, the major difficulty in making useful (*i.e.* calibrated) astronomical observations at sub-mm wavelengths is due to the presence of the atmosphere. The steady improvement in receiver sensitivity will only serve to accentuate this problem.

The atmosphere leads to three sources of difficulty: (a) it effectively contributes to the receiver noise by radiation; (b) signals from astronomical sources are diminished by their passage through it; and (c) its effects are both temporally and spatially variable. Local masses of wetter air lead to 'anomalous refraction' [2], which affect the direction from which the radio waves come and hence the observed signal strength, and rapid changes in optical depth lead to considerable difficulty in obtaining reliably calibrated signals.

The main variable component of the atmosphere is water vapour, which has a considerable number of transitions of various strengths throughout the sub-mm window, and which even from the highest mountain sites effectively reduces the atmospheric transmission to zero for wavelengths shorter than about 300μm. For example, from Mauna Kea, at an altitude of 4200m, the minimum observed precipitable water vapour pressure is about 0.4mm. The best transmission which can be obtained at $\lambda350\mu$m is about 40% and this is achieved on only very few nights; 'useful' nights (when the transmission is perhaps 20% at these wavelengths) occur about 30% of the time, and even then, in no very predictable way (not at least in the past year or two). In addition, our experience suggests that the driest and clearest weather is also likely to be the most variable. For instance, results from a recently installed atmospheric monitor operating at 225 GHz at the Caltech Submillimeter Observatory show that changes in opacity of a factor of two or more in less than one hour can occur relatively often.

One counters the difficulty imposed by a variable atmosphere by a combination of observing techniques. For broad-band detection in particular, the ability to switch rapidly (at several Hz) between the source and a nearby reference position is essential; the resulting difference signal is then synchronously detected. Frequent calibration and pointing measurements on sources of known strengths are required, and a system of continuously providing an internal relative calibration, say, via a chopper wheel, is essential, if meaningful observations are to be made. Similar techniques, with less rapid switching, have to be used in observing spectral lines. Despite this, we are only just beginning to obtain an understanding of both the atmosphere and our calibrators, and much work remains to be done.

Advances in Telescope Technology

Without a doubt, the greatest advance in sub-mm astronomy in the past few years has been the inception of a number of large ground-based telescopes which are intended for, or which can be operated at, sub-mm wavelengths. Essential parameters of these are summarized in Table 1. The goal is to achieve the best surface accuracy over the largest possible area, with the results that the telescope beam (HPBW) is well-defined to high frequencies. In addition, since the atmosphere is a major limiting factor, placing the telescope at altitude is also desirable. The latter brings with it the additional expense and logistical concerns of operating complex devices in remote and occasionally hostile environments. In this table, we have somewhat arbitrarily defined a shortest wavelength at which each telescope

Table 1: Operating† Sub-mm-wave Telescopes

	Diameter (m)	Surface rms (μm)	Minimum λ (mm)	HBPW ($''$)	Altitude (m)	Location and ref.
IRAM	30	85	0.8	8	3300	Pico Valeta [3]
JCMT	15	30	0.3	15	4000	Mauna Kea [12]
SEST	15	75	0.6	15	2400	La Silla [6]
CSO	10.4	35	0.35	23	4000	Mauna Kea
NRAO	12	65‡	0.7	20	1900	Kitt Peak
KOSMA	3	30	0.35	70	3150	Gornergrat [32]
SMT†	10	15	0.3	23	2800	Mt. Graham [4]

† Using a shaped secondary mirror (subreflector)
‡ The SMT is under construction; these are design goals

may usefully be operated. This is primarily a reflection of the surface rms accuracy. There comes a point, as wavelength decreases, at which the proportion of power in the main peak (the 'beam') of the response pattern of the antenna becomes so small, and additionally, the beam shape less well defined, that the antenna is no longer useful for observations.

The rms surface accuracy pertains to the complete aperture of the antenna. The typical surface accuracy of individual panels is considerably better; for instance the panels on the JCMT have rms accuracies of about 10–15μm. The panels in this case are fabricated of aluminum honeycomb structures, with a final very thin layer of pure aluminum providing the reflecting surface[12] .

For the moment, setting the panels in the correct positions, and maintaining the correct settings provides the major challenge, rather than panel accuracy itself. The basic structure of most modern altazimuth telescopes rests on homologous design principles, in which deformations under gravity result in predictable changes as a function of elevation. Use of these construction principles allows one to exceed the normal expectations of steel construction techniques. Thermal variations appear to provide the most critical barrier to further improvements in telescope surface accuracy and pointing control. Keeping such variations to less than 1°C requires either the controlled circulation of air around an enclosed surface support structure (as in the case of IRAM[3] and SEST[6]), or the enclosure of the complete telescope structure, as in the case of CSO. The JCMT is also closed across the viewing aperture by means of a large removable membrane essentially transparent to submillimeter waves; this has the added advantage that the JCMT can observe the Sun directly without heat damage to the subreflector. The panels and backing structure of the SMT[4] will be constructed of CFRP (carbon-fiber-reinforced plastic), which has better performance under both gravity and thermal loading.

Ideally, one would like to be able to set the telescope surface to one-twentieth of the shortest operating wavelength. Measurements of telescope surface deformations are usually done by one variant or another of a holographic technique, in which the amplitude and phase over the beam pattern is mapped. The JCMT uses maps of the beam pattern made at two focus settings, of a millimeter-wave source placed in the near field, while the CSO uses a shearing Fourier Transform Spectrometer[22] to observe planets. The holographic technique is flexible, and has a number of advantages amongst which is that it requires no physical contact with the surface. In principle, the measurement accuracy achievable is also unlimited; this is especially important at the higher frequencies. For instance, at 800 GHz, improving the surface accuracy from 35 to 25μm results in a factor two improvement in telescope efficiency.

Progress in Receiver Technology

The major advance in the last few years has been the development of receivers in which the key element is the superconducting tunnel junction. Such SIS receivers are almost common-place at millimeter wavelengths, and are now available at the low end of the submillimeter range. They have the advantages that they are low noise, robust, and compact, and require low local oscillator power for effective mixing. Of most interest to the astronomer is that receivers built using SIS junctions have a low internal noise contribution. In terms of equivalent black-body temperature, the 'receiver temperature' may be as low as 300 K or better at 345 GHz, which is to be compared with numbers which can be several times greater for Schottky diode receivers. Thus the internal noise of the receiver itself is not greatly different from that of the contribution from the sky to the total system noise, which means that receiver performance in such cases is approaching the point at which additional gains will be difficult, at least for ground-based systems.

Considerable effort is also going into extending the frequency range over which SIS technology is applicable. This is being tackled on two fronts. First, experimentation with new junction materials has shown tremendous promise. Lead alloys and niobium nitride[27] would seem to offer possibilities for operation throughout the submillimeter region. Size is also an important consideration; some junctions are less than one square micron in area, which has ramifications for the fabrication technique. Second, efforts to extend the frequency range of other receiver system elements are meeting with considerable success. One example is the use of direct optical coupling of the submillimeter radiation to the junction via a lens; a particular configuration called a 'bowtie' SIS junction[18] offers good efficiency up to perhaps 1000 GHz. Also, the use of a quasi-optical planar log spiral antenna promises similar performance. Local oscillator sources have been developed[21] which use a Gunn oscillator in combination with a $1 - 8$ times multiplier to provide enough power for mixing for SIS receivers between 100 to 1000 GHz. In addition, local oscillator frequencies can be generated using laser lines of various gases; almost complete coverage from about 600 to 2000 GHz is now possible.

Astronomy at Submillimeter Wavelengths

Spectral Surveys

The low densities prevalent in interstellar clouds, coupled with extremely long timescales and shielding from dissociating radiation leads to a chemistry which is very different from that which is familiar on Earth. Transient species which are difficult or impossible to create in the laboratory may be observed at leisure by radio telescopes. The value of surveys covering wide frequency ranges is that they not only tell us about the excitation properties of known molecules through the observations of multiple transitions, but that they also reveal previously unknown lines of new species. In the specific case of sub-mm astronomy, only a few such surveys have been made, and they are not especially rewarding in terms of new species, although they have been helpful in understanding the excitation of molecules in 'warm' (say, with a kinetic temperature of 30 K) clouds. The next generation of receivers currently under development may well see better progress in the detection of new species.

Spectroscopic surveys at mm and sub-mm wavelengths have concentrated for the most part on three objects, the Orion Molecular Cloud OMC-1, the giant molecular cloud close to the Galactic Center Sgr B2, and the carbon-rich envelope of the late-type star IRC+10216. Both OMC-1 and Sgr B2 are similar in that they are regions of intense star-forming activity. In OMC-1 a strong outflow component (the 'plateau') possesses a different chemistry from the more quiescent material in the region (*e.g.*, [11]). Jewell *et. al.*[14] have surveyed with the NRAO 12-m telescope the region from 330.5 to 360.1 GHz in OMC-1, with a yield of 180 lines, 29 of which are unassigned. Stutzki *et. al.*[25] have surveyed a small region around 800 GHz, and very recently at the JCMT the same group have examined a second small region between 673.65 and 674.7 GHz in OMC-1. In Sgr B2 Sutton *et. al.*[26] have surveyed the region between 330 and 355 GHz at two different positions in the source. Approximately 100 lines are observed, and significant excitation and/or chemical differences can be observed between the two parts of the same object.

A few common species (*e.g.* CH_3OH, CH_3CN, and SO_2) yield a disproportionately large number of submillimeter lines in the surveys of OMC-1 and Sgr B2; this trend appears to continue to the higher frequencies. Furthermore, at submillimeter wavelengths a significant fraction (one estimate[25] gives 30 - 40% at around 800 GHz) of the total emission from dense molecular clouds is due to molecular lines.

IRC+10216 presents an entirely different chemistry (see *e.g.* [16]) from the star formation regions, and should be very rewarding in the long term, but sub-mm spectral survey work has not advanced very far in this source to date. So far Avery *et. al.* (priv. comm.) have surveyed a part of the 345 GHz window. It is clear from this work that the spectral line density is much lower than that of either OMC-1 or Sgr B2, but that with the greater sensitivity of the SIS receivers some interesting discoveries are likely to be made.

A few lines of species expected specifically to exist in interstellar clouds have been searched for; for instance MgH should show two triplets of lines at around 355 GHz (Watt, priv. comm.), there should be a line of *para*-H_3O^+ [34] at 365 GHz, and *ortho*-$H_2D^+(1_{10}-1_{11})$ should be seen at 372 GHz. All three species however so far have eluded positive detections. Often the problem is not that the lines are especially weak, but rather that the line density due to unrelated species is so great that the possibility of confusion is very high. An improvement in receiver technology will not help in such cases.

Stellar objects

Large ground-based telescopes are ideal for the study of objects which have small angular size, such as stellar envelopes. Dust (which also provides a good environment for the formation of molecules) is a very common component of stellar systems; stars form out of a collapsing dense cloud, which exhibits a number of interesting observable properties as the young stars begin nucleosynthesis, persists to a greater or lesser degree in the stable stages (main sequence) of a star's life, and may again be important as a dense circumstellar envelope in the later stages of certain types of stars. In the sub-millimeter region perhaps the most striking advances have been made in continuum observations of the dust in all such stages. From a knowledge of the submillimeter spectrum, with some reasonable assumptions, one can derive good estimates of the dust temperature and emissivity law, the total mass of dust, and some knowledge of the grain sizes in the stellar envelope.

Sub-mm continuum observations (*e.g.* [1,29]) of pre-main-sequence stellar objects, which are often embedded in extensive clouds of ambient dense material, show that (a) the dust appears to be distributed in a dense circumstellar disk, (b) the grain sizes of the dust are typically several tens of microns, and (c) the disk masses are in the range 0.05 through 1 solar mass. The latter number is especially interesting, because it suggests that sufficient material exists in such circumstellar disks for planets to be formed.

Through a combination of radiation pressure, stellar winds, and relative motion of the star and progenitor cloud, most main-sequence stars are visible objects. However, even a significant fraction of these possess a dusty circumstellar disk, first discovered by infrared observations. In the submillimeter region, observations[5,8] show that in such disks the grain temperature is between 50 and 140 K, and that grain sizes are of the order of 100μm. It is tempting to relate these circumstellar disks with the Oort cloud postulated to be a component of our own solar system. Some of these stars (β Pic, Vega) are thought to have attendant planets.

For the moment, spectral line observations at sub-mm wavelengths have contributed rather less to the overall picture of stellar envelopes. However, the dusty envelopes and outflows are prime locations for molecules, and once again, the next generation of more sensitive receivers should produce some valuable insights. The key to such discoveries is the high angular resolving power of the new large ground-based telescopes; furthermore, the higher excitation lines of common molecules such as CO, HCN, and CS, for example, at the higher frequencies not only lead to smaller beamwidths, but also the ability to probe the denser, warmer regions of circumstellar envelopes closer to the star itself. Little work has been done with the necessary sensitivity yet, but early work recently begun with the JCMT has examined the CO(6–5) transition within the inner 30″ or so of the envelope of the prototypical carbon star IRC+10216 (van der Veen and Sahai, priv. comm.). Another recent example is the detection[10] in the CO(3–2) transition at 345 GHz of a high-velocity (±200 km/s) outflow from the carbon-rich object CRL618. This object is especially interesting, in that it is thought to be an example of the short-lived transition stage between the red-giant phase of steady mass outflow and the onset of a fast wind, which, in setting up a shocked and compressed shell of material, ultimately leads to the formation of a planetary nebula. Finally, in another unexpected result, in the peculiar young stellar object MWC349, a number of high-excitation hydrogen recombination lines have been observed to exhibit maser action at mm wavelengths, and more recently, variability. Masing recombination lines have now been seen in the sub-mm region (Thum, priv. comm.), and it will be very interesting to determine the frequency range of this behavior.

Star Formation Regions

Regions of star formation contain a wide range of gas and dust densities, temperatures and excitation levels. The high resolution of large ground-based antennas at the higher frequencies are now beginning to show some interesting new facets of these complex clouds.

Firstly, the higher excitation CO(6–5) and (7–6) transitions at 691 and 800 GHz are powerful diagnostics because they are sensitive to both the higher densities (above about 10^4 cm^{-3}), and higher kinetic temperatures (≥ 50 K). The lines can be very strong; for instance in OMC-1 the total luminosity of the CO(7–6) line alone is estimated to be about ten times that of the Sun. In OMC-1 in particular, the CO observations show that three physically distinct components can be identified: (1) a warm (100–250 K) kinetically-active gas, which is the prevalent form of material in the region; (2) warmer (100–500 K) quiescent gas, which appears to be heated by UV photons from the bright stars in the region, and which is a principal coolant of the material; and (3) cool (20–40 K) quiescent material. The latter appears in absorption against the bright line background of the first two components, and similar effects are seen in other objects (*e.g.* DR21[13]). Still higher densities (in excess of 10^7 cm^{-3}) can be inferred from the detection[24] of the high-excitation HCN(9–8) transition at 797 GHz in OMC-1.

A second area where sub-mm astronomy is making useful additions to our understanding of star-formation regions relates to the photodissociation boundary region between dense molecular cloud and hot ionized material. It is possible to observe the molecular material through the various rotational transitions of CO, neutral gas via two transitions of atomic carbon, and, from the Kuiper Airborne Observatory, the 158-μm line of singly-ionized carbon. In those few instances (*e.g.* M17, OMC-1) which have been studied (*e.g.* [7,23]), in the necessary detail, it is found that a high-temperature, high-density boundary layer exists at this interface. The C$^+$ emission extends much further than might have been expected into the dense molecular cloud. This is interpreted[23] as evidence in favor of the extreme clumpiness of the latter. In addition, it is found that the neutral carbon appears to be distributed in a relatively thin layer at the cloud interface, where the high-excitation CO rotational lines are found also to be the strongest. One may expect that as the higher spatial resolutions come into routine use, this picture of a cloud interface region will become much clearer, and more complex. Already, the first observations of the CO(4–3) and CI $^3P_1$–$^3P_0$ transitions at 461 and 492 GHz have been made (Padman, White, priv. comm.) of OMC-1 and other regions with the JCMT using a 12″ beam. Using the same telescope observations of high-excitation CO lines (including the first detection of the ^{13}CO(6–5) transition) have been made by Genzel *et. al.* (priv. comm.) with an angular resolution of 8″ or less.

Solar System Observations

It is now possible to detect with current continuum bolometer instrumentation using the JCMT all asteroids of size greater than about 150 km. Asteroids are thermal bodies; however, the measurements show a decrease of brightness temperature with increasing frequency. This has been interpreted[20] as being the result of a dusty surface layer (regolith) containing particles of size around 100μm. In addition, the first sub-mm light curve (the result of rotation) for an asteroid (Vesta) has also been obtained. In the near future, it is expected that it will be possible to infer differences in surface composition from the continuum spectra of asteroids and other solid solar system objects. Comets can also be detected in the sub-mm continuum; Jewitt and Luu (priv.comm.) have detected P/Brorsen-Metcalf at 800μm, from which they infer that a silicate particle model leads to the best fit to the data. Finally, alone amongst the large sub-mm telescopes, the JCMT is capable of carrying out solar observations, and some early results are now available[15] at 850μm which show that chromospheric supergranulation can be readily observed, as can sunspots, although in this case, the contrast is rather low at this wavelength.

Concluding Remarks

It has not been possible to do justice to the burgeoning field that is sub-mm astronomy in a few short pages. I have left out developments in many areas that deserve mention. For instance, I have not discussed extragalactic work, the Galactic Center, or sub-mm observations shortward of 300μm; there are a number of references to recent work in Watt and Webster[28]. I hope however I have conveyed some of the excitement and movement of work in this wavelength range.

The future appears to be bright for sub-mm astronomy. A number of major technical advances are close to fulfillment which should see the realization of many of the original hopes for the field. Beyond these steps, there are a number of other advances which should be made in the next ten years. Amongst these are (a) 'high temperature' SIS junction fabrication, (b) active antenna surface control and improved panel construction, (c) the advent of multi-feed line and continuum receiver systems (such as the NRAO 8-beam receiver[18], and 'SCUBA' [9]), (d) carbon-fiber-reinforced plastic antenna construction, (e) interferometry[30], and (f) the launch of space telescopes (*e.g.* [17]) for sub-mm astronomy.

I thank my many colleagues, too numerous to mention by name, who kindly gave their time and effort to tell me about their latest work, sent me preprints, and agreed to let me say a few words about it, sometimes even before they had thought about it themselves.

References

[1] Adams, F.C., Emerson, J.P., Fuller, G.A.; 1989, *Astrophys. J.*, in press

[2] Altenhoff, W.A., Baars, J.W.M., Downes, D., Wink, J.E.; 1987, *Astron. Astrophys.* **184**, 387

[3] Baars, J.M.W., Greve, A., Hooghoudt, B.G., Penalver, J.; 1988, *Astron. Astrophys.* **195**, 364

[4] Baars, J.W.M., Krügel, E., Martin, R.N.; 1990, in Watt and Webster, p. 45

[5] Becklin, E.E., Zuckerman, B.; in Watt and Webster, p. 147

[6] Booth, R.S., Delgado, G., Hagström, M., Johansson, L.E.B., Murphy, D.C., Olberg, M., Whyborn, N.D., Greve, A., Hanson, B., Lindström, C.O., Rydberg, A.; 1989, *Astron. Astrophys.* **216**, 315

[7] Boreiko, R.T., Betz, A.L., Zmuidzinas, J.; 1990, *Astrophys. J.*, in press

[8] Chini, R., Krügel. E., Kreysa, E.; 1989, *Astron. Astrophys.*, in press

[9] Duncan, W.D.; 1990, in Watt and Webster, p. 51

[10] Gammie, C.F., Knapp, G.R., Young, K., Phillips, T.G., Falgarone, E.; 1989, *Astrophys. J. (Letters)* **345**, L87

[11] Genzel, R., Stutzki, J.; 1989, *Ann. Rev. Astron. Astrophys.* **27**, 41

[12] Hills, R.E.; 1985, in Proceedings of the ESO-IRAM-Onsala Workshop on "(Sub)millimeter Astronomy", p. 63

[13] Jaffe, D.T., Genzel, R., Harris, A.I., Lugten, J.B., Stacey, G.J., Stutzki, J.; 1988, *Astrophys. J.* **344**, 265

[14] Jewell, P.R., Hollis, J.M., Lovas, F.J., Snyder, L.E.; 1989, *Astrophys. J. Suppl.* **70**, 833

[15] Lindsey, C.A., Yee, S., Roellig, T.L., Brock, D.R., Duncan, W.D., Watt, G.D., Webster, A.S., Hills, R., Jeffries, J.T.; 1990, preprint

[16] Lucas, R., Guélin, M.; 1990, in Watt and Webster, p. 97

[17] Melnick, G.J.; 1990, in Watt and Webster, p. 93

[18] Payne, J.M.; 1989, *Proc. I.E.E.E.* **77**, 993

[19] Phillips, T.G.; 1988, in Wolstencroft and Burton, p. 1

[20] Redman, R.O., Feldman, P.A., Halliday, I., Matthews, H.E.; 1990, *"Asteroids, Comets, Meteors III"*, eds. C.-I. Lagerkvist, H. Rickman, B.A. Lindblad, M. Lindgren; p. 163

[21] Rothermal, H., Phillips, T.G., Keene, J.; 1989, *Int. J. IR. Mm. Waves* **10**, 83

[22] Serabyn, E., Masson, C.R., Phillips, T.G., 1990, in Watt and Webster, p. 41

[23] Stutzki, J., Stacey, G.J., Genzel, R., Harris, A.I., Jaffe, D.T., Lugten, J.B.; 1988a, *Astrophys. J.* **332**, 379

[24] Stutzki, J., Genzel, R., Harris, A.I., Herman, J., Jaffe, D.T.; 1988b, *Astrophys. J. (Letters)* **330**, L125

[25] Stutzki, J., Genzel, R., Graf, U.U., Harris, A.I., Jaffe, D.T.; 1989, *Astrophys. J. (Letters)* **340**, L37

[26] Sutton, E.C., Jaminet, P.A., Danchi, W.C., Masson, C.R., Blake, G.A.; 1990, in Watt and Webster, p. 105

[27] Vaneldik, J.F., Routledge, D., Brett, M.; in Watt and Webster, p. 75

[28] Watt, G.D., Webster, A.S.; 1990, editors, *"Submillimetre Astronomy"*, Astrophysics and Space Science Library, 151, Kluwer Academic Publishers

[29] Weintraub, D.A., Sandell, G., Duncan, W.D.; 1989, *Astrophys. J. (Letters)* **340**, L69

[30] Welch, W.J.; 1990, in Watt and Webster, p. 81

[31] Winnewisser, G., Armstrong, J.T.; 1989, editors, *"The Physics and Chemistry of Interstellar Molecular Clouds"*, Lecture Notes in Physics, 331, Springer-Verlag

[32] Winnewisser, G., Zimmermann, P., Hernichel, J., Miller, M., Schieder, R., Ungerechts, H.; 1990, *Astron. Astrophys.*, in press

[33] Wolstencroft, R.D., Burton, W.B.; 1988, editors, *"Millimetre and Submillimetre Astronomy"*, Kluwer Academic Publishers

[34] Wootten, A., Boulanger, F., Zhou, S., Combes, F., Encrenaz, P., Gerin, M., Bogey, M.; 1990, in Watt and Webster, p. 107

Observational Constraints for the Chemistry of Dark Clouds

S. R. Federman[1,2], W. T. Huntress[1], and S. S. Prasad[3]

[1] *Jet Propulsion Laboratory, MS 183-601, 4800 Oak Grove Drive, Pasadena, CA 91109, U. S. A.*
[2] *Department of Physics and Astronomy, University of Toledo, Toledo, OH 43606, U. S. A.*
[3] *Departments of Physics and Astronomy, University of Southern California, University Park,*
MC-1341, Los Angeles, CA 90089-1341, U. S. A.

A search for correlations arising from molecular line data has been
made in order to place constraints on the chemical models of dense
interstellar clouds. Observations of CO, H_2CO, NH_3, HC_3N, and HC_5N
have been compiled. The results for CO place constraints on the
strength of the UV field where self-shielding is important. These
results also indicate that the CO abundance increases with A_V for
directions with $A_V \leq 4$ mag. For H_2CO, a quadratic relationship is
obtained in plots versus H_2 column density. Ammonia shows no
correlation with H_2, $C^{18}O$, HC_3N, or HC_5N, because for dark clouds
the column density of NH_3 is nearly constant at 10^{15} cm^{-2}. A strong
correlation is found between HC_3N and HC_5N, indicating the chemical
link between the cyanopolyynes.

INTRODUCTION

Molecular line data for dark clouds has been compiled in order to constrain
chemical models through the use of observed trends between species. When plotted
in a log-log fashion, the slope and the intercept have been used as an aid in
understanding the chemical kinetics, physical conditions, and spatial correspond-
ences within a cloud. Relations of the form
$$\log N(A) = \alpha + \beta \log N(B)$$
have been sought, with N(X) the column density of species X. Several restrictions
have been placed on the compilation in order to acquire a self-consistent set of
data. These restrictions include: (1) exclusion of clouds with substantial star
forming activity (e.g., ρ Oph Cloud), (2) incorporation of data in directions with
$A_V \geq 1$ mag to minimize contamination from nearby dense clumps, (3) exclusion of
species with saturated lines, and (4) averaging different observations for a
specific molecule at the same position. The column density for CO has been derived
from multiplying the column density $C^{18}O$ by 500. For positions with $A_V \leq 4 - 5$ mag,
the standard relationship between $N(H_2)$ and A_V has been used (6).

Details of the compilation, including the sources of the molecular data, are
given in Federman et al. (2). In the present contribution, several results from
the compilation are highlighted.

RESULTS

The analysis of the data for oxygen-bearing species resulted in a clearer
understanding of the chemistry for these species. When $C^{18}O$ data are converted to
CO through the use of a constant ratio of 500, N(CO) is found to increase more
rapidly than linearly with $N(H_2)$ for A_V between 1 and 4 mag. The increase, however, is not as steep as the increase seen in the data for diffuse clouds. Least-
squares fits to these data yield a slope of ~ 1.5 for the dark cloud data and a

slope of ~2 for the diffuse cloud data. A slope greater than one for the dark
cloud data indicates that the abundance of CO increases with depth in the outer
regions of the clouds, where C^+ is processed into the stable CO molecule. Such an
increase is consistent with the most recent theoretical calculations (8). The fits
also reveal that at 10^{21} H_2 cm^{-2}, N(CO) is a factor of 6 greater in the dark clouds.
When examined in light of the theoretical results (8), this result implies that the
strength of the UV field where CO shields itself from dissociation is one-half the
strength of the average Galactic field. This decrease in field strength is not
unexpected because the diffuse outer portions of a cloud attenuate some of the
radiation.

In contrast with the case for the chemistry of CO in the outer portions of
clouds, the chemistry of H_2CO is poorly understood in these regions. All attempts
to model the H_2CO chemistry produce column densities far too small. The compilation
of 6 cm measurements of formaldehyde shows that $N(H_2CO)$ increases quadratically
with $N(H_2)$ for $A_V \leq 4$ mag and appears to remain constant with $N(H_2)$ when $A_V \gtrsim 4$ mag.
These trends may aid in clarifying the chemical steps involved in the production
of H_2CO.

The data on nitrogen compounds reveal information applicable to cloud inter-
iors. Ammonia does not correlate with any of the other species that were compiled.
The lack of correlations between ammonia and the cyanopolyynes, in particular, may
be the result of an observational peculiarity: A nearly constant value of ~ 10^{15}
cm^{-2} for $N(NH_3)$ is found for all the cloud cores sampled. This result is illustrated in Figure 1, where log $N(HC_5N)$ is plotted against log $N(NH_3)$. The chemistry
of HC_3N and HC_5N may be coupled to NH_3, but because of ammonia's peculiar chemistry

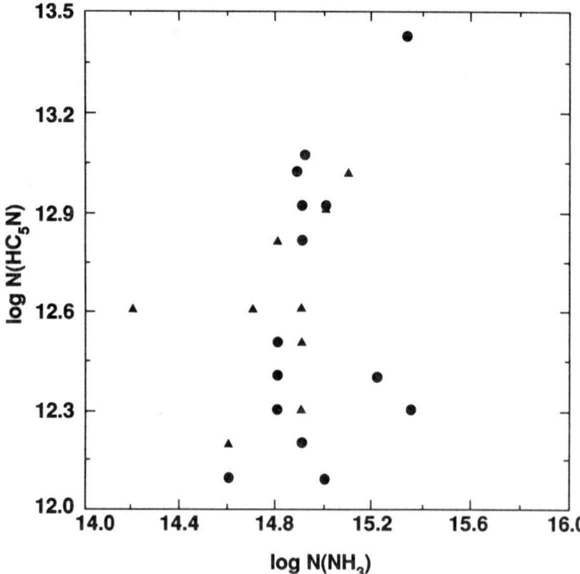

Figure 1 – Log $N(HC_5N)$
versus log $N(NH_3)$. The
different symbols represent
different sources for the
data, as discussed in (2),
where the original figure
appeared.

or excitation, any revealing relationships are masked. The strides recently made
in quantifying the chemistry of NH_3 (3,4,5,9) indicate that the constant value for
$N(NH_3)$ arises from special conditions for excitation.

The results for the carbon-chain molecules indicate trends of use to chemical modeling. A highly significant correlation between $N(HC_5N)$ and $N(HC_3N)$ with a slope of unity strongly suggests that the chemistry of these two cyanopolyynes is connected. Furthermore, the data for several dark clouds indicate that the value of ~ 3 for the ratio $N(HC_3N)/N(HC_5N)$ is universal and not specific to TMC-1. The presence of a universal ratio was noted previously (1,7), but for a smaller sample of clouds. The weakness of the correlation between the cyanopolyynes and $C^{18}O$ favors the conclusion that these molecules are spatially distinct within clouds. Thus, observations of $C^{18}O$ may not sample all the core material of an interstellar cloud.

CONCLUSIONS

Correlations provide important information for modeling interstellar clouds. The strength of the correlation and the parameters of the least-squares fit, such as slope and intercept, provide clues to the underlying chemical network and to the physical model for the cloud. Through the use of this concept, greater insights into the chemistry and physics involving interstellar species have been obtained.

REFERENCES

(1) Bujarrabal, V., Guelin, M., Morris, M., and Thaddeus, P. 1981, Astr. Ap., 99, 239.
(2) Federman, S.R., Huntress, W.T., and Prasad, S.S. 1990, Ap. J., in press.
(3) Galloway, E.T., and Herbst, E. 1989, Astr. Ap., 211, 413.
(4) Herbst, E., Defrees, D.J., and McLean, A.D. 1987, Ap. J., 321, 898.
(5) Langer, W.D., and Graedel, T.E. 1989, Ap. J. Suppl., 69, 241.
(6) Myers, P.C., Linke, R.A., and Benson, P.J. 1983, Ap. J., 264, 517.
(7) Snell, R.L., Schloerb, F.P., Young, J.S., Hjalmarson, A., and Friberg, P. 1981, Ap. J., 244, 45.
(8) van Dishoeck, E.F., and Black, J.H. 1988, Ap. J., 334, 771.
(9) Yee, J.H., Lepp, S., and Dalgarno, A. 1987, M.N.R.A.S., 227, 461.

Evolution of H$_2$O Ice in Dark Clouds

Masuo TANAKA[1], Shuji SATO[2], Tetsuya NAGATA[3], and Tetsuo YAMAMOTO[4]

[1] Institute of Astronomy, University of Tokyo, Mitaka, Tokyo 181, Japan
[2] National Astronomical Observatory, Mitaka, Tokyo 181, Japan
[3] Department of Physics, Kyoto University, Kyoto 606, Japan
[4] Institute of Space and Astronautical Science, Yoshinodai, Sagamihara 229, Japan

Evolution of H$_2$O ice in dark clouds is discussed, based on our infrared spectroscopic observations of the ρ Ophiuchi sources at IRTF with the CGAS in Mauna Kea (Fig. 1).

Water is believed to be the most abundant molecule after H$_2$ and CO. For the overall understanding of the interstellar chemistry, it is especially needed to connect gas phase and solid phase chemistry. The infrared observations of volatile molecules on grain surfaces in dark clouds play this role.

Our main results are summarized as follows.

(1) Figure 2 shows a relation of column densities between dust grains and water ice molecules on grain surfaces. The distribution of the data shows that H$_2$O ice mantles of dust grains are formed in a dense region where UV radiation is shielded. The slope of the line (Fig. 2) indicates a thin icy mantle of grain and the column density of H$_2$O molecules in the ice mantle is $\simeq 0.5 \times 10^{18}\ \tau_{ice}$.

(2) The non-zero τ_{ice} values of the emission line stars (maybe T Tauri stars) in spite of their small extinction indicate that once the H$_2$O ice mantle is formed, it can survive against strong UV radiation up to the stage of T Tauri stars. Furthermore, the existence of H$_2$O ice around T Tauri stars suggests that the majority of H$_2$O molecules in dark clouds is in a solid phase, not in a gas phase, as is expected from the fact that the condensation temperature of ~150 K is much higher than cloud temperatures.

(3) Figure 3 shows relative depths at 3.4 μm. From intracloud to ptotostars, the absorption keeps the same profile, whereas from protostars to T Tauri stars the profile changes. This suggests alteration of the material which is responsible for the absorption. The absorption at 3.4 μm may be attributed to C-H bonding. If so, $\tau_{3.4}/\tau_{ice}$ might be a measure of CH relative to OH. Some materials, which cannot be specified now, are probably responsible for the 3.2 - 3.7 μm absorption and must become abundant relative to H$_2$O or optically activated through the evolution from protostars to T Tauri stars.

Please refer the following paper for a more complete discussion.
"THREE-MICRON ICE-BAND FEATURES IN THE ρ OPHIUCHI SOURCES"
by M. TANAKA, S.SATO, T.NAGATA, AND T.YAMAMOTO, 1990, ApJ, 352, 724

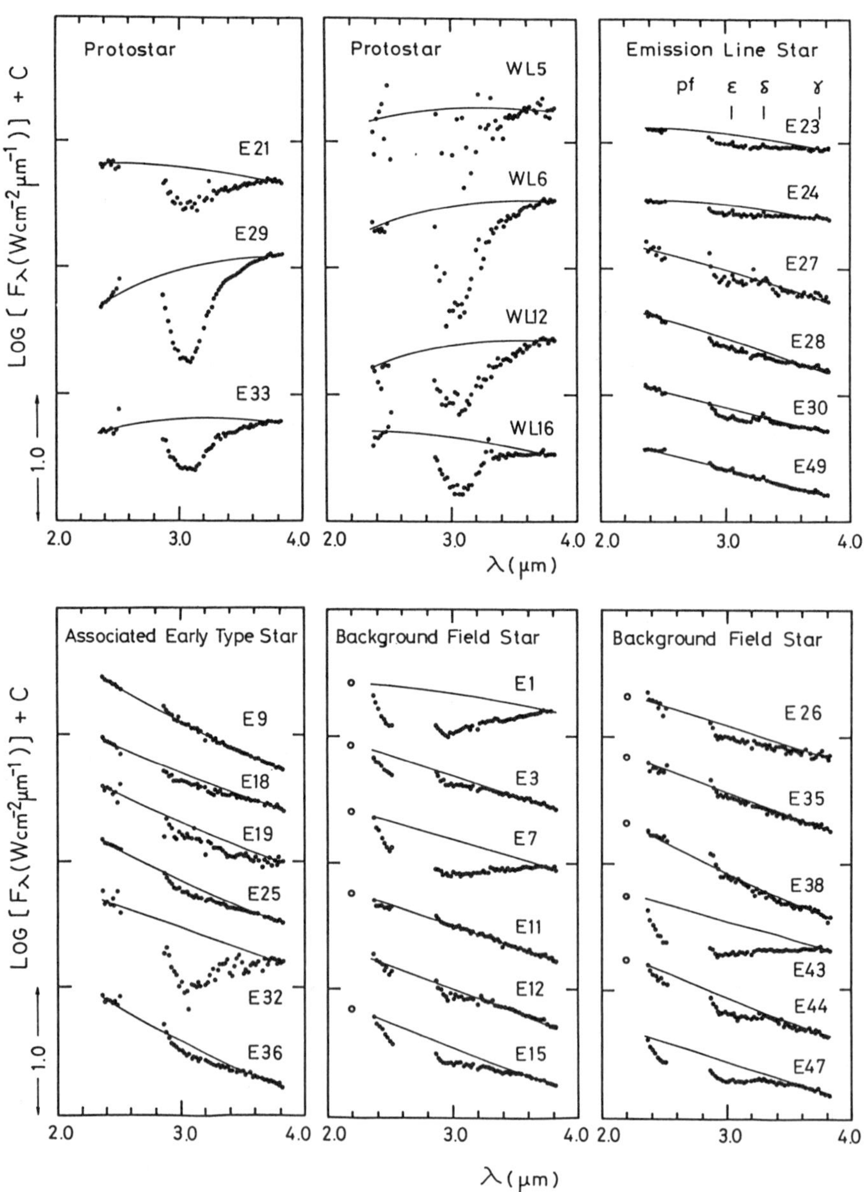

Figure 1   2-4 μm spectra of the ρ Oph sources.

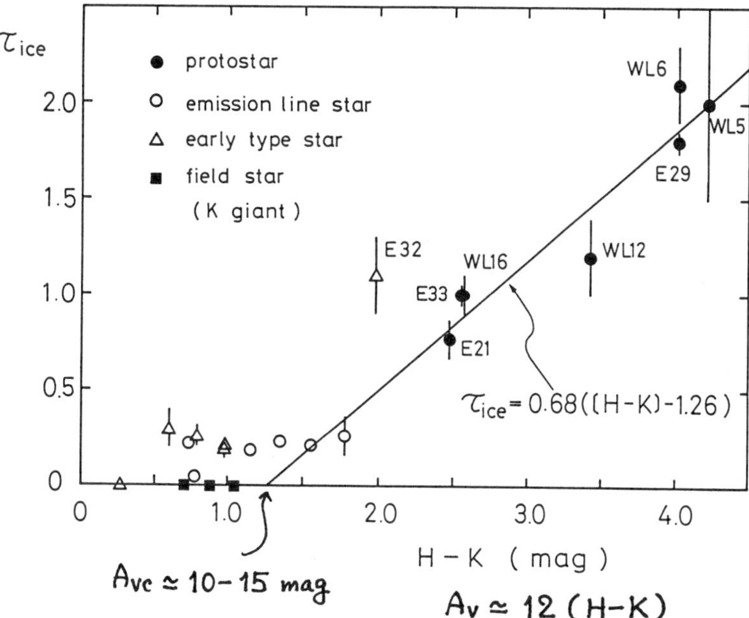

Figure 2   The relation between τ_{ice} and an infrared color H-K.

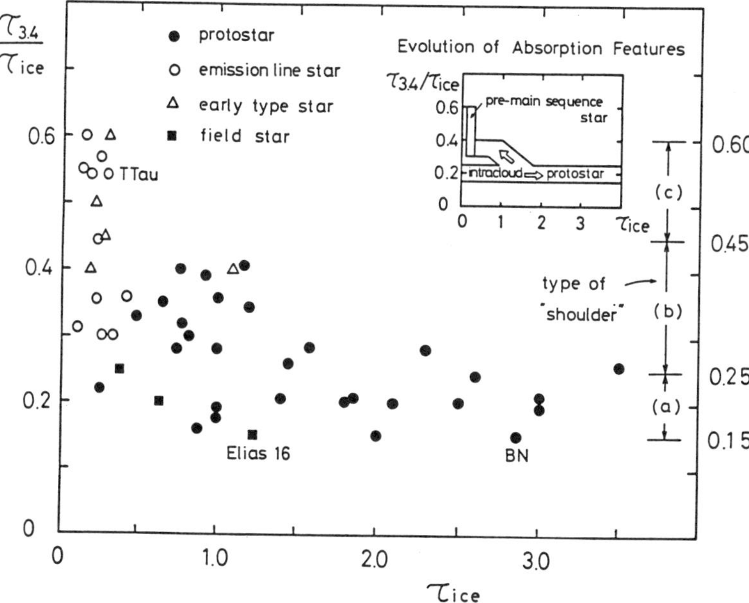

Figure 3   Evolution of the absorption profile from intracloud to
            T Tauri stars through protostars.

IR Images of the Dissipating Molecular Cloud: Excitation of Hydrogen at the Work Surface

Saeko S. HAYASHI[1], Tetsuo HASEGAWA[2], Masuo TANAKA[2], Masahiko HAYASHI[3],
C. ASPIN[1], I. S. MCLEAN[4], P. W. J. L. BRAND[5], and I. GATLEY[6]

[1] *Joint Astronomy Centre, Hilo, Hawaii, U. S. A.*

[2] *Institute of Astronomy, University of Tokyo, Mitaka, Tokyo 181, Japan*

[3] *Department of Astronomy, University of Tokyo, Bunkyo-ku, Tokyo 113, Japan*

[4] *University of California at Los Angeles, U. S. A.*

[5] *Department of Astronomy, University of Edinburgh, U. K.*

[6] *National Optical Astronomy Observatories, U. S. A.*

In the *interstellar laboratory*, the vibrationally excited molecular hydrogen emission arises practically from a mixture of fluorescent and dynamical excitation. Recent measurements of the ionized and the molecular hydrogen emission with a high spatial-resolution instrument resolved the detailed structure in the star-forming region S106. The displacement between the ionized peaks and the bright molecular spots corresponds to approximately 5″ and the molecular peaks are further from the central star. The high density derived with the low excitation temperature suggests that the origin of the *thermal* component in the H_2 emission is either *high-density fluorescence* in photodissociation regions formed on the dense molecular clumps, or it is from *slow shocks* driven by the stellar wind propagating into the dense clumps. These observational approaches directly look into the region where the placental molecular cloud is dispersing after the formation of the star.

I. MEASUREMENTS

The abundant lines of ionized and molecular hydrogen in the infrared wavelengths are handy tools in probing the regions of recent star formation. The vibrationally excited molecular hydrogen emission comes from *thermal excitation* in dynamical shocks (*e.g.* 1) and/or the fluorescent emission due to *radiative excitation (e.g.* 2, 3, 4). In the interstellar laboratory, the lines are generally found to be a mixture of these two mechanisms and the proportion varies among the objects (5).

S106 is a bipolar HII region with a young massive star, S106IR, at the center. It is already *visible* and in its final stage of formation, emerging from the placental molecular cloud. The measurements of Br γ and H_2 v=1-0 S(1) lines were made in August 1987, with the infrared array system IRCAM on the 3.8 m United Kingdom Infrared Telescope, Mauna Kea, Hawaii. The pixel-defined spatial resolution was 0.″62, corresponding to $1.8 \ 10^{-3}$ pc, or $5.6 \ 10^{15}$ cm for this object. A Fabry-Perot etalon with a spectral resolution of 130 km s^{-1} was placed in front of the camera window, and a cold wide-band K filter in series with a cold narrow band Br γ or H_2 v=1-0 S(1) line filter was used.

The most conspicuous difference between the distribution of Br γ and vibrationally excited H_2 emission is that the H_2 originates slightly farther from S106IR. Each bright peak is seen in both Br γ and H_2 with an almost perfect one-to-one correspondence, but the peaks of H_2 are located just outside the Br γ by 3″ − 7″. These offsets are definitely larger than the positional uncertainties of the H_2 and Br γ images.

II. DISCUSSION

The excitation of molecular hydrogen in this region was first noted by Longmore *et al.* (6) that all of the observed H_2 emission could be produced either by shock excitation or by fluorescence by UV radiation from S106IR. In the latter case, the observations can be understood in terms of a model of photodissociation regions (*e.g.* 7), in which the surface of a dense molecular cloud is illuminated by an intense far-ultraviolet radiation field. The overall configuration of the S106 region is illustrated in Figure 1 based on the current high resolution images. A detailed discussion concerning the excitation and the structure of the S106 region is given by Hayashi *et al.* (8).

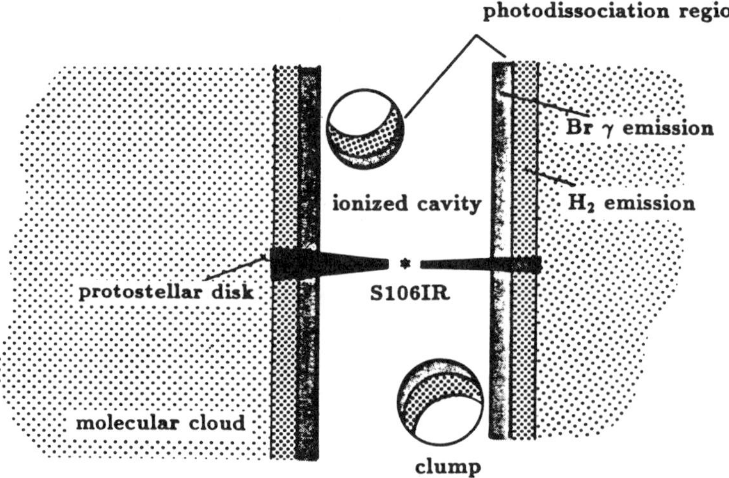

Figure 1. A schematic picture of the cross section of the S106 region, showing the HII region, the H_2 emission region, and the ambient molecular cloud (cited from Ref. 8).

Theories of the fluorescent molecular hydrogen have shown that when the predissociation density is low ($n(H_2) \leq 10^4$ cm^{-3}), the level population distributions of fluorescent H_2, and thus the line intensity ratios, are insensitive to the physical parameters of the photodissociation region (*e.g.* 9, 10). The case of higher predissociation density has recently been considered (11, 12). When $n(H_2) \geq 10^5$ cm^{-3}, collisional de-excitation in the photodissociation region thermalizes the population in the lower vibrational levels of molecular hydrogen at a gas temperature of ~ 1000 K, and so the infrared line ratios no longer adopt the values characteristic of fluorescence in the low density case. Tanaka *et al.* decomposed the molecular hydrogen emission in S106 and indicated that about 65 % of its v=1-0 S(1) emission comes from thermally excited H_2 at an excitation temperature less than 1400 K and that the rest arises from *low density* fluorescence. There are two potential origins for the *thermal* component: collisional de-excitation of fluorescent H_2 (*high-density fluorescence*) or excitation in shocks driven by the expansion of the HII region and the outflow from the protostar.

The shock interpretation has two feasible energy sources, the pressure of the expanding HII region, and the stellar wind from S106IR. The shock will have propagated at a speed ahead of the ionization front equal to the shock speed divided by the compression ratio in the shock, which will be of order ten. In these circumstances, a 10 km s^{-1} shock will have propagated $3\ 10^{16}$ cm in $\sim 10^4$ years. This scale represents the separation between the Br γ and H_2 peaks. A calculation of a 7 km s^{-1} shock propagating into a region of initial density of 10^5 cm^{-3} (13) seems to be consistent with the observation of the H_2 clumps, that is, the surface brightness of 10^{-4} erg s^{-1} sr^{-1} and the temperature of 1200 K.

Now we turn to the high density fluorescence scenario. As the Br γ ridges delineate the ionization fronts, the displacement of $3'' - 7''$ ($2.7 - 6.1\ 10^{16}$ cm) between the Br γ and adjacent H_2 peaks is a measure of the depth of the H_2 emitting zone from the ionization front. If we take $A_v \simeq 2$ mag for the extinction between the ionization front and the H_2 emitting region (7), the total hydrogen density of the photodissociation region is estimated as n(total) $\simeq 10^5$ cm^{-3}. The deduced density is sufficiently high to thermalize the level populations. The resultant thermal gas temperature is predicted to be ~ 1000 K (12) which is reminiscent of the correspondingly low temperature for the thermal component in S106.

The present high resolution images resolves the H_2 peaks for the first time, and gives a good measure of the surface brightness as well. The peak brightness of the H_2 v=1-0 S(1) emission clumps is (1-2) 10^{-4} erg s^{-1} cm^{-2} sr^{-1} with a clump-interclump contrast of 5 : 1 to 10 : 1. Sternberg and Dalgarno (12) calculated the surface brightness of the fluorescent H_2 emission for a grid of the intensity of the incident UV radiation and the density in the photodissociation region. The UV radiation field incident upon the ionization front in S106 is estimated as $\sim 10^4$ times stronger than the general interstellar background. For this radiation field, the observed surface brightness can be reproduced, within an uncertainty caused by various possible geometries of the emitting zone, if the density is n(total) $\geq 10^5$ cm^{-3} in the clumps and n(total) $\simeq 10^4$ cm^{-3} between clumps. These density values agree with the estimates from the millimeter wave observations (*e.g.* 14).

In conclusion, the high density in the photodissociation region in S106 (n(total) $\simeq 10^5$ cm^{-3}) estimated from the present H_2 image suggests either that the *high-density fluorescence* is the major origin for the *thermal* component revealed by Tanaka *et al.* (5), or that this component is produced by *low velocity shocks*. It is this component which comprises the emission in the bright clumps visible of the H_2 image. In the present model, the *fluorescent* component identified in (5) arises naturally from the less dense (n(total) $\simeq 10^4$ cm^{-3}) portions of the neutral envelope around S106.

Acknowledgements. The United Kingdom Infrared Telescope is operated by the Royal Observatory Edinburgh on behalf of the Science and Engineering Research Council of the United Kingdom. This international collaboration program was promoted by Dr. N. Kaifu at Nobeyama Radio Observatory, Japan. S. S. H. has been strongly encouraged by Dr. H. Suzuki.

References

1. Beckwith, S., Persson, S. E., Neugebauer, G., and Becklin, E. E. 1978, *Astrophys. J.*, 223, 464.

2. Hayashi, M., Hasegawa, T., Gatley, I., Garden, R., and Kaifu, N. 1985, *M. N. R. A. S.*, 215, 31p.

3. Gatley, I., Hasegawa, T., Suzuki, H., Garden, R., Brand, P., Lightfoot, J., Glencross, W., Okuda, H., and Nagata, T. 1987, *Astrophys. J. (Letters)*, 318, L73.

4. Hasegawa, T., Gatley, I., Garden, R. P., Brand, P. W. J. L., Ohishi, M., Hayashi, M., and Kaifu, N. 1987, *Astrophys. J. (Letters)*, 318, L77.

5. Tanaka, M., Hasegawa, T., Hayashi, S. S., Brand, P. W. J. L., and Gatley, I. 1989, *Astrophys. J.*, 336, 207.

6. Longmore, A. J., Robson, E. I., and Jameson, R. F. 1986, *M. N. R. A. S.*, 221, 589.

7. Tielens, A. G. G. M., and Hollenbach, D. 1985, *Astrophys. J.*, 291, 722.

8. Hayashi, S. S., Hasegawa, T., Tanaka, M., Hayashi, M., Aspin, C., McLean, I. S., Brand, P. W. J. L., and Gatley, I. 1990, *Astrophys. J.*, in press for May 1 issue.

9. Black, J., and van Dishoeck, E. 1987, *Astrophys. J.*, 322, 412.

10. Sternberg, A. 1988, *Astrophys. J.*, 332, 400.

11. Hollenbach, D. J. 1988, *Astro. Lett. and Communications*, 26, 191.

12. Sternberg, A., and Dalgarno, A. 1989, *Astrophys. J.*, 338, 197.

13. Brand, P. W. J. L., *et al.* 1990, in preparation.

14. Kaifu, N., and Hayashi, S. S. 1987, Proc. of the IAU Symp. No. 115, "Star Forming Regions", eds. M. Peimbert and J. Jugaku (D.Reidel).

Precise Position Measurements of Molecular Maser Sources
(A plan)

Masaki MORIMOTO

Nobeyama Radio Observatory, Nobeyama, Minamisaku, Nagano 384-13, Japan

ABSTRACT

VLBI observations of molecular line maser sources using the 34m antenna of Kashima SpaceC
ommunication Center, Communications Research Laboratories and the 45m telescope of Nobeyama RadioO
bservatory are in progress. They will provide accurate positions of maser spots and so physicala
nd dynamical conditions of extended envelopes of late type variable stars.

Maser Emission from Late Type Variable Stars

Late type variable stars have extended gas envelopes. Gas outflow through an envelope is caused
by the passage of a shock wave associated with the pulsation of the central star.   This gas
ejection is known to be an important supplier of heavy atoms to the interstellar space.

These stars are the sources of strong maser emission of OH(1.6GHz), H_2O(22GHz) and SiO(43GHz).
Because of the difference in energy levels, the sources of emission correspond to different heights
in the envelope. VLBI observations provide a unique opportunity to study the envelope and the mass
ejection phenomenon by accurate position measurements.

For example, measuring only relative positions of maser spots gives the linear size of the
region where the spots are distributed and thus the height of the level where the maser emission
occurs. This information, combined with the radial velocity, will give the excitation and kinematic
conditions in the envelope.

Pioneering observations of the SiO maser were made by Lane (1983), whose results have shown that
v=1 and 2 masers are distributed in an area several times the stellar diameter.

KNIFE, Kashima Nobeyama InterFErometer

We have recently started VLBI observations of H_2O and SiO lines using the 34m telescope at the
Kashima Space Communication Center, Communications Research Laboratories and the 45m telescope of
Nobeyama Radio Observatory (KNIFE, Kashima Nobeyama InterFErometer).

Successful test experiments have shown that this interferometer can provide much higher
positional accuracy and higher sensitivity than previously possible.

At present software to reduce relative positions of different velocity components is being
developed and first serious observations will start October 1990.

Observational plans

a)Differential astrometry of maser spots.

As the first step we will measure relative positions of different velocity components of SiO and
H_2O masers in a regular interval, as the determination of the relative position is relatively easy
with present VLBI technique.

Because of the high sensitivity and positional accuracy, it is possible to observe sources up to 10kpc distance with a useful accuracy.

With observations in a regular interval and simultaneous to H_2O maser, we will directly "see" the movements of the maser spots during the passage of the shock wave at different levels in the envelope.

b)Two v states----possibility of measuring absolute positions

Because of the large bandwidth and the high sensitivity of the system it is also possible to simultaneously observe masers of two (in some cases three) different vibrational states of SiO, and thus directly compare their positions. Comparison of the behavior of the masers at different v states will give a new constraint to the maser mechanism.

More exciting is that through this comparison, "group delay" (a parameter conveniently used in geodetic VLBI experiments for comparing absolute positions) is determined and absolute positions can be determined with an accuracy of at least 10mas or perhaps 10 times better.

This opens up a possibility of comparing positions of Galactic and extragalactic objects very precisely for the first time. For example, a transversal motion of 200km/s at a distance of the Galactic Center can be measured in two years.

c)More antennas

A 10m radio telescope usable to 43GHz is under construction at Mizusawa Geodynamic Observatory, 400km north of Kashima. This will add a NS baseline to KNIFE and more accurate two dimentional positions can be measured very easily. It will also help to measure the size of the maser spots, etc.

d)Other than maser lines

We will also observe active galactic nuclei(AGN) with KNIFE. Spectra of VLBI spots of AGN are not well known at this high frequency, and more interesting is their variation with time. It is also possible to find interesting sources with KNIFE and to help to try more extensive high resolu tion observations using the mm-wave global VLBI.

Reference
Lane A. P., "The Spacial Structure of Silicon Monoxide Masers"   IAU Symposium No110, VLBI and on Compact Radio Sources, D. Reidel Pub., Ed., by R. Fanti, K. Kellerman and G. Seti, p329-330

II
LABORATORY SPECTROSCOPY

Microwave Spectroscopy of Transient Interstellar Molecules

Shuji Saito

Department of Astrophysics, Faculty of Science, Nagoya University, Chikusa-ku, Nagoya 461-01, Japan

Laboratory microwave spectroscopy of transient molecules is discussed from an astronomical point of view. A high sensitivity millimeter and submillimeter wave spectrometer has been developed and used to detect more than twenty transient molecules. Seven transient species among the molecules studied in the laboratory have been identified as new interstellar molecules. Finally, the discussion is directed at the spectroscopically and astronomically important aspects of our latest study on the CH_2CN radical.

I. INTRODUCTION

About eighty different interstellar molecules have been identified and reported in the literature so far. They are mainly made of hydrogen, oxygen, carbon, nitrogen, silicon, and sulfur atoms whose cosmic abundances are relatively high. About one fourth of the interstellar molecules detected are simple inorganic molecules composed of two to five atoms. The remaining three fourths are various kinds of simple organic molecules. It is remarkable that about half of them are transient molecules in terrestrial conditions: molecular ions, free radicals, and unstable molecules, as listed in Table I. Most of the newly identified interstellar molecules are exotic transient species. The recent development of the radio telescope has greatly improved its sensitivity and extended the frequency region covered, especially the high frequency region where stronger signals bearing different kinds of information are expected. As a result, a large number of unidentified lines have been reported and are waiting to be assigned. Most of the unidentified lines may be attributed to new transient molecules.

Table I. Transient Interstellar Molecules Detected by Radio Telescopes

Molecular ions:	SO^+, HCO^+, HCS^+, HN_2^+, HCO_2^+, H_2CN^+
Free radicals	
diatomic:	CH, OH, CN, NO, NS, SO, CSi, CP
polyatomic:	HCO, CCS, cyclic-C_3H, C_3N, CH_2CN,
	C_2H, C_3H, C_4H, C_5H, C_6H
Unstable molecules	
diatomic:	CS, SiO, SiS, PN
polyatomic:	HNO, HNC, C_2Si, H_2CS, C_3O, C_3S, C_3H_2, CH_2NH, C_4Si

Generally the transient molecules are chemically active and difficult to produce in high concentration in the laboratory, so that a great number of interesting transient molecules have not been studied in the laboratory and their detailed transition frequencies are unknown. Laboratory studies of spectroscopically and astronomically unknown transient molecules, especially potential candidates of interstellar molecules, are occasionally essential to their identification in space. Since the spectral lines of a new interstellar molecule provide us with unique information characteristic of its energy-structure and also information about the physical and chemical conditions of the molecule and its surroundings, the number of available lines must be increased for deeper understanding of the physical and chemical processes in space. Therefore, further laboratory high-resolution spectroscopy must be made for candidates of new interstellar molecules and for transient molecules which have not been studied.

II. MICROWAVE SPECTROSCOPY OF TRANSIENT MOLECULES

As noted above, the spectrometer suitable for transient molecule spectroscopy must overcome difficulties originating from their chemical activity in the laboratory. Essential features in the laboratory microwave spectroscopy of transient molecules are summarized into three points as follows (1): (i) how to make the spectrometer more sensitive, (ii) how to produce transient molecules efficiently, and (iii) how to predict transition frequencies accurately. A good combination of these three essential points makes it easy to detect the spectral lines of transient molecules by overcoming their low concentration. However, the sensitivity of the spectrometer is the most important of these three features. If the sensitivity of the spectrometer is much improved, it is possible to search for a transient species even under conditions of nonoptimized production or poorly predicted frequency region. Thus, the high sensitivity may partly cover deficiencies in the other features. This makes possible wide frequency searching in a shorter time. As is well known, the peak absorption coefficient of a rotational transition is in proportion to the third power of the transition frequency for linear or symmetric top molecules or to the second power for asymmetric-top molecules (2). If we are able to overcome various experimental difficulties in high-frequency spectroscopy, we will obtain much higher effective sensitivity to detect a molecular line.

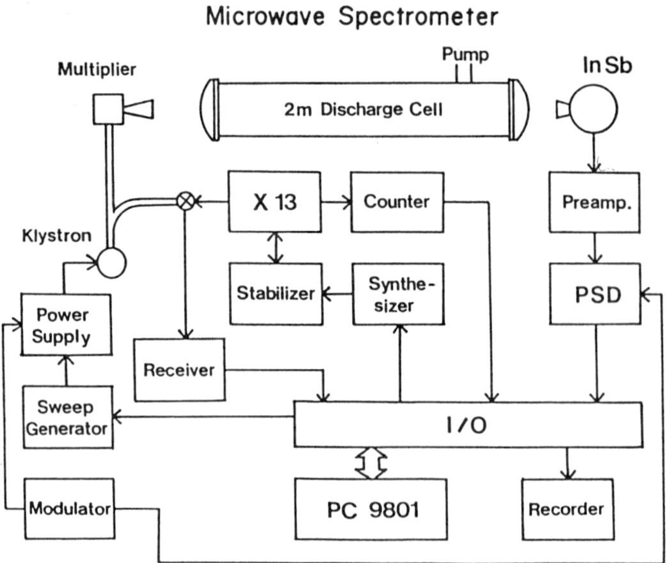

Fig. 1.   A block diagram of the microwave spectrometer for transient molecules (2).

We have developed a millimeter-and submillimeter-wave spectrometer for transient molecules (3).   Figure 1 shows a schematic diagram of the spectrometer for a source modulation system covering the frequency region of 40 to 500 GHz, combined with a free space discharge cell.   The frequency region below 130 GHz is covered by a series of millimeter-wave klystrons.   Microwave power above 130 GHz is supplied by several multipliers.   An InSb photo-conductive detector cooled to liquid helium temperature is used to detect radiation from the power source through the absorption cell. The free space absorption cell, which was designed after Woods (4), is made of a free Pyrex glass tube 2 m in length and 10 cm in outer diameter.   It has a cylindrical stainless steel electrode at each end.   Transient species are directly produced in the cell with a glow discharge or hollow cathode discharge.   The sensitivity of the spectrometer developed is about 2×10^{-9} cm^{-1} expressed by the minimum detectable peak absorption coefficient.   This means that the minimum detectable total number of molecules is about 1.5×10^{11} and the minimum detectable density is 1.1×10^{7} molecules/cm^3.   This high sensitivity is achieved by a combination of low noise and high power source, low noise detector, and low loss cell for the high frequency spectroscopy.

III. NEW INTERSTELLAR MOLECULES

The high sensitivity attained has enabled us to study various new transient molecules in the laboratory.   Table II lists transient molecules studied at Nagoya University in these four years. Among them, seven molecules have been identified as new interstellar molecules.   Identification of the new interstellar molecules has been substantiated by various but close collaborations between laboratory spectroscopy and radio astronomy.   The $C_6H(^2\Pi_i)$ radical was elucidated only from the interstellar unidentified spectral lines (5, 6), and detailed knowledge about the energy structure of the free radical, that is, $^2\Pi_i$, was essential to its identification.   New laboratory spectroscopic detections resulted in the definitive identifications of several new interstellar molecules such as, $CCS(^3\Sigma^-)$ (7), C_3S (8), $CH_2CN(^2B_1)$ (9, 10), and C_4Si (11), from many unidentified lines observed previously by radio telescopes (12, 13).   Furthermore, we have studied several molecules, first in laboratory, as good candidates for interstellar molecules, and then tried to detect their corresponding interstellar lines toward molecular clouds by using radio telescopes.   We identified the cyclic-$C_3H(^2B_2)$ (14) and the vibrationally excited C_4H radical (15).   Very recently Turner, Guélin, and their collaborators identified the CP radical in the circumstellar envelope of the carbon star IRC+10216 (16, 17), which we detected first in the laboratory (18).   Some of the newly detected transient molecules are potential candidates for interstellar molecules: the H_3S^+ ion (19) is an important cornerstone in sulfur chemistry in space, H_2Cl^+ (20) is a unique precursor ion to the interstellar HCl molecule, and the $PS(^2\Pi_r)$ (21) radical is one of the potential candidates in circumstellar envelopes.   The following discussion is limited to spectroscopically and astronomically interesting aspects of the results for the CH_2CN radical.

Table II. New Transient Molecules Studied at Nagoya University[a]

New interstellar molecules:

$C_6H(^2\Pi_i)$, $CCS(^3\Sigma^-)$, C_3S, cyclic-$C_3H(^2B_2)$, $CH_2CN(^2B_1)$, C_4Si, $CP(^2\Sigma^+)$

Candedates for interstellar molecules:

H_3S^+, H_2Cl^+, $PS(^2\Pi_r)$, $H_2CN(^2B_1)$, $HS_2(^2A'')$, NHCN($^2A''$),

MgS, CaS, $AlS(^2\Sigma^+)$, $H_2NS(^2B_1)$, $H_2NO(^2B_1)$, $HSiS(^2A')$

Other molecules:

$NF(^1\Delta)$, HBF^+, $CO(^3\Pi,^3\Sigma)$

---

[a] 1986-1989

IV. $CH_2CN(^2B_1)$

The identification of CH_2CN was made by a collaborative combination of high sensitivity millimeter-wave laboratory spectroscopy and high-resolution centimeter-wave radio astronomy. Thaddeus and his collaborators (13) suggested in 1985 in their paper on the first ring carbene, C_3H_2, that three harmonically related lines, U20120, U80484, and U100602, detected in Sgr B2 and TMC-1 might be due to a linear isomer of C_3H_2, ethynyl methylene, HCCCH. We assigned these three unidentified lines to the rotational transitions of a new free radical, CH_2CN, based on its laboratory microwave spectroscopy (9).

The spectral lines of CH_2CN were detected in a discharge in pure CH_3CN when we examined satellite lines of the HCCN radical (22). We observed several strong paramagnetic lines which could not be assigned to any satellite lines of HCCN, or to any lines of C_3N and C_4H (C_3N and C_4H were expected to exist in the chemical system used). The paramagnetic lines showed an a-type R-branch K structure at about every 20 GHz, modified by the fine structure typical of a doublet radical. They also showed a very complicated hyperfine structure for the $K_{-1}=0$ transitions and a well-resolved triplet structure for a fine structure component of the $K_{-1}=1$ transitions. These facts led us to assign the observed strong paramagnetic lines to the transitions of the $CH_2CN(^2B_1)$ radical, because the ground electronic state of the CH_2CN radical was considered to be $^2B_1$, in analogy to the related radical CH_3. In addition symmetric rotational levels of K_{-1}=even correlate with the symmetric hydrogen nuclear spin function with $I(H)=1$, giving 18 fine and hyperfine components to the even rotational levels, whereas the odd rotational levels pair with the antisymmetric spin function with $I(H)=0$, giving 6 components.

The complicated spectral lines were analyzed by using a conventional Hamiltonian including the hyperfine interaction term of the two nuclei for the doublet radical. We determined 20 molecular constants, by least-squares fitting the observed spectral lines, including the rotational constants, centrifugal distortion constants, spin-rotation coupling constants with their centrifugal distortion terms, and hyperfine coupling constants for hydrogen and nitrogen, as shown in Table III.

Table III. Molecular Constants of $CH_2CN(^2B_1)^a$.

Parameter	Value (MHz)[b]	Parameter	Value (MHz)[b]
A	285752(181)	$^s\Delta_{KN}$	0.0103(42)
B	10246.52(10)	$^s\Delta_K$	0.099(99)
C	9876.25(10)	$a_F(H)$	-60.93(42)
Δ_N	0.004004(27)	$T_{aa}(H)$	-15.74(41)
Δ_{NK}	0.40908(40)	$T_{bb}(H)$	16.3^c
Δ_K	22.0^c	$a_F(N)$	7.22(65)
δ_N	0.000167(33)	$T_{aa}(N)$	-15.445(94)
δ_K	0.148(50)	$T_{bb}(N)$	-14.18(45)
$H_K N$	-0.00116(11)	$\chi_{aa}(N)$	-4.290(112)
ε_{aa}	-658.0(65)	$\chi_{bb}(N)$	2.15^c
ε_{bb}	-24.11(20)		
ε_{cc}	-1.84(20)		

[a]Reference (9).

[b]Values in parentheses denote one standard deviation and apply to the last digits of the constants.

[c]Fixed.

When we observed two a-type R-branch transitions with $K_{-1}=0$ at 80.48 GHz and 100.6 GHz, we found that their frequencies corresponded well with those of U80484 and U100602 described above. Moreover, after we finished the analysis of the low-N, R-branch transitions and obtained a set of the fine and hyperfine coupling constants for both the hydrogen and nitrogen nuclei, we definitely identified other astronomically observed lines as fine and hyperfine components of the 2_{02}-1_{01} transition, as shown in Figure 2. The spectral pattern calculated from the molecular constants is given in the lower half of the Figure, showing a good agreement between the observed interstellar frequencies and the calculated ones. We also identified eleven lines as components of the 1_{01}-0_{00} transition, detected at 20.120 GHz in TMC-1 with the NRAO 43 m telescope (10). However, the initial analysis was not precise enough to assign even weaker interstellar spectral lines and to derive a reliable column density of CH_2CN in TMC-1. For this purpose the spectral intensity was accurately evaluated for each hyperfine component.

The observed interstellar spectral lines given in Figure 2 are not a simple pattern derived from a combination of two triplet times triplet patterns. The reason for this is that the magnitude of the spin-rotation interaction, and the hyperfine interactions of the hydrogen and nitrogen nuclei are similar in energy, especially for the $K_{-1}=0$ levels. As a result, no coupling scheme of the angular mementa concerned could be good quantum numbers to the effective component level. For an example, the strong line observed at 40247.59 MHz is assigned to the 2_{02}-1_{01}, $J=3/2$-$3/2$, $F_1=5/2$-$3/2$, $F=7/2$-$5/2$ transition. However, the effective wave function for the upper level is given by

$$\Psi_u = 0.779\Psi(2_{02}, 3/2, 5/2, 7/2) - 0.625\Psi(2_{02}, 5/2, 5/2, 7/2) + 0.0464\Psi(2_{02}, 5/2, 7/2, 7/2),$$

and the effective wave function for the lower level by

$$\Psi_l = 0.729\Psi(1_{01}, 3/2, 3/2, 5/2) - 0.682\Psi(1_{01}, 1/2, 3/2, 5/2) - 0.0564\Psi(1_{01}, 3/2, 5/2, 5/2),$$

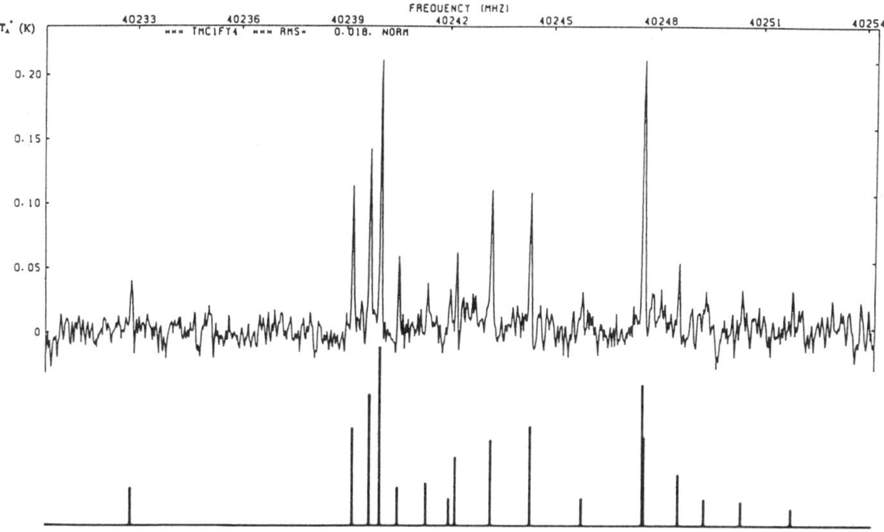

Fig. 2. The observed spectrum of the 2_{02}-1_{01} transition of CH_2CN in TMC-1, shown with the calculated frequencies and relative intensities (vertical bars).

where $\Psi(J_{K_{-1}K_{+1}}, J, F_1, F)$.   Therefore, the spectral intesity is not simply calculated from the square of a general formula for the matrix element of the dipole moment,

$$< N'_\tau, S, J', I_H, F'_1, I_N, F' \parallel \mu_a \parallel N_\tau, S, J, I_H, F_1, I_N, F >$$

$$= (-1)^{N'+S+J+J'+I_H+F_1+F'_1+I_N+F+1} \times [(2J'+1)(2J+1)(2F'_1+1)(2F_1+1)(2F'+1)(2F+1)]^{1/2}$$

$$\times \begin{Bmatrix} J & J' & 1 \\ N' & N & S \end{Bmatrix} \begin{Bmatrix} F_1 & F'_1 & 1 \\ J' & J & I_H \end{Bmatrix} \begin{Bmatrix} F & F' & 1 \\ F'_1 & F_1 & I_N \end{Bmatrix} < N'_\tau \parallel \mu_a \parallel N_\tau > \qquad (1)$$

but it must be calculated by taking into account the mixing of wave functions.   The spectral line intensity for the components of the 2_{02}-1_{01} and 1_{01}-0_{00} transitions was calculated from Eq.(1) by using the effective wave functions transformed from the pure wave functions defined by the coupling scheme of the angular momenta, $J=N+S$, $F_1=J+I_H$, and $F=F_1+I_N$. The calculated results for the 2_{02}-1_{01} transition are shown in the lower half of the Figure 2.   A good correspondence was obtained between the relative intensities of the observed and calculated spectral lines.   Furthermore, the detailed intensity analysis made it possible to assign 16 observed weak lines to the fine and hyperfine components of the 2_{02}-1_{01} transition.

Since the intensity of each component relative to the total intensity of the rotational transition is now known, a total column density of the CH_2CN radical in TMC-1 may be estimated from observed specral line intensities on the assumption that the spectral lines used are optically thin, and the level populations are in local thermal equilibrium with the excitation temperature of 10K. The result is 1.0×10^{14} cm^{-2} for TMC-1.   The column density of CH_2CN in TMC-1 may be compared with that of CH_3CN: 1.0×10^{13} cm^{-2} in TMC-1 (23). CH_2CN and CH_3CN are considered to be produced from the same precursor molecular ion, CH_3CNH^+, so that the production of CH_2CN and CH_3CN depends on the branching ratio of the dissociative recombination reaction of CH_3CNH^+. The large ratio of CH_2CN to CH_3CN observed in TMC-1 suggests a possibility that the precursor ion to CH_2CN is different from that of CH_3CN, CH_2CNH^+.   The chemistry of CH_2CN and its related molecules is discussed in a separate paper (24).

The detailed analysis of the spectral lines observed in the laboratory as well as in interstellar space has established that CH_2CN is a unique interstellar free radical, showing the most complicated spectral pattern among those interstellar molecules detected so far.   Our result also shows that extensive knowledge of the spectroscopic details of the CH_2CN radical is essential to full understanding of its interstellar spectral lines.

V. CONCLUSIONS

Several new interstellar molecules have been discussed from the standpoints of laboratory microwave spectroscopy and radio astronomy.   Vast, cold, and low-density space environment is a unique molecular laboratory, whose physical and chemical conditions are rarely attained in the laboratory on Earth.   This unique space laboratory is favorable to the existence of transient molecules, most of which are very exotic and non-terrestrial.   These exotic transient molecules are generally difficult and challenging problems for laboratory spectroscopy.   In other words, laboratory spectroscopy may be enriched by astronomical studies on non-terrestrial transient species which represent new development in high-resolution molecular spectroscopy.   On the other hand, detailed knowledge about new transient molecules obtained by laboratory spectroscopy is essential to a deeper understanding of physical and chemical processes in space.   The rotational and vibrational energy structure, the line strength of transitions, the dipole moment, and the chemical behavior of each molecule obtained by laboratory spectroscopy is inevitably necessary to analyses and deductions of its excitation mechanism, its column density determination, its kinetic behavior in clouds, and its interstellar chemistry in space.   Furthermore, the mutually beneficial aspect between laboratory spectroscopy and radio astronomy has been now accelerated by recent technical progress in both fields.   Mutually stimulating collaborations herald remarkable new progress in both radio astronomy and laboratory spectroscopy.

ACKNOWLEDGMENTS

Most of the results discussed in the paper were obtained by collaborations with several astronomers and spectroscopists, the late H. Suzuki, M. Ohishi, and N. Kaifu of Nobeyama Radio Observatory, K. Kawaguchi of the Institute for Molecular Science, S. Yamamoto of Nagoya University, and W. M. Irvine of the University of Massachusetts, to whom the author is very grateful.

This study was supported by Grants-in-Aid from the Ministry of Education, Science, and Culture (Nos. 62470015, 62740277, 63632514, 63540192, and 01632513).

REFERENCES

(1) Saito, S. 1978, **Pure Appl. Chem., 50,** 1239.
(2) Townes, C. H. and Schowlow, A. L. 1955, **Microwave Spectroscopy,**McGraw-Hill, New York.
(3) Yamamoto, S. and Saito, S. 1988, **J. Chem. Phys., 89,** 1936.
(4) Woods, R. C. 1973, **Rev. Sci. Instr., 44,** 282.
(5) Suzuki, H., Ohishi, M., Kaifu, N., Ishikawa, S., Kasuga, T., Saito, S., and Kawaguchi, K. 1986, **Publ. Astrom. Soc. Japan, 38,** 911.
(6) Saito, S., Kawaguchi, K., Suzuki, H., Ohishi, M., Kaifu, N., and Ishikawa, S. 1987, **Publ. Astrom. Soc. Japan, 39,** 193.
(7) Saito, S., Kawaguchi, K., Yamamoto, S., Ohishi, M., Suzuki, H., and Kaifu, K. 1987, **Astrophys. J., 317,** L115.
(8) Yamamoto, S., Saito, S., Kawaguchi, K., Kaifu, N., Suzuki, H., and Ohishi, M. 1987, **Astrophys. J., 317,** L119.
(9) Saito, S., Yamamoto, S., Irvine, W. M., Ziurys, L. M., Suzuki, H., Ohishi, M., and Kaifu, K. 1988, **Astrophys. J., 334,** L113.
(10) Irvine, W. M., Friberg, P., Hjalmarson, A., Ishikawa, S., Kaifu, N., Kawaguchi, K., Madden, S. C., Matthews, H. E., Ohishi, M., Saito, S., Suzuki, H., Thaddeus, P., Yamamoto, S., and Ziurys, L. M. 1988, **Astrophys. J., 334,** L107.
(11) Ohishi, M , Kaifu, N., Kawaguchi, K., Murakami, A., Saito, S., Yamamoto, S., Ishikawa, S., Fujita, Y., Shiratori, Y., and Irvine, W. M. 1990, **Astrophys. J., 345,** L83.
(12) Kaifu, N , Suzuki, H., Ohishi, M., Miyaji, T., Ishikawa, S., Kasuga, T., Morimoto, M., and Saito, S. 1987, **Astrophys. J., 317,** L111.
(13) Thaddeus, P., Vrtilek, J. M., and Gotlieb, C. A. 1985, **Astrophys. J., 299,** L63.
(14) Yamamoto, S., Saito, S., Ohishi, M., Suzuki, H., Ishikawa, S., Kaifu, N., and Murakami, A. 1987, **Astrophys. J., 322,** L55.
(15) Yamamoto, S., Saito, S., Guelin, M., Cernicharo, J., Suzuki, H., and Ohishi, M. 1987, **Astrophys. J., 323,** L149.
(16) Turner, B. E., Tsuji, T., Bally, J., Guelin, M., and Cernicharo, J. 1990, **Nature,** in preparation.
(17) Guelin, M., Cernicharo, J., Paubert, G., and Turner, B. E. 1990, **Astron. Astrophys.,** in press.
(18) Saito, S., Yamamoto, S., Kawaguchi, K., Ohishi, M., Suzuki, H , Ishikawa, S., and Kaifu, K. 1989, **Astrophys. J., 341,** 1114.
(19) Saito, S., Yamamoto, S., and Amano, T. 1987, **Astrophys. J., 314,** L27.
(20) Saito, S., Yamamoto, S., and Kawaguchi, K. 1988, **J. Chem. Phys., 88,** 2281.
(21) Ohishi, M., Yamamoto, S., Saito, S., Kawaguchi, K., Suzuki, H , Kaifu, K., Ishikawa, S., Takano, S., Tsuji, T., and Unno, W. 1988, **Astrophys. J., 329,** 52.
(22) Saito, S., Endo, Y., and Hirota, E. 1984, **J. Chem. Phys., 80,** 1427.
(23) Irvine, W. M. and Schloerb, F. P. 1984, **Astrophys. J., 282,** 516.
(24) Turner, B. E., Friberg, P., Irvine, W. M., Saito, S., and Yamamoto, S. 1990, **Astrophys. J.,** in press.

Microwave Spectrum of CCS Radical in Vibrationally Excited States

Mitsutoshi TANIMOTO[1], Shuji SAITO[2], Satoshi YAMAMOTO[2], and
Kentarou KAWAGUCHI[3]

[1] *Department of Chemistry, Faculty of Science, Shizuoka University, Ohya, Shizuoka 422, Japan*
[2] *Department of Astrophysics, Nagoya University, Chikusa-ku, Nagoya 464, Japan*
[3] *Nobeyama Radio Observatory, National Astronomical Observatory, Nobeyama, Minamimaki, Nagano 384-13, Japan*

Microwave spectra of the interstellar molecule CCS have been observed in three vibrationally excited states $v_1=1$, $v_2=1$ and $v_3=1$ to obtain information on the vibrational frequencies. The spectrum of the v_1 and v_3 excited states (stretching vibrations) were analyzed using a Hamiltonian for the $^3\Sigma$ vibronic state, whereas the spectrum of the v_2 excited state (bending) was analyzed with Hamiltonian matrix elements for the $^3\Pi$ vibronic state. For the v_2 bending excited state it was necessary to include a small spin-orbit interaction constant A. The frequency of the bending vibration was estimated from the l-type doubling constant q and was found to be rather low when compared with that of OCS.

INTRODUCTION

In 1984 Dr. Suzuki and collaborators detected a strong unidentified line U45379 in the Taurus Molecular Cloud 1 (1,2). It took several years before this unidentified line was finally assigned through a laboratory spectroscopic study to the CCS radical (3). This was the first observation that carbon chain molecules bearing a sulfur atom exist in substantial amount in interstellar space. Before this identification, sulfur-containing carbon chain molecules had not been known at all in the fields of both spectroscopy and astronomy. This is in strong contrast to the oxygen analogue CCO, which was extensively studied with ultraviolet and infrared spectroscopy (4,5,6) and its vibrational frequencies precisely determined. There is, however, no information on the vibrational frequencies of the CCS radical. Hence the microwave spectrum of the CCS radical in vibrationally excited states was measured to gain some information on the vibrational frequencies of CCS.

EXPERIMENTAL AND ANALYSIS

The spectrum of the CCS radical was observed using a source modulation spectrometer with a glow discharge cell (7), which was 2 m in length and 90 mm in inner diameter. The experimental condition was optimized by observing the ground state lines of the radical. A mixture of 20 mTorr CS_2 and 30 mTorr He was flowed through the cell and glow discharged with 200 mA of current. The cell was cooled down to about −30 C, to make the absorption lines stronger.

We searched for the spectral lines around 285 GHz, where Saito et al. (3) observed the ground-state transition N=22-21. We picked up candidate lines whose intensity was stronger than 5% (1/20) of the intensity of the ground state lines, which is sufficient to detect vibrationally excited lines.

We observed many relatively stronger lines on the lower frequency side of the ground state lines, whereas we observed weaker lines on the higher frequency side. When a degenerate bending motion is excited, a pair of lines with equal intensity would be expected as a result of the l-type doubling. Hence, if the lines on the lower frequency side should be due to excited states of bending vibration, they must have counterparts on the higher frequency side. However, we could not observe any corresponding lines of similar intensity on the higher frequency side. Accordingly

these strong lines could not be assigned to the lines of the bending vibration.
While we were picking up the candidate lines, we also noted the behaviour of each
absorption line for the Zeeman effect. The Zeeman effect on the two components F_1
and F_3 among the three lines for the triplet state molecules is expected to be large,
whereas the Zeeman effect on the central F_2 line is small. The central F_2 component
would show only some broadening and did not collapse significantly on application of
the magnetic field. Referring to the Zeeman behaviour and to the fine structure
splitting expected from those in the ground state, we selected the F_1, F_2 and F_3
components. We compared the vibration-rotation interaction constants o with that of
OCS and assigned the stronger lines on the low frequency side to the excited states
of the v_1 and v_3 stretching vibrations. Two set of three lines on the higher
frequency side were subsequently ascribed to the l-type doubled v_2 states. The fine
structure splittings were rather different from those in the ground state. From the
crude rotational constants obtained in this way we estimated the transition
frequencies down to 230 GHz and those for 100 GHz region (N=10-9...8-7). The v_1 and
v_3 states belong to the same vibronic $^3\Sigma^-$ state as that of the ground state. The
analysis of the spectrum was analyzed by using the Hamiltonian for a $^3\Sigma$ state (8).
On the other hand, the v_2 states belong to the different vibronic state of $^3\Pi$
symmetry, and thus the spectrum was analyzed by using the effective Hamiltonian for
the $^3\Pi$ vibronic state (9).

RESULTS AND DISCUSSION

    The molecular constants for the vibrationally excited states are listed in Table
1. The centrifugal distortion constant D is related to the vibrational frequency and
the observed constants yield a vibrational frequency of about 830 cm^{-1}. This
vibration will probably be the lower of the two stretching vibrations, i. e. the C=S
stretching vibration. This stretching vibration is nearly equal to that for OCS. On
the other hand, the l-type doubling constant q is correlated with the vibrational
frequency v_2. This information has been used to estimate the v_2 bending frequency to
be about 300 cm^{-1}. The bending vibrational frequency is about two thirds of the OCS
bending frequncy. This corresponds well to a large difference in the rotational
constant difference between the ground and the v_2 excited states in CCS. Although we
were not able to draw any information on the highest stretching vibration v_3 from the
present microwave spectroscopic study, the information on vibrational frequencies of
the CCS radical will be helpful in observing the infrared spectrum of the molecule.

References
(1) Suzuki, H., Kaifu, N., Miyaji, T., Morimoto, M., Ohishi, M., and Saito, S. 1984,
    Astrophys. J., **282**, 197.
(2) Kaifu, N., Suzuki, H., Ohishi, M., Miyaji, T., Ishikawa, S., Kasuga, T.,
    Morimoto, M., and Saito, S. 1987, Astrophys. J. (Letters), **317**, L111.
(3) Saito, S., Kawaguchi, K., Yamamoto, S., Ohishi, M., Suzuki, H. and Kaifu, N.
    1987, Astrophys. J. (Letters), **317**, L115
(4) Jacox, M. E., Milligan, D. E., Moll, N. G., and Thompson, W. E. 1965, J. Chem.
    Phys., 43, 3734.
(5) Devillers, C., and Ramsay, D. A. 1971, Canadian J. Phys., **49**, 2839.
(6) Yamada, C., Kanamori, H., Horiguchi, H., Tsuchiya, S., and Hirota, E. 1986, J.
    Chem. Phys., **84**, 2573.
(7) Yamamoto, S., and Saito, S. 1988, J. Chem. Phys., 89, 1936.
(8) Gordy, W., and Cook, R. L. 1984 Microwave Molecular Spectra (New York: John
    Wiley), chap. 4.
(9) Effantin, C., Michaud, F., Roux, F., d'Incan, J., and Verges, J. 1982, J. Mol.
    Spectrosc., **92**, 349.

Vibrationally Excited Carbon-Chain Molecules in the Laboratory and in Space; C_3N and C_4H

Hitomi MIKAMI

Department of Astrophysics, Faculty of Science, Nagoya University, Chikusa-ku, Nagoya 464-01, Japan

The rotational transitions of the C_3N ($\tilde{X}^2\Sigma^+$) radical in the lowest excited bending mode, ν_5, have been observed in the laboratory, and have been least-squares analyzed to determine the molecular constants. A number of the measured or calculated line frequencies have been surveyed in the carbon star envelope IRC+10216; none has been detected. The line intensity of C_3N in the ν_5 state is less than 1/15 - 1/20 of that in the ground vibrational state. The small line intensity ratio between the ν_5 state and ground state of C_3N contrasts with the large corresponding ratios observed for the two lowest bending states, ν_7 and $2\nu_7$, of C_4H in this source. A sharp contrast between the populations in the vibrationally excited states for C_3N and C_4H in the envelope of IRC+10216 suggests that the high efficiency for the vibrationally excitation of C_4H is brought by a radiative pumping via its low-lying electronically excited state.

The rotational spectra of several molecules in vibrationally excited states have been detected in the carbon star envelope IRC+10216 by using radio telescopes- HCN(1,2), CS(3), SiS(4,5), HC_3N(6), and C_4H(4,7). The observed line shapes indicate that the first four molecules are excited in the inner part of the envelope, and only C_4H is excited in the outer part. Furthermore, the line intensities of the species excited in the inner part are typically 20 times or more weaker in their vibrationally excited states than in ground vibrational states, whereas the lines of two lowest bending states of C_4H, ν_7 and $2\nu_7$, have intensities as large as 1/3 and 1/6 of those of the ground state, respectively. Since the ν_7 and $2\nu_7$ states of C_4H lie at about 200 K and 400 K above the ground state, respectively(7), the large intensity ratios for two lowest bending states of C_4H suggest an efficient excitation of C_4H to the vibrationally excited states in the low density and low temperature region. In order to obtain further information on the excitation mechanism of C_4H in the envelope of IRC+10216, we studied other vibrationally excited carbon-chain radicals with laboratory microwave spectroscopy, in order to conduct astronomical searches with radio telescopes. We chose the C_3N radical, because C_3N is isoelectronic with C_4H, and the vibrational frequency of its lowest bending mode, ν_5, is close to that of the ν_7 state of C_4H. Further, the line intensities of C_3N and C_4H in their ground states are almost the same in the outer envelope of IRC+10216.

In the laboratory experiment, we used a source-modulation spectrometer and a 2 m long free space absorption cell cooled to about 100 K (8). The C_3N radical was produced by using dc glow discharge in a flowing mixture of C_2H_2, N_2, and He. The optimum conditions were 9 mTorr for C_2H_2, 7 mTorr for N_2, 6 mTorr for He, and 200 mA for the discharge current. The microwave source is a frequency doubler or tripler powered by millimeter-wave klystrons. We observed ten rotational transitions of C_3N in the ν_5 state in the frequency region of 188 GHz to 278 GHz. We analyzed the measured transition frequencies by a least-squares method, and determined the molecular constants for C_3N in the ν_5 state. The determined molecular constants are given in Table I, with those for C_4H in the ν_7 state for comparison.

On the basis of the laboratory result, we examined the frequency region covered by the spectral survey of IRC+10216 with the IRAM 30 m radio telescope(6). No lines were detected, the most significant upper limits being $T_A^* < 0.03$ - 0.04 K (e.g. Figure 1). This means that the line intensity of the ν_5 state of C_3N is less than 1/15 - 1/20 of that of the ground state. This result is similar to the cases of HCN, CS, SiS, and HC_3N in the vibrationally excited states in IRC+10216. The small intensity ratios of these molecules contrast with the large values for the ν_7 and $2\nu_7$ states of C_4H. Therefore, the vibrational excitation of C_4H suggests the presence of a particular excitation mechanism for C_4H.

C_4H in the ν_7 state has a spin-orbit coupling constant, A_{so}, of about -90000 MHz(7) (see Table I). On the other hand, the A_{so} value of C_3N in the ν_5 state is about -5300 MHz. The A_{so} value in a $^2\Sigma$ electronic state is considered to indicate a degree of the vibronic mixing between the ground electronic state, $\tilde{X}^2\Sigma^+$, and the first electronically excited state, $\tilde{A}^2\Pi_i$. The A_{so} constants found for the lowest bending states of C_3N and C_4H suggest a large difference in energy between the first electronically excited states of the two molecules. C_2H has an A_{so} value of about -10000 MHz(9) in the ν_2 state, and has the first electronically excited state at about 3600 cm^{-1} above the ground electronic state(10). The A_{so} constant of C_4H in the ν_7 state is much larger than that of C_2H in the ν_2 state. Hence, the energy of the first electronically excited state for C_4H is expected to be smaller than 3600 cm^{-1}. On the other hand, A_{so} of C_3N in the ν_5 state is smaller than that of C_2H in the ν_2 state, so that the $\tilde{A}$ state of C_3N must lie at a level of higher than 3600 cm^{-1}.

These facts led us to the conclusion that C_4H is radiatively excited to the first electronically excited state, $\tilde{A}^2\Pi_i$, and the population in the ν_7 and $2\nu_7$ states is supplied by the ensuing downwards cascading process from the $\tilde{A}$ state. Infrared radiation from the central star IRC+10216, which has the peak flux at 10 μ(11), is responsible for the efficient pumping of C_4H to the first electronically excited state. On the other hand, C_3N is not effectively pumped to the first electronically excited state, because the $\tilde{A}$ state of C_3N is located at an energy far from the peak-power radiation from IRC+10216. The excitation mechanism is shown schematically in Figure 2. In the figure, the solid lines imply the allowed electronic transitions, while the dashed lines represent the weak vibronic transitions.

We also studied spectral lines of C_3H(12) and C_5H(13) in their vibrationally excited states in the laboratory, but we did not recognized any features of them in IRC+10216. This result consistently suggests that C_3H and C_5H could not be excited to the first electronically excited state in the envelope of IRC+10216, because the energy differences between their two lowest electronic states correspond to radiation in the near-infrared or visible.

As a summary, we found that the infrared radiation field of the carbon star IRC+10216 has a strong interaction with the first electronically excited state of C_4H, but not with those of other carbon-chain molecules, and showed that the spectral lines of vibrationally excited C_4H can be one of the powerful tools probing the radiation field in the circumstellar envelope.

REFERENCES

1. Ziurys, L. M. and Turner, B. E. 1986, Ap. J. (Letters), **300**, L19.
2. Lucas, R., Omont, A., Guilloteau, S., and Nguyen-Q-Rieu 1986, Astr. Ap. (Letters), **154**, L12.
3. Turner, B. E. 1987a, Astr. Ap. (Letters), **182**, L15.
4. Guelin, M., Cernicharo, J., Navarro, S., Woodward, D. R., Gottlieb, C. A., and Thaddeus, P. 1987, Astr. Ap. (letters), **182**, L37.
5. Turner, B. E. 1987b, Astr. Ap. (Letters), **183**, L23.
6. Cernicharo, J., Guelin, M., and Kahane, C. 1990, in preparation.
7. Yamamoto, S., Saito, S., Guelin, M., Cernicharo, J., Suzuki, H., and Ohishi, M. 1987, Ap. J. (letters), **323**, L149.
8. Yamamoto, S, and Saito, S. 1988, J. Chem. Phys., **89**, 1936.
9. Woodward, D. R., Pearson, J. C., Gottlieb, C. A., Guelin, M., and Thaddeus, P. 1987, Astr. Ap. (letters), **186**, L14.
10. Curl, R. F., Carrick, P. G., and Merer, A. J. 1985, J. Chem. Phys., **82**, 3479.
11. Campbel, M. F., Elias, J. H., Gezari, D. Y., Harvey, P. M., Hoffmann, W. F., Hudson, H. S., Neugebauer, G., Soifer, B. T., Werner, M. W., and Westbrook, W. E. 1976, Ap, J., **208**, 396.
12. Yamamoto, S., Saito, S., Suzuki, H., Deguchi, S., Kaifu, N., Ishikawa, S., and Ohishi, M. 1990, Ap. J., **348**, 363.
13. Takano, S., Yamamoto, S., and Saito, S. 1990, in preparation.

Table I. Molecular constants of C$_3$N and C$_4$H
in vibrationally excited states (MHz)[a]

	C$_3$N(ν_5)	C$_4$H(ν_7)[b]
B_v	4967.5600(14)	4762.8472(15)
α_i	−19.9392(18)	−4.1915(17)
D_v	0.0008263(11)	0.00089118(98)
A_{so}	−5295.(42)	−89799.(18)
γ_v	−17.80(12)	−37.92(20)
p	1.916(54)	17.834(52)
q	11.9376(27)	14.9674(30)
q_D	−0.0000613(22)	−0.0001183(20)

[a]The numbers in parentheses denote three standard deviations in units of the last significant digits.
[b]Ref. (7)

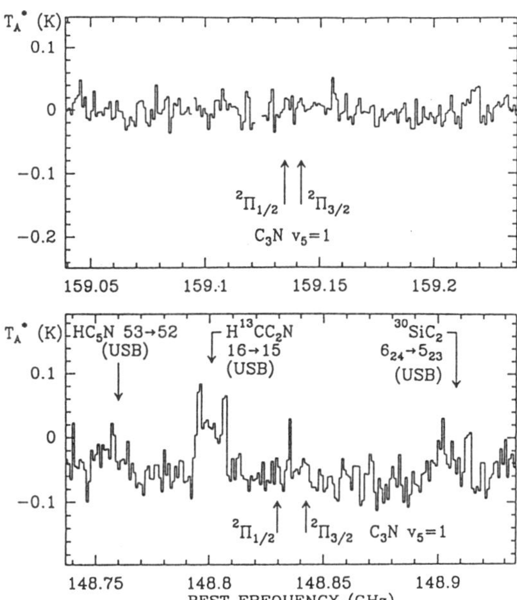

Figure 1. Two spectra centered at the frequencies of the $J = 31/2 \rightarrow 29/2$ ($^2\Pi_{1/2}$) and $33/2 \rightarrow 31/2$ ($^2\Pi_{3/2}$) (upper spectrum), and $J = 29/2 \rightarrow 27/2$ ($^2\Pi_{1/2}$) and $31/2 \rightarrow 29/2$ ($^2\Pi_{3/2}$) (lower spectrum) rotational transitions of C$_3$N in the ν_5 state, observed in the direction of the carbon star envelope IRC+10216 with the IRAM 30 m radio telescope. The upper spectrum has been observed in the single sideband mode, the lower in double sideband mode. The lines labelled USB lie in the receiver upper sideband.

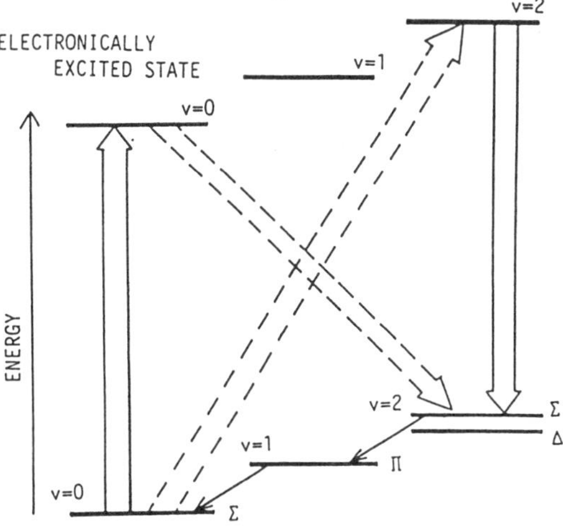

Figure 2. The excitation mechanism of C$_4$H in the envelope of IRC+10216. The solid lines indicate the allowed electronic transitions, and the dashed lines represent weak vibronic transitions. C$_4$H is excited up to the first electronically excited state by infrared radiation from the central star IRC+10216, and subsequently, falls into the the vibrational levels in the ground electronic state.

The Structure and Chemistry of the Carbon-Chain Molecule, C_3H

Mitsuka KANADA*

Department of Astrophysics, Nagoya University, Chikusa-ku, Nagoya 464-01, Japan

The rotational spectra of three ^{13}C isotopic species of the C_3H radical were studied with laboratory microwave spectroscopy. The detailed molecular constants for the three ^{13}C species were determined by least-squares analyses of the observed ^{13}C hyperfine-resolved spectral lines. The molecular structure of C_3H was determined from the rotational constants for the ^{13}C, D, and normal species. The spin density of unpaired electrons was derived from the hyperfine coupling constants. The molecular structure and spin density obtained indicate delocalization of electrons including the unpaired electrons. Astronomical identifications of the ^{13}C-C_3H species are discussed on the basis of the interstellar chemistry of C_3H and its related molecules.

The linear carbon-chain molecules C_nH are specific to cold dark clouds. The spectral lines of CH to C_6H have been identified in the circumstellar envelope IRC+10216 and the molecular cloud TMC-1, in the years 1973 to 1986(1-7). The ground electronic states of even-numbered carbon-chains (C_2H, C_4H) are $^2\Sigma$, whereas those of odd-numbered chains (CH, C_3H, C_5H) are $^2\Pi$. However, the electronic structure of C_6H was found to be $^2\Pi$, though it has an even-number of carbon atoms. The electronic structure of the carbon-chain molecules is more difficult to predict for longer chain members. Another peculiarity of carbon-chain molecules is their abundances in astronomical sources. C_nH (n=2-6) in IRC+10216 and TMC-1 are found to have different abundances between odd and even-numbered chains(6,7). Odd-numbered chains(C_3H and C_5H) are much more deficient than even-numbered ones(C_2H and C_4H).

From a high-resolution spectroscopic point of view the peculiarities of the carbon-chain molecules must reflect their detailed molecular structures. Exact knowledge about the molecular structure of each member of C_nH may allow us obtain information about longer members of C_nH, for example C_7H and C_8H. The CH and C_2H radicals have been extensively studied with high-resolution spectroscopic methods(8,9) and their molecular structure are known. So far no molecular structure has been reported for C_3H.

We studied rotational spectra of three ^{13}C isotopic species of C_3H with laboratory microwave spectroscopy(10). The ^{13}C-C_3H radicals were produced by discharging a mixture of C_2H_2, ^{13}CO, and He in free space cell cooled to liquid-nitrogen temperature. We observed the rotational spectra of ^{13}C-C_3H in the ground state, $^2\Pi_r$, and also in the vibrationally excited state of the CCH bending mode v_4, $^2\Sigma^\mu$. Most observed spectral lines show hyperfine structure due to the ^{13}C nucleus. Several low-J transitions show further splittings due to the hydrogen nucleus. We observed 9 rotational transitions for the C1, C2, and C3 species in the ground state ($^2\Pi_r$), and 6 rotational transitions in the vibrationally excited state ($^2\Sigma^\mu$), where C1, C2, and C3 denote $^{13}CCCH$, $C^{13}CCH$, and $CC^{13}CH$, respectively. The line intensities of the three species were not remarkably different from one another. The observed spectral lines were least-squares analyzed to determine the molecular constants, including the hyperfine coupling constants for the ^{13}C and H nuclei. In the analysis the vibronic coupling between the ground and vibrationally excited states was taken into account. The molecular constants are given in Table I.

We determined the molecular structure of C_3H from the rotational constants for the ^{13}C, deuterated, and normal species, and obtained the spin density of unpaired electrons from the observed hyperfine coupling constants for the three ^{13}C nuclei. The structure determined is given in Table II, where the

---

*Present address: NTT LSI Laboratories, 3-1 Morinosato Wakamiya, Atsugi-shi, Kanagawa, Japan.

structure obtained by an ab initio calculation at the MCSCF(6-31G)(11) level is also given for comparison. The bond lengths obtained in the present study agree well with those given by the ab initio calculation. The C_1-C_2 bond length was determined to be 1.326 Å and the C_2-C_3 bond length to be 1.254 Å, which can be compared with that for the normal C-C bond length of 1.504 Å and the C≡C bond length of 1.203 Å, respectively. The large deviations from the normal bond lengths are considered to be caused by delocalization of π electrons including unpaired electrons. The C-H bond length is shorter than the normal C-H bond length as found in C_2H_2. The short C-H bond may correspond to a vibrationally averaged bond length along the molecular axis. As a result, the molecular structure of the C_3H radical in the ground state is considered to be a hybrid between two classical structures: an acetylene form and an allene form as shown in Figure 1. The picture of the structure obtained is consistent with the spin density of the unpaired p-electrons on each carbon atom estimated from the hyperfine coupling constants, d+c/3(12), as shown in Table III. The spin density is largest at C_1 and smallest at C_2. This is schematically shown in Figure 2. The molecular orbital of the unpaired electrons has a node at the central carbon atom C_2. ·

Since the transition frequencies of the three ^{13}C species for C_3H are now available, we discuss the possibility of their astronomical identification. According to Adams and Smith(13), the linear C_3H radical as well as the cyclic C_3H radical and the cyclic C_3H_2 are produced by the dissociative recombination of the cyclic or linear $C_3H_3^+$ ion.

$$\left.\begin{array}{l} \text{l-}C_3H_3^+ \\ \\ \text{c-}C_3H_3^+ \end{array}\right\} + e \quad \rightarrow \quad \left\{\begin{array}{l} \text{l-}C_3H,\ \text{c-}C_3H \\ \\ \text{c-}C_3H_2 \end{array}\right.$$

where l- means the linear form and c- the cyclic form. In 1988 Bell and his collaborators reported N(c-C_3HD)/N(c-C_3H_2)~0.08, and N(c-^{13}C-C_3H_2)/N(c-C_3H_2)~0.025 in TMC-1(14). If l-C_3H is produced by the reaction scheme given above, the following relations are derived assuming statistical D/H arrangements;

$$N(C_3D)/N(C_3H) \sim 0.04,$$
and
$$N(\text{l-}^{13}C\text{-}C_3H)/N(\text{normal-}C_3H) \sim 0.012.$$

These relations can be used as one of several crucial tests for the precursors suggested above if the column densities for the D and ^{13}C species of C_3H are obtained for some astronomical sources. However, the spectral intensity of the J=3/2-1/2 transition for $C_3H(^2\Pi_{1/2})$ was reported to be 0.35 K in TMC-1(3). This means that spectral lines of the D and ^{13}C species of C_3H could hardly be detected even with modern high-sensitivity radio telescopes. Rapid progress of radio-telescope techniques will give a chance to detect these species in space in future.

References
(1) Ziurys, L. M., and Turner, B. E. 1985, Ap. J., <u>292</u>, L25.
(2) Tucker, K. D., Kutner, M. L., and Thaddeus, P., 1974, Ap. J., <u>193</u>, L115.
(3) Thaddeus, P., Gottlieb, C. A., Hjalmarson, A., Johansson, L. E. B., Irvine, W. M., Friberg, P., and Linke, R. A., 1985, Ap. J. <u>294</u>, L49.
(4) Guelin, M., Cernicharo, J., Navarro, S., Woodward, D. R., Gottlieb, C. A., and Thaddeus, P., 1987, Astron. Astrophys., <u>182</u>, L37.
(5) Cernicharo, J., Kahane, C., Gomez-Gonzalez, J., and Guelin, M., 1986, Astron. Astrophys., <u>194</u>, L1.
(6) Suzuki, H., Ohishi, M., Kaifu, N., Ishikawa, S., Kasuga, T., Saito, S., and Kawaguchi, K., 1986, Publ. Astron. Soc. Japan, <u>38</u>, 911.
(7) Saito, S., Kawaguchi, K., Suzuki, H., Ohishi, M., Kaifu, N., and Ishikawa, S., 1987, Publ. Astron. Soc. Japan, <u>39</u>, 193.
(8) Lubic, K. G., and Amano, T. 1984, J. Chem. Phys., <u>81</u>, 1655.
(9) Yan, W., Dane, C. B., Zeitz, D., Hall, J. L., and Curl, R. F., 1987, J. Mol. Spectrosc. <u>123</u>, 486.
(10) Yamamoto, S., and Saito, S. 1988, J. Chem. Phys., <u>89</u>, 1936.
(11) Osamura, Y. 1988, Private communication.
(12) Frosch, R. A., and Foley, H. M. 1952, Phys. Rev., <u>88</u>, 1337.
(13) Adams, N. G., and Smith, D. 1987, Ap. J., <u>317</u>, L25.
(14) Bell, M. B., Avery, L. W., Matthews, H. E., Feldman, P. A., Watson, J. K. G., Madden, S. C., and Irvine, W. M., 1988, Ap. J., <u>326</u>, 924.

Table I.   Molecular Constants of the $^{13}C_3H$ radical   (MHz)[a].

Constants	^{13}CCCH		C^{13}CCH		CC^{13}CH	
Ground State ($^2\Pi r$)						
B	10755.821	(32)	11187.907	(40)	10842.577	(36)
D	0.0047834	(58)	0.0051136	(80)	0.0047888	(68)
A_{SO}	432418	(22)	432279	(26)	431463	(31)
γ	−51.6	(53)	−51.07	(69)	−47.15	(91)
P	−6.87	(11)	−7.48	(11)	−7.96	(16)
q	−11.907	(53)	−12.837	(63)	−12.314	(55)
a(^{13}C)	62.6	(61)	−6.7	(64)	44.1	(69)
b	61.7	(49)	−70.0	(63)	137.0	(53)
c	−80	(25)	98	(31)	−74	(28)
d	125.8	(55)	0.0[b]		90.8	(55)
b+c/3	35.0	(97)	−37	(12)	112	(54)
d+c/3	99	(10)	33	(10)	66	(56)
a(H)			16.2	(76)	17.1	(99)
d					14.6	(99)
v_4 State ($^2\Sigma$)						
B	10778.590	(67)	11211.150	(80)	10863.109	(75)
D	0.004643	(19)	0.004979	(22)	0.004673	(25)
γ	−34.73	(31)	−35.97	(19)	−35.58	(21)
γ_D	0.00094	(60)				
ES	608160	(1358)	604970	(1380)	594624	(1210)
b(^{13}C)					161.1	(18)
β	−1202.4	(89)	−1253	(10)	−1219	(92)
Standard Deviation	0.0246		0.0304		0.0332	

[a] 3σ

[b] Fixed.

Table II.   Bond Length of C_3H (Å)

	$r(C_1-C_2)$	$r(C_2-C_3)$	$r(C_3-H)$
This work	1.326	1.254	1.017
MCSCF(6-31G)	1.366	1.247	1.055

Table Ⅲ.　　Spin Densities of C_3H. (%)

	C1	C2	C3	total
s character	0.927	0.988	2.97	4.89
p character	36.9	12.2	24.6	73.7
total	37.8	13.2	27.6	78.6

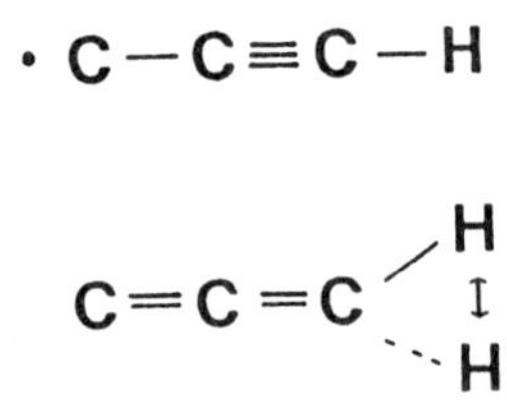

Figure 1. Molecular structure

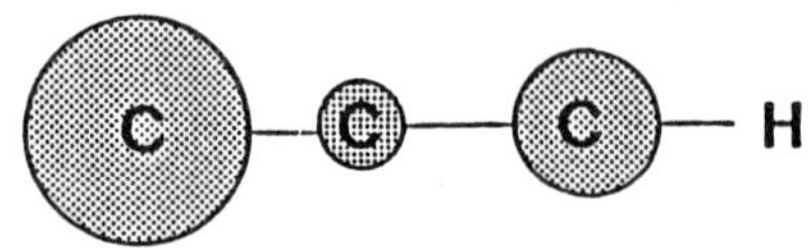

Figure 2. Spin densities

The Dissociative Recombination of H_3^+ with Electrons is not Very Slow

Takayoshi AMANO

Herzberg Institute of Astrophysics, National Research Council, Ottawa, K1A 0R6, Canada

The dissociative recombination rate coefficients for H_3^+ in the ground vibrational state is measured for the five rotatinal levels by monitoring the decay of the infrared absorption signals as a function of time. The rate coefficients are 1.8, 2.5, and 4.4 in units of 10^{-7} cm^{3}s^{-1} for H_3^+ at 273 K, 210 K, and 110 K, respectively. These values agree very well with those obtained using the stationary afterglow or the merged beam techniques, but they disagree with that obtained by Smith and coworkers($\leq 2 \times 10^{-8}$cm^{3}s^{-1}) using the flowing afterglow/Langmuir probe(FALP) method. The temperature dependence obtained here is consistent with those obtained previously using the stationary afterglow and the merged beam techniques. The measurements are extended to HN_2^+ and HCO$^+$, and the rate coefficients obtained agree well with the values determined previously.

INTRODUCTION

The process which we are going to discuss is a very simple but very fundamental one

$$H_3^+ + e^- \rightarrow \text{products.} \tag{1}$$

As H_3^+ plays a key role in laboratory discharge plasmas, in planetary atmospheres, and in interstellar space, its dissociative recombination rate coefficient has been measured repeatedly using various techniques. However, there is a controversy about the dissociative recombination rate of this process.

In 1984, Adams, Smith, and Alge(1) claimed that the dissociative recombination of H_3^+ with electrons was immeasurably slow, and set the upper limit of the rate coefficient to be 2×10^{-8} cm^{3}s^{-1}. More recently Adams and Smith(2) revised the upper limit to be even smaller, less than 10^{-11}cm^{3}s^{-1}. They ascribed the previous larger values of the rate coefficients of the order of 10^{-7}cm^{3}s^{-1} to vibrationally excited H_3^+, referring to a theoretical calculation by Michels and Hobbs(3) who showed that H_3^+ in $v \leq 2$ states should have negligibly small dissociative recombination rate coefficients due to the absence of the crossings of the dissociative potential surfaces of H_3. Also recently Hus *et al.*(4) repeated the merged beam experiments and obtained a cross section which is about an order of magnitude smaller than the previous value determined using a similar technique(5). They suggested that the ions in the experiment by Auerbach *et al.*(5) were predominantly vibrationally excited and the rate coefficient of the order of 10^{-7} cm^{3}s^{-1} obtained was not the rate for H_3^+ in the ground state (see also a paper by Mitchell *et al.*(6)). Macdonald, Biondi, and Johnsen(7) confirmed in 1984 the results obtained by Leu, Biondi, and Johnsen(8). Johnsen, however, suggested in his recent review article(9) that the rate coefficients obtained previously(7,8) were likely to be flawed by an impurity ion CH_5^+.

Smith and Adams present their view on the current situation in one of their recent review articles(10). In it, they write, "The most recent attempt to determine α_e for H_3^+, which used infrared absorption spectroscopy indicated that the α_e for H_3^+ ($v = 0$) is in excess of 10^{-7} cm^{3}s^{-1}, but it is now recognised that, at the high electron and H_3^+ number densities used in the experiment, the process being observed was collisional-radiative recombination (and not two-body dissociative recombination) which cannot occur under the very low charge number density conditions of interstellar clouds." The description about "the most recent attempt"(11) is very biased and misleading. There is no experimental evidence to support the statement that "it is now recognised that the process being observed was collisional- radiative recombination.", and it is the main objective of this presentation to resolve this misinterpretation. Since my preliminary result(11), I have extended my measurements to more vibration-rotation transitions at three different temperatures i.e. liquid nitrogen, dry-ice, and ice temperatures. Also the measurements were extended to include the determination of the dissociative recombination rate coefficients of HN_2^+ and HCO^+. A full paper describing the details of these results will be published elsewhere(12), and therefore here I shall present only the summary of the results.

EXPERIMENTAL TECHNIQUE

The experimental technique used in this work is very simple and straightforward. Figure 1 shows a schematic diagram of the experimental setup. Ions were generated in a cooled hollow cathode by applying pulse discharge and the concentration of the ion was monitored by observing the decay of the infrared absorption signals as a fuction of time. The cathode was made of stainless steel tube of 80 cm long. We used two cathodes of different diameter, one 38 mm and the other 60 mm to examine the effect of diffusion loss. Copper tubing of 1/4 inch diameter was wound around the cathode to allow suitable coolant to flow through to cool the discharge. The frequency of the infrared laser radiation from a difference frequency laser system was held fixed at the center of a particular vibration-rotation transition line. The laser beam was sent through a hollow cathode discharge cell fitted with a set of multi-traversal mirrors. The typical pathlength used in this work was 28.8 m. The transmitted laser radiation was received by a liquid nitrogen cooled InSb photovoltaic detector and the signal was amplified with a preamplifier(Infrared Associates, Model PPA-15-IS). The bandwidth of the detector and the preamplifier system was estimated to be about 10 MHz. The decay of the absorption intensity after pulse discharge was recorded with a digital storage oscilloscope(HITACHI VC6050). The stored signals were transferred to a microcomputer(HP 9816) for analysis.

Figure 2 shows an example of the time dependent signal of absorption of H_3^+. The vertical axis represents the signal intensity from the preamplifier in units of mV. This example is for the 2 0 $_1$ $\leftarrow$ 1 0 transition of the ν_2 fundamental band of H_3^+(13,14) at 2725.898 cm^{-1} recorded at liquid nitrogen temperature. The peak intensity corresponds to 62 % absorption. The second channel shows the current waveform recorded by monitoring the voltage across a 10 Ω resister inserted between the cathode and the ground. The peak current corresponds to 2.5 A. The pulse width was chosen to be long enough for the intensity to reach the steady state(almost).

The hydrogen gas was obtained from a commercial cylinder(Matheson or Linde, Ultra-high-pure grade, 99.999 %) and was introduced into the cell through a liquid nitrogen trap to eliminate condensable impurities.

ANALYSIS AND RESULTS

The decay of an ionic species due to dissociative recombination is described by

$$dN_+/dt = -k_e N_+ N_e \tag{2}$$

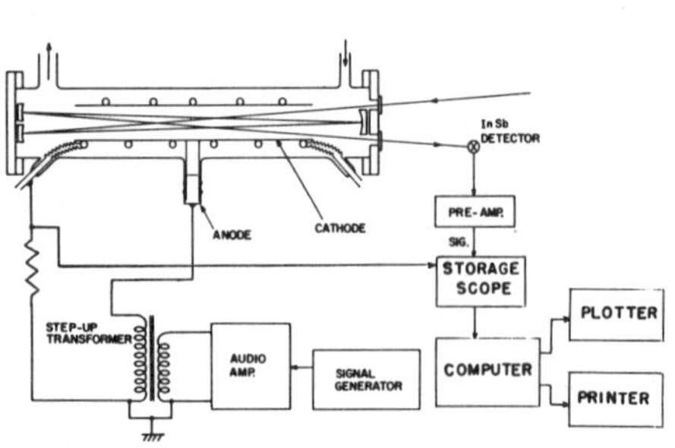

Figure 1. Schematic diagram of the measurement system.

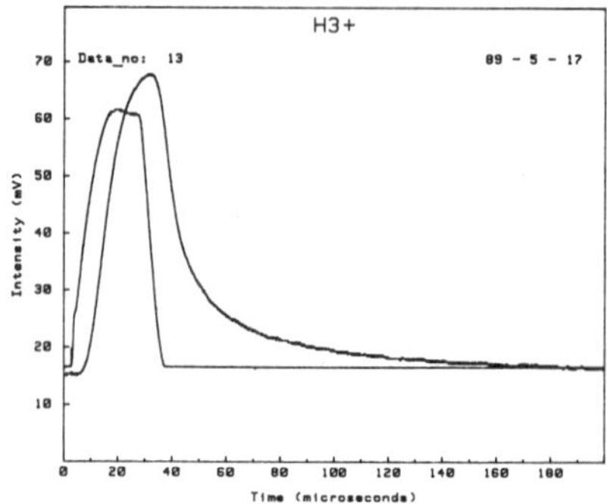

Figure 2. Transient absorption signal of the 2 0 $_1$ $\leftarrow$ 1 0 transition of the ν_2 fundamental band and the discharge current recorded at liquid nitrogen temperature.

where N_+ and N_e denote the abundances of ions and electrons respectively. Because the reaction between H_2^+ and H_2 is very rapid(15,16), the abundance of H_2^+ is negligibly small, and H_3^+ is by far the dominant positive ion, when the discharge reaches the steady state. It is also known that the relative abundances of H^+ and H_5^+ are less than a few percent of the total positive ions in the discharge through H_2(15,16). Thus we can assume $N_+ \sim N_e$ in very good approximation. Equation (2) is then written as

$$dN_+/dt = -k_e N_+^2, \tag{3}$$

and the solution is given as

$$1/N_+(t) = 1/N_+(0) + k_e t. \tag{4}$$

As the rotational relaxation time is of the order of 10 ns and the dissociative recombination decay time of the order of 10 μs under the pressure range of our measurements, we can assume that the population of an individual level, N_{vJ}, is always given by $f_v f_J N_+(t)$ with time-independent $f_v f_J$, where $f_v f_J$ is the fractional population of the level. Under this assumption, eq.(4) can be rewritten as

$$1/N_{vJ}(t) = 1/N_{vJ}(0) + (k_e/f_v f_J)t. \tag{5}$$

This formulation implicitly assumes that the rate coefficient k_e has no or little rotational dependence. However, this equation will apply even when k_e has a rotational dependence, as long as the rotational relaxation is much faster than the dissociative recombination decay, in which case k_e should be interpreted to be a rotationally averaged value.

The infrared absorption is given in terms of the absorption coefficient as

$$I(\omega, t) = I_c e^{-\alpha(\omega,t)l} \tag{6}$$

where I_c is the infrared power without absorption and l is the path length. The absorption coefficient is given by

$$\alpha(\omega, t) = \frac{8\pi^2 \, \bar{\mu}^2 \omega}{3ch}(N_1(t) - N_2(t))\phi(\omega), \tag{7}$$

where $\bar{\mu}$ is the dipole matrix element, N_1 and N_2 are the populations of the lower and the upper levels, respectively, and $\phi(\omega)$ is a line shape function. For infrared absorption lines in the 3 - 4 μm region with reasonably low vibrational temperature as is the case in the present measurements, N_2 is negligible, and N_1 is given by $N_1 = f_v f_J N_+$. The abundance is thus calculated from the peak intensity as

$$N_1(t) = \Gamma^{-1}\ln(I_c/I(\omega_0, t)) \tag{8}$$

where Γ is given by

$$\Gamma = \frac{8\pi^2 \, \bar{\mu}^2 \omega_0}{3ch}\phi(\omega_0) \cdot l. \tag{9}$$

The transition dipole moment for H_3^+ was calculated from the line strength $S(f - i)$ given in Table 2 of Miller and Tennyson's paper(17). For HN_2^+ and HCO^+ the transition dipole moments calculated by Botschwina(18,19) were used, and the J dependent part of the matrix element was as given in Townes and Schawlow's book(20).

To derive the abundance of the ion from the absorption intensity using eqs.(6)-(9), the translational(kinetic), vibrational, and rotational temperatures in addition to the transition dipole moment should be known. The translational temperature was found to be very close to the cell temperature from the linewidth measurements. The rotational temperature was determined from the relative intensities among the several vibration-rotation transitions. The vibrational temperature was not determined precisely, but, from non-observartion of the hot-band from the $v_2 = 1$ state, f_v was assumed to be 1.0. Detailed discussions about determination of these characteristic temperatures are found in ref.(12). The electron temperature was not measured in this work. As discussed in ref.(12), we assumed that the electron temperature is equal to the translational temperature of the ions.

Figure 3(a) is the plot of log $(N_+(t))$ which clearly indicates that the decay is not exponential. Figure 3(b) shows an example of the plot of $1/N_+(t)$ obtained from the measured decay curve for the $2\,0_1 \leftarrow 1\,0$ transition recorded at liquid nitrogen temperature. The origin of the time($t = 0$) in these figures was chosen at the point where the discharge current diminished completely to zero. This very good linear relationship shows that the decay process is indeed the dissociative recombination. The slope, $(4.294\pm0.006) \times 10^{-7}$ cm^{3}s^{-1}, and the intercept in this example are typical of many runs repeated over the past several months. From this intercept the ion abundance at $t = 0$ is obtained to be 5×10^{11}cm^{-3}. The abundance varied from one run to another depending on discharge conditions, but this example is a typical one of many runs. The statistical uncertainty of the slope is far smaller than the fluctuations observed from one run to another which usually amounted to several %.

As it was found that the recombination rate coefficient had no measurable pressure dependence in the pressure range of 0.1-1 Torr, most measurements were carried out in the 200-600 mTorr pressure range. Table I summarizes the rate coefficients thus determined for the five rotational states at three different temperatures. These values were obtained by taking the average of several runs(typically five). The standard deviation amounts to several to 10 % which is caused probably by fluctuations of the laser power.

Figure 4 shows the temperature dependence of the rate coefficients together with the values for HN_2^+ and HCO^+. The temperature dependences obtained previously(7,8,21,22) are indicated by broken lines. The temperature dependence measured by Macdonald, Biondi, and Johnsen(7) shows a curious deviation from the T^{-1} dependence below about 400 K of electron temperature. Our data suggest that the T^{-1} dependence may extend further toward the lower temperature region. Our data for HCO^+ are consistent with the temperature dependence, $T_e^{-0.69}$, obtained by Ganguli *et al.*(21) in the range of $295 < T_e < 5500$ K, and suggest that it can be extrapolated to 100 K. Our measured points for HN_2^+ fall on the line obtained by Mul and McGowan(22).

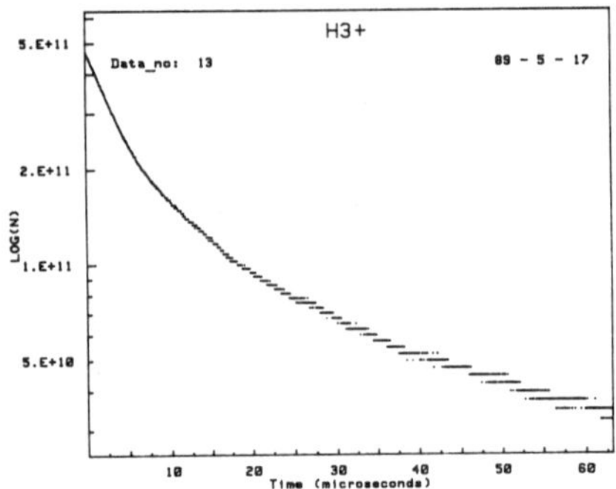

Figure 3(a). The $\log(N_+(t))$ vs. t plot of the decay signal given in Fig.2. The origin of the time is chosen at the point where the current falls to zero completely.

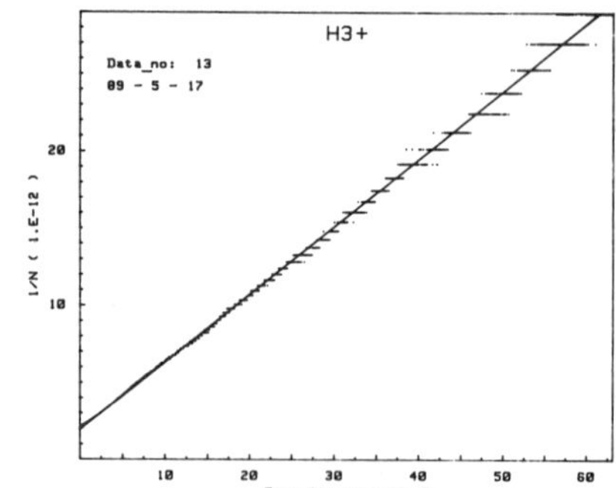

Figure 3(b). The $1/N_+(t)$ vs. t plot of the same signal as in Figure 3(a).

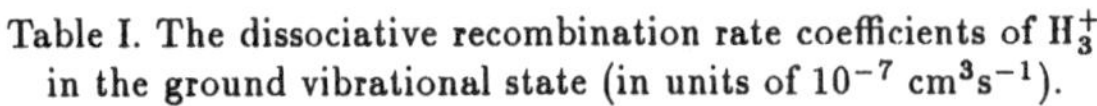

Table I. The dissociative recombination rate coefficients of H_3^+ in the ground vibrational state (in units of 10^{-7} cm^{3}s^{-1}).

J, K	110 K	210 K	273 K
1, 0	4.1(2)[a]	2.5(1)	1.72(5)
1, 1	4.1(1)	2.7(2)	1.77(10)
2, 2	4.6(4)	2.4(2)	1.85(6)
3, 3	4.5(5)	2.6(2)	1.91(7)
4, 4	------	2.2(2)	1.9(4)

[a] standard deviation in units of the last quoted digits.

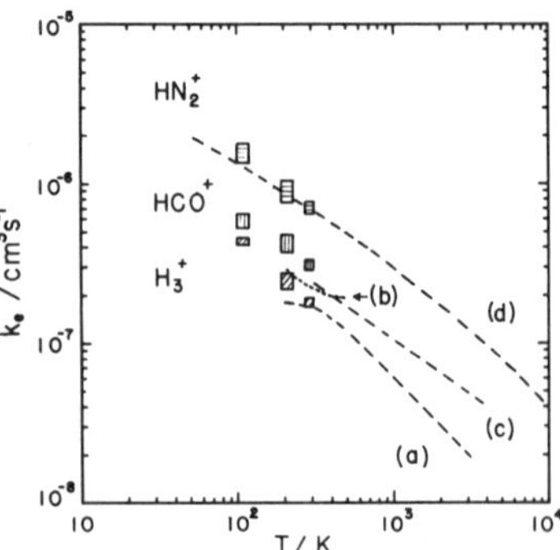

Figure 4. The temperature dependence of the dissociative recombination rate coefficients for H_3^+, HN_2^+, and HCO^+. Each rectangle represents the measured value, and the size reflects the estimated uncertainty. The dashed and dotted lines are: (a) ref.(7), (b) ref.(8), (c) ref.(21), and (d) ref.(22).

DISCUSSION

Since the value did not agree with the most recent result obtained by Smith and coworkers which seemed to be gaining general acceptance, we carefully examined possible causes which might have made our result different from theirs. One may argue that the decay process observed in our previous experiment(11) as well as in this experiment is not due to the dissociative recombination process but other processes such as ambipolar diffusion, formation of H_5^+, reactions with impurities, or collisional-radiative recombination process(23,24). I am going to discuss these possibilities one by one below.

The ambipolar diffusion loss rate, $D_a p/\Lambda^2$, is estimated to be 50 s^{-1}Torr by assuming $D_a p \sim 500$ cm^{2}s^{-1}Torr(25) and $\Lambda \sim 3$ cm(radius of the cathode). So the ambipolar diffusion loss is negligibly small in the pressure range used in this work. Experimentally it is also clear that the ambipolar diffusion is not a dominant process, since no pressure dependence of the decay rate is observed and $N_+(t)$ did not obey the exponential decay.

H_3^+ may be lost by the following three body reaction to lead the formation of H_5^+,

$$H_3^+ + 2H_2 \rightarrow H_5^+ + H_2. \tag{10}$$

The rate coefficient for this process was measured by Hiraoka and Keberle (26) to be $4.4 \times 10^{-24}T^{-2.3}$cm^{6}s^{-1}, where T is the temperature in K. When $[H_2]$ is 1×10^{16}cm$^{-3}(\sim 300$ mTorr), the loss rate of H_3^+ is calculated to be of the order of 100 μs at 100 K which is about an order of magnitude slower than the rate we obtained in our experiment. Also it is clear experimentally, because, if this process is responsible for the decay we observed, the decay should be exponential and should exhibit conspicuous pressure and temperature dependences.

The possibility of impurities as a cause of faster decay is ruled out immediately by examining the time dependence of the decay signal (see Figs.3(a) and 3(b)). If constant leak or back diffusion of impurity gases from the pumping system caused the depletion of H_3^+, the decay should be exponential. This was demonstrated in ref.(12) by mixing a small amount of N_2 intentionally to hydrogen discharge.

The ion density (and also electron density) in the hollow cathode plasma employed in this work was typically $N_+ \sim 5 \times 10^{11}$ cm^{-3} at $t = 0$ for H_3^+. (There is a typographical error in the caption to Figure 3 of ref.(11). The ion abundance at $t = 0$ is 3×10^{11}cm^{-3} rather than 3×10^{12}cm^{-3}). If we assume $T = 110$ K and $[e]=3 \times 10^{11}$cm^{-3}, the collisional-radiative recombination rate coefficient is calculated to be 7.4×10^{-7}cm^{3}s^{-1} using eq.(29) in a paper by Stevefeld, Boulmer, and Delpech(24) derived for transitions between hydrogenic energy levels of high principal quantum number. One may conclude that this mechanism explains the process observed in this experiment. However, this is not the case for two reasons. First, the collisional-radiative recombination rate coefficient has a distinctively different temperature dependence. At $T = 210$ K, the rate coefficient would be 4.0×10^{-8}cm^{3}s^{-1} and at $T = 275$ K it would be 1.2×10^{-8}cm^{3}s^{-1}, which disagree with the experimental results. Second, if this process is dominant, $1/N_+(t)$ cannot be linear with t. As given in Fig.4 as an example of the signals taken at $T = 110$ K, $1/N_+(t)$ shows a linear dependence on t over an order of magnitude of $N_+(t)$. From these experimental observations, we can conclude that this is not a dominant process of the ion loss in our experiments.

A remaining question is why the FALP measurement yielded such a slow "immeasurable" dissociative recombination rate. Figure 2 of the paper by Adams, Smith, and Alge(1) clearly shows an initial decrease of the electron density in the H_3^+ recombination measurements. They discarded this portion as being caused by the vibrationally excited H_3^+. As discussed in ref.(12), our observation implies that the vibrational relaxation is very fast as expected from the rate coefficient, 3×10^{-10} cm^{3}s^{-1} given by Kim, Theard, and Huntress(27). Adams, Smith, and Alge(1) cited the detection of CH_3^+ with CH_4 added to H_3^+ flow as an evidence of the existence of the vibrationally excited H_3^+. Their observation indicated that less than 5 % of the initial products from the reaction between H_3^+ and CH_4 was CH_3^+, suggesting that vibrationally excited H_3^+ could not be the dominant constituent. This observation merely indicates that vibrationally excited H_3^+ existed at the early stage of the reaction, but does not guarantee the persistence of hot H_3^+ further downstream of the flow tube. Although an implication of the argument given above is that the initial decay observed in their experiment is indeed mainly due to H_3^+ in the ground state, there remains a question. The decay became slow somewhat abruptly at the distance $z \sim 50$ cm (see Fig.2 of ref.(1)) and turned to be diffusion limited afterward. To unravel this puzzle, one should repeat the FALP experiment or conduct similar measurements.

ACKNOWLEDGMENTS

I thank Takako Amano for implementing the data acquisition system and writing an analysis program. I also gratefully acknowledge R. H. Schwendeman and H. Sasada for providing their Voigt line profile fitting program.

REFERENCES

(1) Adams, N. G., Smith, D., and Alge, E., 1984, J. Chem. Phys., **81**, 1778.
(2) Adams, N. G., and Smith, D., 1987, *Astrochemistry. IAU Symposium No. 120* (Dordrecht: Reidel).
(3) Michels, H. H. and Hobbs, R. H., 1984, Ap. J.(Letters), **286**, L27.
(4) Hus, H., Youssif, F., Sen, A., and Mitchell, J. B. A., 1988, Phys. Rev. A, **37**, 2543.
(5) Auerbach, D., Cacak, R., Caudano, R., Gaily, T. D., Keyser, C. J., McGowan, J .Wm., Mitchell, J. B. A., and Wilk, S. F. J., 1977, J. Phys. B:Atom. Molec. Phys., **10**, 3797.
(6) Mitchell, J. B. A., Ng, C. T., Forand, L., Janssen, R., and McGowan, J. Wm., 1984, J. Phys. B: At. Mol. Phys. **17**, L909.
(7) Macdonald, J. A., Biondi, M. A., and Johnsen, R., 1984, Planet Space Sci. **32**, 651.
(8) Leu, M. T., Biondi, M. A., and Johnsen, R., 1973, Phys. Rev. A.,**8**, 413.
(9) Johnsen, R., 1987, Int. J. Mass Spectrom. Ion Proc.**81**, 67.
(10) Smith, D. and Adams, N. G., 1989, J. Chem. Soc. Faraday Trans. 2, **85**, 1613.
(11) Amano, T., 1988, Astrophys. J. Lett.**329**, L121.
(12) Amano, T., 1990, J. Chem. Phys. to be published.
(13) Oka, T., 1980, Phys. Rev. Lett.,**45**,531.
(14) Watson, J. K. G., Foster, S. C., McKellar, A. R. W., Bernath, P. F., Amano, T., Pan, F. S., Crofton, M. W., Altman, R. S., and Oka, T., 1984, Can. J. Phys.,**62**,1875.
(15) Saporoschenko, M., 1965, Phys. Rev.,**139**,A349.
(16) Albritton, D. L., Miller, T. M., Martin, D. W., and McDaniel, E. W., 1968, Phys. Rev., **171**, 94.
(17) Miller, S., and Tennyson, J., 1988, Astrophys. J. **335**, 486.
(18) Botschwina, P., 1984, Chem. Phys. Lett. **107**, 535.
(19) Botschwina, P., 1989, *Ion and Cluster Ion Spectroscopy and Structure*, pp. 59-108. ed. J. P. Maier(Elsevier, Amsterdam).
(20) Townes, C. H., and Schawlow, A. L., 1955, *Microwave Spectroscopy*,(McGraw Hill, New York).
(21) Ganguli, B., Biondi, M. A., Johnsen, R., and Dulaney, J. L., 1988, Phys. Rev. A. **37**, 2543.
(22) Mul, P. M., and McGowan, J. Wm., 1979, Astrophys. J. Lett. **227**,L157.
(23) Bates, D. R., Kingston, A. E., and McWhirter, R. W. P., 1962, Proc. Roy. Soc. A**267**, 297.
(24) Stevefelt, J., Boulmer, J., and Delpech, J-F., 1975, Phys. Rev. A. **12**, 1246.
(25) von Engel, A., 1983, *Electric Plasmas: Their Nature and Uses*, (Taylor and Francis, New York).
(26) Hiraoka, K., and Kebarle, P., 1975, J. Chem. Phys., **63**, 746.
(27) Kim, J. K., Theard, L. P., and Huntress, W. T., 1974, Int. J. Mass Spectrom. Ion Phys., **15**, 223.

Infrared Spectroscopy of Transient Species

Kentarou KAWAGUCHI

Nobeyama Radio Observatory, Minamimaki, Minamisaku, Nagano 384-13, Japan

High-resolution infrared spectroscopy of transient species such as free radicals and molecular ions is discussed with reference to astrochemical interest. The infrared diode laser and Fourier transform studies of the C_3 molecule are described in connection with its detection in IRC+10216; possibility of the observation of far infrared transitions in molecular clouds is discussed.

INTRODUCTION

Detection and identification of new molecular lines from space are very exciting, and useful for investigating chemical evolution and monitoring physical conditions of interstellar clouds and circumstellar envelopes. In the last three years the following species were identified ; C_6H, CCS, C_3S, cyclic-C_3H, CH_2CN, CP, CSi, C_4Si, and HCCCHO(propynal), and CH_3COCH_3(acetone). These species except for CP, $HCCCHO$, and CH_3COCH_3 were new in the laboratory as well as in space. Therefore, laboratory experiments and ab initio calculations were indispensable to assign these species. Other interstellar molecules so far detected are summarized, for example, by Irvine et al.(1).

Through molecular line survey observation, more than one hundred unidentified lines were detected in objects such as TMC-1, Ori-KL, Sgr-B2, and IRC+10216 and remained to be assigned. These lines may be due to hitherto unknown species which may have short lifetime in the laboratory. To identify new species, a highly sensitive detection method is necessary in the laboratory. Microwave spectroscopy is suitable for such a purpose, in fact it contributed to the identification of the new species, as mentioned in Saito's paper of this volume. On the other hand, modern high resolution infrared spectroscopy can be applicable to the detection of short lived molecules such as free radicals and molecular ions. The molecular constants derived from the analysis of Doppler-limited spectra can predict pure rotational transition frequencies in the microwave region within a few MHz accuracy. Especially in molecular ion studies, infrared detection is usually easier than microwave detection. The $HCNH^+$ ion was detected for the first time by using a difference frequency laser system(2), and microwave spectroscopy was then applied to determine more accurate transition frequencies(3); the astronomical detection was successful in Sgr-B2(4). The protonated cyanoacetylene ion HC_3NH^+ was also detected by a difference frequency system(5), however no microwave spectra in laboratory were reported. The astronomical search was carried out, using the infrared data, to derive the upper limit of the ion(6, 7). The infrared study of negative ions is a growing area; however, so far no laboratory microwave and astronomical detections have been reported.

Laboratory infrared data become also useful for identifying spectral lines observed by FIR and infrared astronomy, which is a rapidly developing field. Since non-polar molecules and light molecules such as hydrides have no proper transitions in the microwave region, infrared detection has advantage for monitoring these species. In the present paper we represent recent result of laser spectroscopy and Fourier transform spectroscopy on some transient species, especially concerning C_3 in relation to astronomical observations.

INFRARED STUDY OF C_3

The optical spectrum of C_3 from comets was recognized in the 19th century; however, the identification of the carrier was not done until 1951 by Douglas(8). A search for C_3 through its electronic transition was carried out in interstellar diffuse clouds by Snow et al.(9), to give the upper limit for the column density of 10^{10} cm^{-2} or about 10^{-11} in fractional abundance with respect to hydrogen. In dense clouds the C_3 molecule has been thought to be a precursor to more complex carbon chain molecules.

The molecular structure was believed to be linear from the analysis of the optical spectrum. However ab initio calculations predict a quasi-linear structure with a potential barrier of 21 cm^{-1} (10). The vibrational frequencies were derived to be 1230, 63, and 2040 cm^{-1} for the ν_1, ν_2, and ν_3 vibrations. It should be noticed that the ν_2 frequency is extremely low. The low frequency may be partly due to vibronic interaction from the excited Π state whose ν_2 frequency is 308 cm^{-1}. The direct observation of the vibration-rotation transition has not been reported in the gas phase. We applied the diode laser spectroscopic method to this species(11).

An excimer laser beam of 193 nm was used to photolyze a parent molecule and to produce transient species. The infrared absorption of the transient species was detected by a tunable infrared diode laser in a multi-reflection cell. The C_3 species was produced by the photolysis of diacetylene. The same spectrum was obtained by the photolysis of acetylene and also deuterated acetylene and allene. This means that the species is due to carbon compounds. The spectral pattern with a 1.6 cm^{-1} interval allows us to assign the species to C_3, the interval 1.6 cm^{-1} corresponds to four times the rotational constant of C_3. Since the ν_2 bending frequency of C_3 is quite low, many spectral lines from the excited bending state were observed and assigned. The allene molecule gave the strongest C_3 signal; however, in the 1900 cm^{-1} region allene itself gave strong absorption. Therefore the discharge method was employed, where the C_3 molecule was produced by ac discharge in various kind of carbon compounds.

The strong ν_3 band intensity made it possible to detect the combination $2\nu_2+\nu_3$ band. This observation was very important to derive information about the bending frequency. The observed spectrum was analyzed by using the energy level expression for a linear molecule, and the molecular constants were determined for each vibrational level(12). The large centrifugal distortion constant and also the negative vibration-rotation constant indicate that the C_3 molecule has a quasi-linear structure. This observation agreed with the prediction by the ab initio calculation(10). Figure 1 shows the vibrational energy level structure of C_3 determined by the infrared studies. The energy interval between (020) and (000) was determined accurately. However the important ν_2 fundamental frequency was not derived because of the selection rule.

CARBON AND SILICON COMPOUNDS IN SPACE

Hinkle et al. observed the ν_3 band of C_3 toward the carbon star IRC+10216 by using 4-m telescope at Kitt Peak. Laboratory data mentioned above contributed to the confirmation of the species. The column density

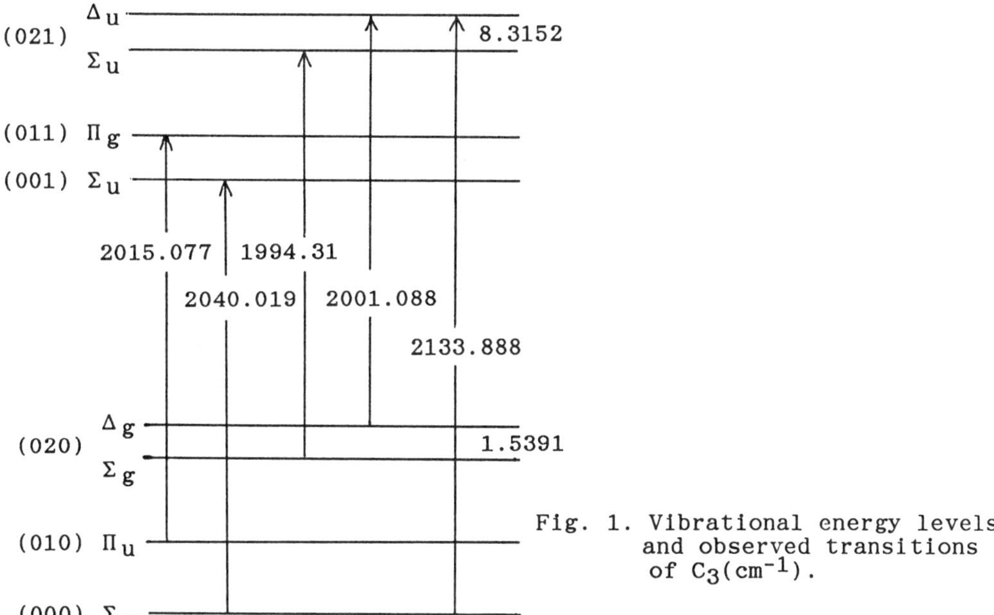

Fig. 1. Vibrational energy levels and observed transitions of $C_3(cm^{-1})$.

of C_3 was determined to be 10^{15} cm^{-2} and the fractional abundance was about 10^{-6} of hydrogen(13). The linear C_5 molecule was also identified in IRC+10216(14) as well as in the laboratory(15, 16). Table I summarizes the abundances of the carbon and related silicon compounds in IRC+10216. The SiC_4 molecule was recently detected at Nobeyama Radio Observatory(17). The C_4 and C_3Si molecules were not identified in the laboratory (gas phase) as well as in space because of their high reactivity. These compounds may have much relation to dust components; that is, graphite is thought to be a carrier of the 2200 Å band and the SiC bond in solid phase gives a 12 micron infrared band. It is an interesting future problem to understand how the small molecules correlate with the large molecules such as C_{60} which may be a dust component.

Table I. Molecular Abundances in IRC+10216

Molecule	CC	CCC	C_5	CCH	C_4H	CO	CH_4
N^a	–	1.0(15)	1.3(14)	3.0(15)	5.0(15)	2.3(19)	2.5(17)

Molecule	CSi	CCSi	C_4Si	HCSi	HC_3Si	SiO	SiH_4
N^a	6.0(13)	1.5(14)	7.0(12)	–	–	2.0(15)	2(15)

a. Column density in cm^{-2}, where a(b) signifies a x 10^b.

In Table I we compare the column densities of carbon compounds and the silicon substituted compounds. The abundance ratio of carbon and silicon compounds is [CCSi]/[CCC] = 0.1 and [C_4Si]/[C_5] = 0.07. These ratios are very close to the atomic abundance ratio of carbon and silicon [Si]/[C] = 0.083. On the other hand the abundance ratio [SiH_4]/[CH_4] is 0.008, which is an order of magnitude smaller than the atomic abundance ratio. Although we have no definite explanation about the production mechanism of these compounds in circumstellar envelopes, we can point out the possibility of a similar adsorption mechanism between carbon and silicon chain compounds to the surfaces of grains. Since CCH and C_4H are fairly abundant species in the circumstellar envelope, HSiC and HC_3Si may have detectable abundances if the same abundance ratio as [Si]/[C] is assumed. The molecular constants of HCSi and HC_3Si have not been determined experimentally.

In an ion-molecular reaction scheme, Suzuki predicted the abundance of C_3 in interstellar clouds to be about 5×10^{-7} of H_2(18), which is almost the same as that of IRC+10216. However in usual molecular clouds the ν_3 band may not be suitable for monitoring, because a strong $5\mu m$ background emission source is necessary for observation of the absorption. On the other hand the ν_2 bending vibrational band may be more suitable. The absorption intensity of the ν_2 band is estimated to be as small as 0.01 times of the ν_3 band. But if we consider an emission line for monitoring, we can estimate a brightness temperature of 1.1 K for strong lines in the optically thin clouds, where the column density, excitation temperature and dipole moment are assumed to be 10^{15} cm^{-2}, 100 K, and 0.19 Debye respectively. According to the molecular constants derived from infrared spectroscopy, the relative line position of the ν_2 band is very well known, but the absolute frequency error is about 0.1 cm^{-1}. Recently the direct observation of the C_3 ν_2 band was carried out by using a tunable far-infrared laser spectrometer(19). The tunable radiation was produced by a mixing of microwave and optically pumped far infrared laser radiation on a Schottky barrier diode. The C_3 molecules were produced by laser vaporization of a graphite rod and cooled in a supersonic expansion. The lowest vibrational frequency of the C_5 molecule is predicted to be 119 cm^{-1}. This band is also a good candidate for sub-mm observation.

HIGH-RESOLUTION INFRARED FOURIER TRANSFORM SPECTROSCOPY

Recently, a high-resolution Fourier transform(FT) infrared spectrometer was installed at the Nobeyama Radio Observatory. The spectrometer(Bruker FTS 120 HR) covers from 10 cm^{-1} to 10000 cm^{-1} with a maximum resolution of 0.002 cm^{-1}. A discharge cell was set outside for observations of transient species. Infrared emission measurements were carried out by using two type of discharge cell. One was a hollow cathode cell with a three-layer structure which made it possible to cool down the cell to liquid nitrogen temperature. Another was a simple microwave discharge cell for electrodeless discharge. The emitted light was focused on the input iris of the spectrometer.

So far we have observed the following atomic and molecular spectra; (1) atomic chlorine and helium spectra in 2000 cm^{-1} region, (2) the molecular nitrogen $^1\Delta_u$-$^1\Pi_g$ electronic transition in the 2700 cm^{-1} region, (3) ArH$^+$ v=1-0, 2-1, 3-2, 4-3,and 5-4 bands by a microwave discharge in Ar and H_2 mixture, (4) the OH radical v=1-0, 2-1 bands. These emissions were observed with 0.02 cm^{-1} resolution and 1-2 hours integration. The N_2 spectrum was also observed with 0.006 cm^{-1} resolution to resolve the Λ-type doubling.

An absorption experiment was carried out by using a White type multi-reflection absorption cell, where the effective path length was set to be about 20m. The sensitivity of the FT spectrometer was checked and compared with that of the laser spectrometer, because it is the most important factor in the detection of less abundant transient species. A source emitting power P, and loss of input power is detected after an absorption cell by a detector which has a noise equivalent power P_n. Then the minimum detectable change ΔP_{min} is given as follows:

$$\Delta P_{min} = 2(P\ P_n)^{1/2}. \qquad (1)$$

In laser spectroscopy the source power is several hundred microwatts in a very narrow frequency region($< 0.0003 cm^{-1}$). Therefore typically a 10^{-4} change of input power can be detected with conditions of $P_n=10^{-12}W/(Hz)^{-1/2}$ and time constant of 0.3 sec. On the other hand in an FT spectrometer, classical sources such as glober lamps or tungsten lamps are used, and the power is very small compared with laser sources. Therefore in general the sensitivity is low. In laser spectroscopy we need a long time for obtaining the whole spectrum of a molecule. For example if the scan is done with a speed of 0.005 cm^{-1}/sec, we need 33 hours for a 100 cm^{-1} survey. In an FT spectrometer 33 hours integration gives $\Delta P_{min}/P=0.0004$, where the frequency coverage is 2000-2500 cm^{-1} with 0.0048 cm^{-1} resolution. The sensitivity is only four times worse than that of laser spectrometer. In lower frequency regions such as mid-infrared and far infrared, the sensitivity is lower than this value. Therefore we must improve the power of the source or use a detector with low noise equivalent power. It is one advantage of an FT spectrometer to observe the whole spectral range at once, and the spectral pattern may be useful for identification of the species.

Figure 2 shows the observed C_3 spectrum by the FT spectrometer in the 2000 cm^{-1} region. The C_3 molecule was produced by a discharge in a mixture of allene and He. The integration time was about 2 hours. The signal strength corresponds to about 4 % absorption. This absorption spectroscopy was also applied for the observations of the CH_3 and NO_3 radicals in the 600 cm^{-1} and 2500 cm^{-1} regions, respectively.

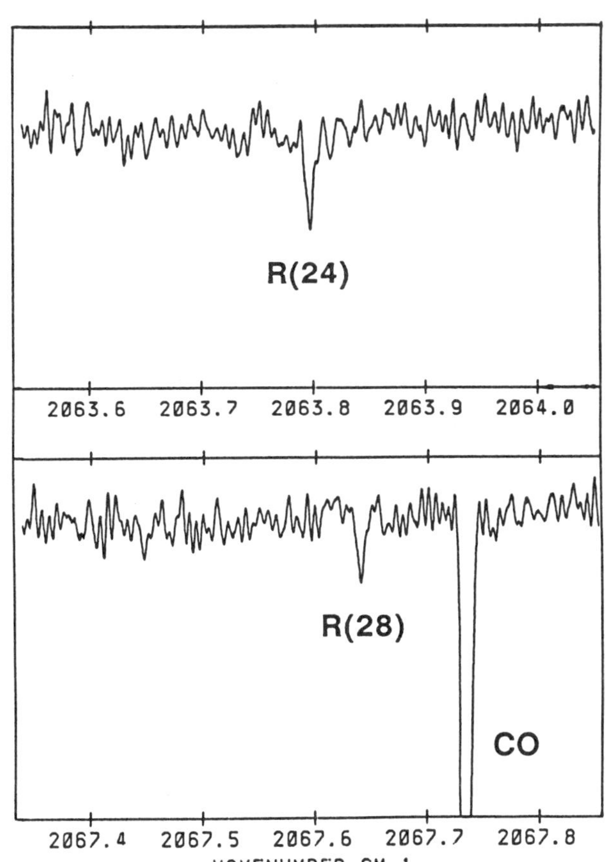

Fig. 2. Observed spectrum of the C_3 ν_3 band.

Acknowledgements

The present author thanks K. Matsumura, H. Kanamori, and E. Hirota for their collaboration in the infrared study of C_3. He is also grateful to N. Kaifu and M. Morimoto for their support in the installation of the FTIR spectrometer.

References

(1) Irvine, W.M., Goldsmith, P.F., and Hjalmarson, Å. 1987, in Interstellar Processes, ed. D.J. Hollenbach and H.A. Thronson(Dordrecht: D. Reidel), p561.

(2) Altman, R.S., Crofton, M.W., and Oka, T. 1984, J. Chem. Phys. **80**, 3911.

(3) Bogey, M., Demuynck, C., and Destombes, J.L. 1985, J. Chem. Phys. **83**, 3703.

(4) Ziurys, L.M. and Turner, B.E. 1986, Ap. J. (Letters), **302**, L31.

(5) Lee, S.K. and Amano, T. 1987, A, Ap. J. (Letters), **323**, L145.

(6) Turner, B.E., Amano, T., and Feldman, P.A. 1990, Ap. J. **349**, 376.

(7) Amano, T., Saito, S., Yamamoto, S., Kawaguchi, K., Suzuki, H., Ohishi, M., and Kaifu, N. 1990, Ap. J. **351**, 500.

(8) Douglas, A.E. 1951, Ap. J. **114**, 466.

(9) Snow, T.P., Seab, C.G., and Joseph, C.L. 1988, Ap. J. **335**, 185.

(10) Kraemer, W.P., Bunker, P.R., and Yoshimine, M. 1984, J. Mol. Spectrosc. **107**, 191.

(11) Matsumura, K., Kanamori, H., Kawaguchi, K., and Hirota, E. 1988. J. Chem. Phys. **89**, 3491.

(12) Kawaguchi, K., Matsumura, K., Kanamori, H., and Hirota, E. 1989, J. Chem. Phys. **91**, 1953.

(13) Hinkle, K.W., Keady, J.J., and Bernath, P.F. 1988, Science, **241**, 1319.

(14) Bernath, P.F., Hinkle, K.H., and Keady, J.J. 1989, Science, **244**, 562.

(15) Heath, J.R., Cooksy, A.L., Gruebele, M.H.W., Schmuttenmaer, C.A., Saykally, R.J. 1989, Science, **244**, 564.

(16) Moazzen-Ahmadi, M., McKellar, A.R.W., Amano, T. 1989, Chem. Phys. Lett. **157**, 1.

(17) Ohishi, M., Kaifu, N., Kawaguchi, K., Murakami, A., Saito, S., Yamamoto, S., Ishikawa, S., Fujita, Y., Shiratori, Y., and Irvine, W.M. 1989, Ap. J. (Letters), **345**, L83.

(18) Suzuki, H. 1979, Prog. Theo. Phys. **62**, 936.

(19) Schmuttenmaer, C.A., Cohen, R.C., Puglianom N., Heath, J. R., Cooksy, A.L., Busarow, K.L., and Saykally, R.J. 1990, Science, **249**, 897.

Interstellar Polycyclic Aromatic Hydrocarbons

L. J. ALLAMANDOLA

NASA-Ames Research Center, MS 245-6, Moffett Field, CA 94035, U. S. A.

The infrared evidence which supports the PAH hypothesis is briefly summarized.
Rather than presenting a general discussion of the assignments, this paper focuses
on the spectroscopic constraints placed on the molecular shape, size and structure
of interstellar PAHs. It is pointed out that ionized, gas phase, symmetric PAHs
containing 20–50 carbon atoms are the dominant contributors to the narrow interstellar infrared emission bands while larger species contribute to the broader
components. These apparently ubiquitous, complex, organic ring molecules are more
abundant than the other interstellar polyatomic molecules known. This points to a
unique chemical history of the interstellar PAHs.

1. INTRODUCTION

The interstellar emission bands at 3050, 1610, "1310", 1160 and 885 cm^{-1} (3.28, 6.2,
"7.7", 8.6 and 11.3 microns) provide very detailed information about an apparently
ubiquitous and abundant component of the interstellar medium. The bands are associated with a large number of different types of astronomical object (1,2,3, and
references therein) implying that the carrier is surprisingly widespread and
extremely stable. While it has been recognized for some time now that the emission
was most likely due to IR fluorescence from molecule sized species pumped by UV
photons (4), the identity of the carriers has remained enigmatic since their discovery in 1973 by Gillett, Forrest and Merrill (5). Current thinking that polycyclic aromatic hydrocarbons (PAHs) and related materials are the long sought after
carriers can be traced to the suggestion of Duley and Williams (1981) that they
arise from the vibrations of chemical subgroups attached to the aromatic moieties at
the surfaces of small (<0.01 μm), amorphous carbon particles (6). Subsequently
Leger and Puget (1984) and Allamandola, Tielens and Barker (1985) proposed that
individual molecule-sized (<.001 μm) PAHs and PAH related materials were the carriers (7,8). The principal reason for this assignment was the resemblance of the
interstellar emission spectra with the vibrational spectra of these materials.
Related observations support an aromatic carrier as well. The fraction of total IR
luminosity radiated in the "1310" cm^{-1} (7.7 μm) feature, which is by far the most
intense of the bands, is strongly correlated with the C/O ratio in planetary nebulae
(9). As the carriers must be produced under harsh conditions in these objects, they
must be extremely stable and carbon rich. Furthermore, although there is some
variation among the relative band intensities, the bands are correlated implying
that a single class of chemical species is responsible (9). Taken together, these
observations are completely consistent with PAHs which are a family of extremely
stable, planar hydrocarbons with a chicken wire like structure.

Interstellar PAHs are thought to be very abundant, moreso in fact than all of the
known interstellar, gaseous, polyatomic molecules combined (7,8). The proposed
ubiquity and high abundance of interstellar PAHs implies that their influence on
various processes and phenomena can be profound (10). For example they may be

responsible for the infrared cirrus discovered by the IRAS satellite (8,11), some of
the diffuse interstellar bands in the visible (12,13,14), maintaining interstellar
cloud temperatures (15), moderating interstellar cloud chemistry (16), and contributing to the deuterium enrichments found in interplanetary dust particles and
meteorites (17,18).

While definitive spectroscopic evidence for PAHs is lacking in the ultraviolet and
visible spectral regions(19,20), the case in the infrared is now compelling.
Several predictions have been borne out (21,22) and the theory continues to be
capable of accommodating new observational results (23). In this paper the spectroscopic evidence for PAHs will be briefly reviewed. This will be followed by a
discussion of the constraints the infrared spectra place on interstellar PAH size,
structure and geometry. An extensive review of these and many other aspects of the
PAH hypothesis can be found in reference 18.

2. INTERSTELLAR SPECTRA AND THE SPECTRA OF PAHs

The interstellar emission spectra consist of strong, well-defined bands, weak bands,
and broad features. To date, the relative intensities of the strong bands appear to
be reasonably well correlated. This is not the case for the weak bands and broad
features. Any model which attempts to explain the IR emission phenomenon should
address all of these characteristics. With one exception, the proposals skirt most
of the issues of spectroscopic detail. There are several interpretations of the
emission phenomenon based on PAHs. Duley and Williams (24,25), and Bussoletti,
Colangeli and collegues (26), while not specifically addressing the weak bands and
broad features, envision that the well-defined bands arise from PAHs and sidegroups
on PAHs which comprise hydrogenated amorphous carbon particles. Sakata and colleagues (27) suggest that the well defined bands originate in components which
comprise a quenched carbonaceous composite (QCC) while Papoular and co-workers
propose the components in a coal (28) are responsible. Leger and d'Hendecourt (29)
and Leger et al. (20), again principally addressing the well-defined bands, envision
that they arise from free, molecule-sized PAHs containing between 50-100 carbon
atoms and not PAHs which are part of carbonaceous particles. Allamandola, Tielens
and Barker (18,30) envision that smaller, free, ionized PAHs containing between 20-
50 carbon atoms are responsible for both the well defined and weak bands while
larger PAHs, PAH clusters, and/or amorphous carbon particles (containing on the
order of 400 carbon atoms) contribute to the broad features. Whether the PAHs are
free or part of a larger particle is, at present, controversial. For the emission
to originate from the aromatic structural units of a particle (HAC, QCC, coal, etc.)
or PAH cluster requires that vibrational energy created by the absorption of the
exciting photon remain localized in a specific PAH unit long enough for it to radiatively relax in the IR ($\approx 10^{-2}$s). Our principle objection to this is that the
timescale for normal nonradiative relaxation in a solid is usually many orders of
magnitude faster. While one might imagine that some individual PAH units may be
very loosely coupled to the remainder of the carbon particle, this must be demonstrated experimentally before it can be accepted. On the other hand, IR emission
from highly vibrationally excited, free molecules is expected and has been
measured. UV-pumped IR emission from the bicyclic aromatic molecule azulene has
been reported by Cherchneff and Barker (31).

In view of the limited space, this article will not deal with the differences
between the various models but rather focus on the underlying, unifying theme that
PAHs and PAH-like species are responsible for the emission. All of the IR emission
features and assignments are summarized in Table I and the emission spectrum from
position 4 in the Orion Bar is compared with the absorption spectra of several PAHs
in Figure 1.

TABLE 1: EMISSION COMPONENTS: PROPERTIES AND ASSIGNMENTS[1]

ν (cm^{-1})	λ (Microns)	FWHH (cm^{-1})	ASSIGNMENT[2]
		THE MAJOR BANDS	
3040	3.29	30	Aromatic $C - H$ stretch ($v=1 \rightarrow v=0$)
1615	6.2	30	Aromatic $C - C$ stretch
1315-1250	7.6-8.0	70-200	Blending of several strong aromatic $C - C$ stretching bands
1150	8.7	–	Aromatic $C - H$ in-plane bend
885	11.3	30	Aromatic $C - H$ out-of-plane bend for non-adjacent, peripheral H atoms
		THE MINOR FEATURES	
3085	3.24	–	Overtone and/or combination involving fundamentals in the 1810-1050 cm^{-1} (5.52-9.52 μm) range
2995	3.34	–	Overtone and/or combination involving fundamentals in the 1810-1050 cm^{-1} (5.52-9.52 μm) range
2940	3.4	"20"	Aromatic CH stretch ($v=2 \rightarrow v=1$)
2890	3.46	–	Overtone/combination band involving fundamentals in the 1810-1050 cm^{-1} (5.52-9.52 μm) range, aromatic CH stretch (high v), aliphatic CH stretch,?
2850	3.51	–	Aromatic CH stretch ($v=3 \rightarrow v=2$), aliphatic CH stretch, overtone/combination band involving fundamentals in the 1810-1050 cm^{-1} (5.52-9.52 μm) range
2810	3.56	–	Aromatic CH stretch (high v), aldehydic CH stretch, overtone/combination band involving fundamentals in the 1810-1050 cm^{-1} (5.52-9.52 μm) range
1960-1890	5.1-5.3	30	Combination of CH out-of-plane and in-plane bend, ?
1785-1755	5.6-5.7	40	Overtone of 885 cm^{-1} (11.3 μm) band; Aromatic $C - C$ stretch; Carbonyl $C = 0$ stretch, ?
1470-1450	6.8-6.9	30	Aromatic $C - C$ stretch, aliphatic CH deformation
840	11.9	–	$C - H$ out-of-plane bend for doubly adjacent H atoms
790	12.7	–	$C - H$ out-of-plane bend for triply adjacent H atoms
		THE BROAD COMPONENTS	
2940 3115-2740[†]	3.5 3.21-3.65[†]	"300"	Overlap of $C - H$ stretching modes, shifted by anharmonic effects, with overtones and combinations of $C - C$ stretch fundamentals in the 1670-1250 cm^{-1} (6-8 μm) region, aliphatic CH stretch?,?
~ 1200 1810-1050[†]	~ 8.5 5.52-9.52[†]	"400"	Blending of many weak aromatic $C - C$ stretching bands
880 950-740[†]	12 10.5-13.5[†]	"160"	Overlap of many aromatic $C - H$ out-of-plane bending modes for non-adjacent as well as doubly and triply adjacent peripheral H-atoms
Red-Near IR Continuum			Electronic transitions between low-lying levels in ionized and complexed PAHs and amorphous carbon particles
Mid-IR Continuum			Quasi-continuum formed by overlapping overtone and combination bands

" " Value estimated from several published spectra.   [†] Rough limits of the feature.

1: When the assignment is not clear, several possible explanations are listed. The first seems most likely.

2: This table is extensively discussed in Allamandola, Tielens, and Barker, 1989.

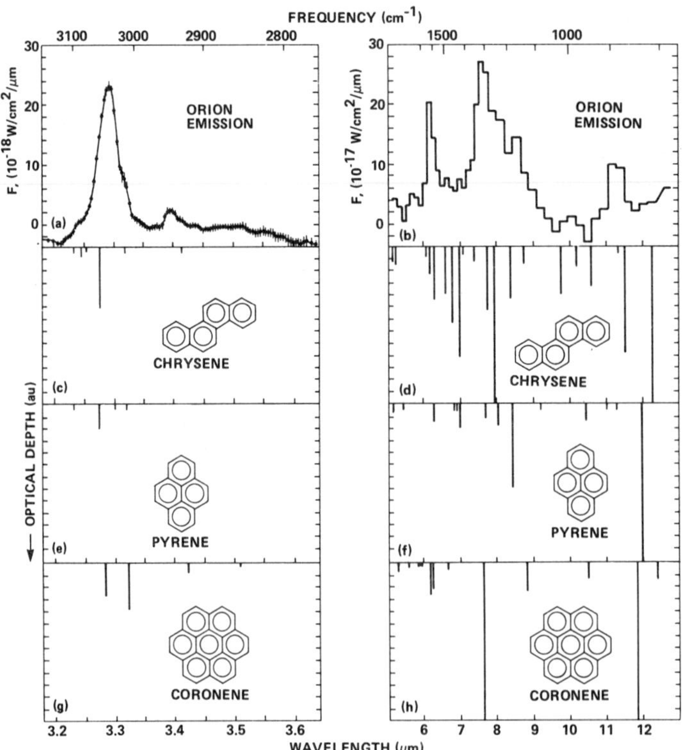

Figure 1.   The 3-13 micron emission spectrum from the Orion Bar compared with the absorption spectra of the PAHs chrysene, pyrene and coronene suspended in KBr pellets. (Orion, 32; Chrysene, 33; Pyrene, 34; Coronene, 35,36. Figure reproduced from 30.

2.1   INTERSTELLAR PAH GEOMETRY

Although the interstellar spectra resemble the laboratory data, they do not match in detail.   The principal challenge here is to select a mixture of PAHs consistent with the constraints posed by the interstellar medium which can reproduce the positions and profiles of the interstellar features.   In the interstellar medium less stable PAHs will be weeded out, leaving a mixture dominated by the most stable forms which are symmetric and condensed (12,14).   One example of a mixture comprised largely of the most stable molecular forms is the soot formed in the high temperature combustion of hydrocarbons.   Figure 2 shows a comparison of the IR emission spectrum in the 1670-1250 cm^{-1} (5.25-10 μm) region from Orion with the Raman spectrum of auto soot.

The Raman spectrum of soot principally probes the aromatic carbon-carbon stretching vibrational frequencies because the Raman scattering cross section for these bonds is very large and they are the most dominant type of bond in the mixture.   This comparison dramatically shows that the C-C vibrational frequencies in a mixture of the more stable PAHs coincides quite well with the frequency range of the interstellar features.   The confident assignment of the interstellar features primarily to symmetric interstellar PAHs is further supported by the good match of the major emission bands in the 2000 to 1000 cm^{-1} (5-10 μm) region from NGC 2023 with the calculated infrared emission spectra from 4 symmetric PAHs (29).

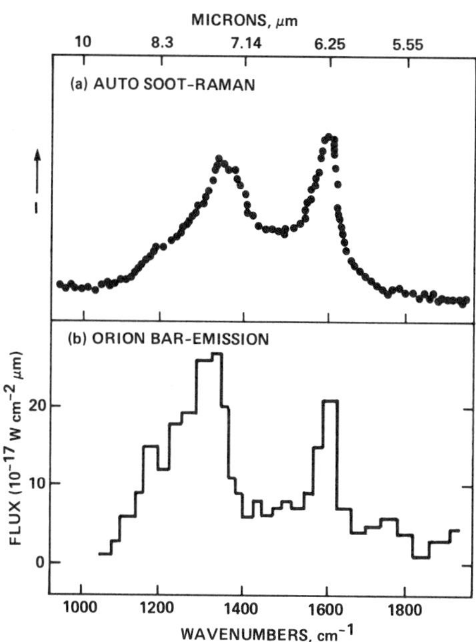

Figure 2. Comparison of the 5 to 10 micron Raman spectrum of auto soot (a form of
amorphous carbon which is rich in the more stable PAHs) with the emission
from Orion (soot spectrum, adapted from 37; Orion, 32). Figure
reproduced from 8.

Another apparent mismatch between the interstellar spectra and the laboratory
spectra of PAHs is evident in the 900-500 cm^{-1} (11-20 μm) region. (Figure 1) The
spectra of PAHs studied in the laboratory have more than one H atom per edge ring.
These bands have long been used by chemists as a diagnostic of the number of
adjacent hydrogen atoms on the edge rings of aromatic molecules. One difficulty in
assigning the emission in this region to PAHs arose because most interstellar
spectra appeared to have only one band in this region (885 cm^{-1}, 11.3 μm) which
implies one hydrogen per ring. This was taken to indicate severe dehydrogenation in
spite of the fact that hydrogen is greater than 10^{7} times more abundant than the
proposed PAHs.

This problem was resolved when it was shown that IRAS spectra of objects emitting
the bands also have a broad plateau of emission from about 950-770 cm^{-1} (10.5-13 μm)
(38). This relieved the difficulties associated with invoking partial dehydrogenation. This also implied that the interstellar PAHs responsible had edge rings with
single, as well as doubly and triply adjacent H atoms, but not four or five, placing
strong constraints on the molecular structures possible and again indicating that
the most stable, symmetric forms are dominant in the interstellar medium. Recent
observations have shown that there is resolvable structure on this plateau
(22,39). The new features are just where one would expect to find them for PAHs
with doubly and triply adjacent H atoms per edge-ring: in the 840-830 cm^{-1} (11.9-12
μm) and 790-775 cm^{-1} (12.6-12.9 μm) regions.

2.2 INTERSTELLAR PAH SIZE

Figure 3 schematically shows the excitation-emission process for an ionized PAH.
Essentially all of the energy deposited in the ion is quickly converted to vibrational energy of the ground state D_0. The intensities in the various vibrational
fundamentals can be calculated as described in detail elsewhere (18). One of the

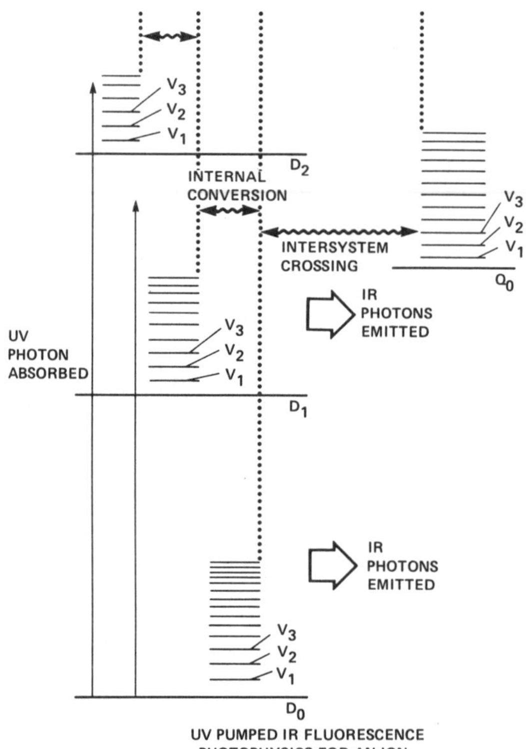

Figure 3.   Schematic energy level diagram for an ionized PAH showing the various
            radiataive and non-radiative relaxation channels possible. Note that in
            the case of an ionized PAH, most of the energy will be emitted in the
            near and mid-IR.

many important applications of this analysis is the estimation of the size and
abundance of the interstellar PAHs from the relative intensities of the emission
features. This analysis shows that PAHs containing between 20-50 carbon atoms are
the major contributors to the narrow interstellar emission bands, that they are more
abundant than all of the known interstellar polyatomic molecules, and that they
contain on the order of 5% of the carbon available in the interstellar medium. This
high abundance of such complex species points to a unique chemical history, one
quite different from gas-phase ion molecule reactions. We have suggested that
interstellar PAHs are produced in carbon-rich circumstellar shells (18).

3.   INTERSTELLAR PAH IDENTIFICATION

Summarizing the above, we conclude that the PAHs which contribute most strongly to
the narrow interstellar infrared emission bands:

    1)   contain between 20-30 carbon atoms, and
    2)   have edge rings with single, as well as doubly and triply adjacent H
          atoms, not four or five.

These conclusions permit us to narrow down the number of likely neutral, isolated
candidates to a handful because the number of symmetric PAHs containing 20-30
carbon atoms is not large. By eliminating those with edge rings containing four or
five hydrogen atoms leaves only a few molecules. When the spectra of these neutral
molecules as well as of their ionized and partially dehydrogenated counterparts

become available, it should be possible to identify the smallest members of the complex interstellar PAH family. This family of important, abundant, carbon-bearing, cyclic species includes free molecules, PAH clusters and amorphous carbon particles.

References

1. Bregman, J.D. 1989, in Interstellar Dust, eds. Allamandola, L.J. and Tielens, A.G.G.M. (Kluwer, Dordrecht) p. 109

2. Roche, P.F. 1989, in Interstellar Dust, eds. Allamandola, L.J. and Tielens, A.G.G.M. (Kluwer, Dordrecht) p. 303

3. Sellgren, K. 1989, in Interstellar Dust, eds. Allamandola, L.J. and Tielens, A.G.G.M. (Kluwer, Dordrecht) p. 103

4. Allamandola, L.J., Greenberg, J.M. and Norman, C.A. 1979, Astron. Astrophys. **77**, 66

5. Gillett, F.C., Forrest, W.J., and Merrill, K.M. 1973, Ap. J. **183**, 87

6. Duley, W.W. and Williams, D.A. 1981, Mon. Not. R. Astro. Soc. **196**, 269

7. Leger, A. and Puget, J.L. 1984, Astro. Astrophys. **137**, L5

8. Allamandola, L.J., Tielens, A.G.G.M. and Barker, J.R. 1985, Ap. J. (Letters) **290**, L25

9. Cohen, M., Allamandola, L.J., Tielens, A.G.G.M., Bregman, J.D., Simpson, J.P., Witteborn, F.C., Wooden, D. and Rank, D. 1986, Ap.. J. **302**, 737

10. Omont, A. 1986, Astron. Astrophys. **164**, 159

11. Puget, J.L., Leger, A. and Boulanger, F. 1985, Astron. Astrophys. **142**, L19

12. van der Zwet, G.P. and Allamandola, L.J. 1985, Astron. Astrophys. **146**, 76

13. Leger, A. and d'Hendecourt, L.B. 1985, Astron. Astrophys. **146**, 81

14. Crawford, M.K., Tielens, A.G.G.M. and Allamandola, L.J. 1985, Ap. J. (Letters) **293**, L45

15. d'Hendecourt, L.B. and Leger, A. 1987, Astron. Astrophys. **180**, L9

16. Lepp, S. and Dalgarno, A. 1988, Ap. J. **324**, 553

17. Allamandola, L.J., Sandford, S.A., and Wopenka, B. 1987, Science **237**, 56

18. Allamandola, L.J., Tielens, A.G.G.M. and Barker, J.R. 1989, Ap. J. Suppl., **71**, Dec.

19. Donn, B.D., Allen, Jr., J.E., and Kahanna, R.K. 1989, in Interstellar Dust, eds Allamandola, L.J . and Tielens, A.G.G.M. (Kluwer, Dordrecht) 181

20. Leger, A., Verstraete, L., d'Hendecourt, L.B., Defourneau, D., Dutuit, O. Schmidt, 1989 in Interstellar Dust, eds. Allamandola, L.J. and Tielens, A.G.G.M. (Kluwer, Dordrecht) p. 173

21. Allamandola, L.J., Bregman, J., Sandford, S.A., Witteborn, F.C. and Wooden, D.H. 1989, Ap. J. (Letters), L59

22. Witteborn, F.C., Sandford, S.A., Bregman, J.D., Allamandola, L.J., Cohen, M. and Wooden, D. 1989, Ap. J., **341**, 270

23. Geballe, T.R., Tielens, A.G.G.M., Allamandola, L.J., Morehouse, A. and Brand, P.W.J.L. 1989, Ap. J., **341**, 278

24. Duley, W.W. and Williams, D.A. 1988, Mon. Not. R. Astro. Soc. **231**, 969

25. Duley, W.W., 1989, in Interstellar Dust, eds. Allamandola, L.J. and Tielens, A.G.G.M. (Kluwer, Dordrecht) p. 141

26. Bussoletti, E. and Colangelli, L. 1989, in Interstellar Dust, eds. Allamandola, L.J. and Tielens, A.G.G.M. (Kluwer, Dordrecht) p. 147

27. Sakata, A. 1989, in Interstellar Dust, eds. Allamandola, L.J. and Tielens, A.G.G.M. (Kluwen, Dordrecht) p. 191

28. Papoular, R., Conard, J., Guiliano, M., Kister, J. and Mille, G., 1989, Astron, Astrophys, in press

29. Leger, A. and d'Hendecourt, L.B. 1987 in Polycyclic Aromatic Hydrocarbons and Astrophysics, eds. Leger, A., d'Hendecourt, L.B. and Boccara, N. (D. Reidel, Dordrecht) 223

30. Allamandola, L.J., Tielens, A.G.G.M. and Barker, J.R. 1987 in Physical Processes in Interstellar Clouds, eds. Morfill, G.E. and Scholer, M. (D. Reidel, Dordrecht) 305

31. Cherchneff, I. and Barker, J.R. (1989) Ap. J. Letters **341**, L21

32. Bregman, J., Allamandola, L.J., Tielens, A.G.G.M., Rank, D.M. and Geballe, T.R. 1988, Ap. J., **344**, 791

33. Cyvin, B.N., Klaeboe, P., Whitmer, J.C. and Cyvin, S.J. 1982a, Z. Naturforsch, **37a**, 251

34. Cyvin, S.J., Cyvin, B.N., Brunvoll, J., Whitmer, J.C., Klaeboe, P., and Gustavsen, J.E. 1979, Z. naturforsch, **34a**, 876

35. Bakke, A., Cyvin, B.N., Whitmer, J.C., Cyvin, S.J., Gustavsen, J.E. and Klaeboe, P. 1979, Z. Naturforsch, **34a**, 579

36. Cyvin, S.J., Cyvin, B.N., Brunvoll, J., Whitmer, J.C., and Klaeboe, P. 1982b, Z. Naturforsch, **379**, 1359

37. Rosen, H. and Novakov, T. 1978, Atmos. Environment **12**, 923

38. Cohen, M., Tielens, A.G.G.M. and Allamandola, L.J. 1985, Ap. J. Letters **299**, L93

39. Roche, P.F., Aitken, D.K. and Smith, C.H. 1988, Mon. Not. R. Astro. Soc. **236**, 485

IR Emission from PAH Molecules and Interstellar Dust

W. W. Duley

Physics Department, York University, 4700 Keele Street, North York, Ontario, M3J 1P3, Canada

Summary

The population of vibrational levels in PAH emitters is examined using a simple optical pumping model. It is shown that significant pumping effects occur for vibrational states with energies ≤ 200 cm^{-1}. The long radiative lifetime of these states implies that they may exhibit enhanced or inverted population relative to lower levels. The apparent absence of strong discrete PAH emission at $\lambda > 15\mu$m suggests the existence of a hitherto unrecognized quenching mechanism and may provide useful insight into the physical state of these emitters.

Introduction

The association of the unidentified infrared (UIR) emission features with interstellar polycyclic aromatic hydrocarbon (PAH) molecules or molecular groups (Duley and Williams 1981, Leger and Puget 1984, Allamandola, Tielens and Barker 1985) is gaining increasing acceptance. The processes involved in the excitation of these emission features have been the subject of much theoretical (see Allamandola, Tielens and Barker 1987 for a review) and some experimental work (Cherchneff and Barker 1989). Infrared emission by PAH is preceded by the absorption of an ultraviolet photon. The energy of this photon is converted in part to vibrational excitation with the PAH molecules as it is shared between the $3N - 6$ available vibrational modes. IR emission occurs when the molecule makes a transition between two of these vibrational states. The rate constant, A_λ for this emission is of the order of 10–10^2 sec^{-1} at $\lambda = 3.28\mu$m.

While the spectroscopy of the UIR emission is becoming increasingly well understood, many questions still remain concerning the details of excitation/emission mechanisms in these large molecules in an interstellar or nebular radiation field. One fundamental question concerns the possibility that IR emission may occur from PAH groups attached to dust grains (Duley and Williams 1988) rather than from gas-phase molecules (Allamandola 1989). The question of the absence of strong discrete emission features longward of 12.7μm (Bregman 1989) must also be addressed.

In this note the response of a PAH molecule to the interstellar radiation field is discussed in terms of a conventional optical pumping model often applied to terrestrial laser media. I show that, if PAH molecules are 'free-flying' gaseous species then optical pumping should lead to significant population inversion at middle and far-infrared (FIR) wavelengths. The absence of discrete emission lines at wavelengths $\lambda > 12.7\mu$m may then provide a clue to the physical state of the emitters.

Excitation & Emission by PAH

Figure 1 shows a simplified representation of the electronic and vibrational energy levels in a PAH molecule. The cross-section for absorption of an ultraviolet photon for such molecules is σ_λ(cm^2). Absorption leads to an electronic transition between the ground state and a range of vibronic levels of one or more excited states. If the radiative flux is F_λ (photons cm^{-2} sec^1 Å^{-1}) then the rate of pumping of the excited state is

$$P = \int \sigma_\lambda F_\lambda \, d\lambda \tag{1}$$

This can be approximated by average values so that $P \approx \sigma F$ where σ is a typical UV absorption cross-section while F is the average integrated UV radiative flux (photons cm^{-2} sec^{-1}).

Some fraction of the incident photon energy $\beta h\nu$ is retained by the molecule after any rapid electronic deactivation (e.g. luminescence) and is distributed among the $3N - 6$ vibrational modes. The simplest approximation would involve

$$\epsilon_i = \frac{\beta h\nu}{3N - 6} \tag{2}$$

where ϵ_i is the total energy of the ith vibrational mode. If $h\nu_i < \epsilon_i$ where $h\nu_i$ is the 0–1 energy of the ith vibrational frequency, then the molecule can emit one or more IR photons with this energy. Absorption of UV energy therefore leads to the population of the ith vibration level.

If we now take the rate of radiative relaxation of this level to be $A_i(\lambda)$ (sec^{-1}) then a simple rate equation can be written to connect molecular populations in the ground (N_0) and i^{th} vibrational levels (N_i). In equilibrium the solution to this equation yields the following relation

$$\sigma F N_0 = A_i(\lambda)N_i \tag{3}$$

or

$$\frac{N_i}{N_0} = \frac{\sigma F}{A_i(\lambda)} \tag{4}$$

Under normal interstellar conditions, the ratio $N_i/N_0 \ll 1$. However, under conditions of intense UV pumping, as exists for example in UIR source regions, N_i, can become comparable to or can exceed N_0. The condition for this to occur is that $A_i(\lambda) \sim \sigma F$. Under these conditions, which are possible when $h\nu_i$ is small, the molecule may exhibit a vibrational population inversion. The possibility of such a radiatively-driven population inversion will exist as long as the collision frequency $\sigma n_{H_2} v < A_i(\lambda)$. For UIR sources where $\sigma \sim 5 \times 10^{-16}$ cm^2, $n_{H_2} \sim 10^4$ and $v \sim 10^5$ cm sec^{-1}, $A_i(\lambda) > 5 \times 10^{-7}$ sec^{-1}. Since $A_i(\lambda) \propto \lambda^{-3}$ while $A_i(3.3\mu$m$) \sim 10$ sec^{-1} this condition implies $\lambda < 900\mu$m. Observation of enhanced excited state vibrational populations should therefore be possible in UIR sources.

For PAH molecules in UIR sources the UV pumping rate σF is estimated to be 2×10^{-3} sec^1 (Geballe et al 1989). Then $N_i/N_0 = 2 \times 10^{-3}/A_i(\lambda)$, so that $N_i \sim N_0$ when $A_i(\lambda) \leq 2 \times 10^{-3}$ sec^{-1}. With

$$\frac{A_i(\lambda)}{A_i(3.3)} = \left(\frac{3.3}{\lambda}\right)^3 \tag{5}$$

and $A_i(3.3) = 10$ sec^{-1} while $A_i(\lambda) = 2 \times 10^{-3}$; $\lambda = 56.4\mu$m. This is the critical wavelength beyond which population inversions could occur between the $v = 1$ vibrational level and the ground state. Typical PAH molecules such as naphthalene, anthracene and coronene all have vibrational modes at $\lambda > 50\mu$m. These modes involve ring vibrations and are unaffected by hydrogenation.

It is important to note that under conditions where collisional deactivation can be neglected (e.g. $n_{H_2} \sim 10^4$ cm^{-3}) every UV excited molecule will emit at least one IR photon from each IR active mode. Because $A_i(\lambda) \propto \lambda^{-3}$ these photons will be emitted sequentially with those at shortest wavelength (i.e. 3.28μm) being emitted first. The emission spectrum of the molecule will therefore depend on time with 3.28μm photons emitted ~ 0.1 sec after excitation while 11.3μm photons will be emitted ~ 100 sec later.

The long lifetime of vibrational states with $h\nu_i \leq 200$ cm^{-1} compared to UV pumping rates then implies that population inversion is possible between $v_i = 1$ and the ground state, or between excited vibrational states. Under these conditions the interstellar medium may exhibit optical gain at frequency ν_i. The gain coefficient g (cm^{-1}) can be written (Milonni and Eberly 1988)

$$g = \sigma(N_i - N_0) \tag{6}$$

with $\sigma = 10^{-19} - 10^{-20}$ cm^2 and $N_i - N_0 \sim N_0/2 = 10^{-6}n_{H_2}$, $g = (10^{-25} - 10^{-26})n_{H_2}$. This is a small value and suggests that amplified spontaneous emission at PAH FIR frequencies may not be observable. However, even in the absence of amplification, the effective brightness temperature at these ν_i will be large.

This should yield strong, discrete emission lines from PAH at $\lambda \geq 50\mu$m with brightness temperatures far above thermal equilibrium values at these wavelengths. Such emission will be unavoidable under conditions where $\sigma F \sim A_i(\lambda) \gg$ collision rate.

In the near absence of collisional deactivation, a PAH molecule exposed to an intense UV radiation flux will always be vibrationally excited. This implies that there would be little population in the normal ground vibrational state i.e. that the effective ground state of the molecule would be at some excited level i where $h\nu \leq 200$ cm^{-1}. The absorption spectrum of PAH (or PAH$^+$) under these pumping conditions would be different from that for normal (cold) PAH.

Middle IR emission from PAH

Vibrational spectra of PAH molecules contain many features at wavelengths longer than the lowest energy UIR feature at 12.7μm (Neerland et al 1980). Many of these are $C - C - C$ out-of-plane bending modes. IRAS spectra of UIR sources occasionally show evidence for discrete emission lines between 13 and 23μm but strong individual lines are not apparent. A good example is the spectrum of the planetary nebula BD $+30°$ 3639 (Bregman 1989) which is featureless between 12.7μm and 23μm. Spectra obtain of NGC 7027 (McCarthy, Forrest and Houck 1978) between 16 and 38μm also show a featureless continuum while emitting a rich variety of lines between 3.28 and 11.3μm.

One explanation for the absence of discrete emission at long wavelengths is that a multitude of individual lines from a variety of different molecules combine to form a continuum. However, as it appears that emission at shorter wavelengths can be understood in terms of a very limited range of small PAH emitters (Duley and Jones 1990) the formation of such a smooth continuum would not seem to be likely.

An alternative explanation for this effect may be found in the excitation and deactivation of PAH. Since a necessary condition for IR emission from a state at $h\nu_i$ is that $A_i(\lambda) > \tau_d^{-1}$ where τ_d is a deactivation timescale, the absence of emission past a given λ may suggest that some process operating over a timescale τ_d acts to limit this emission. Specifically since there is apparently no line emission at $\lambda \geq 15\mu$m, eqn. 5 taken with $A_i(15) = \tau_d^{-1}$ and $A_i(3.3\mu m) = 10$ sec^{-1} yields $\tau_d \sim 10$ sec.

The origin of τ_d may provide a useful insight into the physical state of the PAH emitters. With $\tau_d \sim 10$ sec some 20 times shorter than $(\sigma F)^{-1}$, the interval between UV absorption events, the limit to emission provided by τ_d cannot be due to photo-destruction or modification (isomerization). In addition, since $\tau_d \ll (\sigma n v)^{-1}$, the collision time, gas phase collisional deactivation is not likely the source of τ_d.

The existence of a deactivation timescale τ_d can, however, be understood if the emitters are in fact not free flying molecules, but are instead in island form on the surface of HAC dust. In the model proposed by Duley and Williams (1988) PAH 'islands' in HAC are conjectured to be sufficiently poorly connected thermally to allow localization of absorbed energy over timescales comparable to IR radiative lifetimes. Such properties have been observed in other amorphous materials (Madan and Shaw 1988), Orenstein, Kastner and Vaninov 1982, Phillips 1982) but have not yet been studied experimentally in HAC. An analysis of such processes in HAC, (Duley and Williams 1988) yields a deactivation timescale that depend on the energy of the bonds connecting an excited PAH island to other islands. In this description, τ_d can be identified with an energy delocalization timescale for PAH in HAC. Recent laboratory data showing the existence of pendant aromatic rings in HAC (Tamor *et al* 1989) suggests that the morphology of these rings is such that energy delocalization may indeed be inefficient.

In the rehydrogenation of HAC in ERE sources (Duley and Williams 1990) it is expected that relatively weakly bonded polymeric chains will be removed first leaving PAH molecules poorly connected thermally to each other. The morphology in this case would be analogous to a lacelike structure with PAH groups able to behave like quasi free molecules while still attached to dust particles.

In conclusion, strong emission is expected from free-flying PAH molecules at wavelengths $\lambda \geq 50\mu$m due to radiative pumping of low-lying vibrational levels. The apparent absence of discrete IR emission features at $\lambda > 15\mu$m would imply quenching of emission over a timescale that is short compared to timescale for both collision and photoabsorption. Such quenching can, however, be qualitatively understood if PAH emitters are attached to dust surfaces.

Acknowledgement

This research was supported by grants from the NSERCC.

References

Allamandola, L.J.: 1989, *International School of Physics "Enrico Fermi" Course CXI Solid State Astrophysics* ed. E. Bussoletti, (in press).

Allamandola, L.J., Tielens, A.G.G.M. & Barker, J.R.: 1985, *Astrophys. J.* **290**, L25.

Allamandola, L.J., Tielens, A.G.G.M. & Barker, J.R.: 1987 *Physical Processes in Interstellar Clouds* eds. Morfill, G.E. & Scholer, M. Reidel, Dordrecht, p. 305.

Bregman, J.D.: 1989 *Interstellar Dust Proc. IAU Symp.* 135, eds. Allamandola, L.J. & Tielens, A.G.G.M., Klumer, Dordrecht, p. 109.

Cherchneff, I. & Barker, J.R.: 1989. *Astrophys. J.* **3412**, L21.

Duley, W.W. & Jones, A.P.: 1990 *Astrophys. J.* (in press).

Duley, W.W. & Williams, D.A.: 1981. *Mon. Not. R. Astr. Soc.,* **196**, 269.

Duley, W.W. & Williams, D.A.: 1988. *Mon. Not. R. Astr. Soc.* **231**, 969.

Duley, W.W. & Williams, D.A.: 1990. *Mon. Not. R. Astr. Soc.* (submitted).

Geballe, T.R., Tielens, A.G.G.M., Allamandola, L.J., Moorhouse, A., & Brand, P.W.J.L.: 1989. *Astrophys. J.* **341**, 278.

Leger, A. & Puget, L.: 1984. *Astron. Astrophys.* **137**, L5.

Madan, A. & Shaw, M.P.: 1988. *The Physics and Applications of Amorphous Semiconductors* Academic Press, New York.

McCarthy, J.F., Forrest, W.J. & Houck, J.R.: 1978. *Astrophys. J.* **224**, 109.

Millonni, P.W. & Eberly, J.H.: 1988. *Lasers* J. Wiley, New York.

Neerland, A., Cyvin,, B.J., Brunvall, J., Cyvin, S.J. & Klaeboe, P.: 1980. *Z. Naturforsch.* **35A**, 1390.

Orenstein, J., Kastner, M.A. & Vaninov, V.: 1982: *Phil. Mag.* **B46**, 23.

Phillips, J.C.: 1982. *Phys. Rev.* **B25**, 1397.

Tamor, M.A., Wu, C.H., Carter, R.O. & Lindsay, N.N.: 1989. *Appl. Phys. Letters* **55**, 1388.

Figure caption

Figure 1: Simplified representation of electronic and vibrational levels in a PAH molecule.

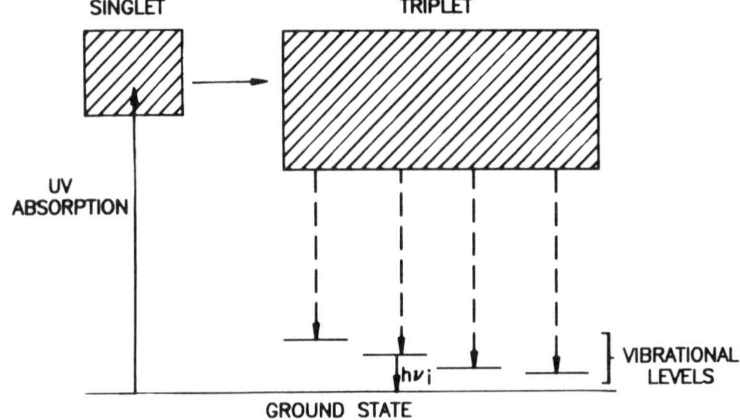

The Time Dependent Structure of Carbonaceous Grains and Mantles in the Interstellar Medium

A. P. Jones[1], W. W. Duley[1], and D. A. Williams[2]

[1] *Department of Physics, York University, 4700 Keele Street, North York, Ontario, M3J 1P3, Canada*
[2] *Department of Mathematics, UMIST, PO Box 88, Manchester, M60 1QD, U. K.*

Interstellar HAC likely contains two major carbon phases; graphitic and polymeric/diamond-like. This structure is the result of carbon species accreting onto interstellar grains as hydrogen-rich polymeric and/or diamond-like materials which are subsequently partially photoprocessed into a graphitic type of material. It is shown that a HAC model for interstellar carbon grains can explain the IR emission features (UIR bands) in terms of small linear PAHs that are an intimate component of interstellar HAC. We propose an evolutionary cycle for HAC which shows that carbon grains are a dynamic component of the interstellar medium.

1 Introduction

Hydrogenated amorphous carbon (HAC) forms in the interstellar medium by the accretion of carbon atoms, ions and radicals onto the surfaces of pre-existing silicate grains (1,2). In the hydrogen rich interstellar medium the initial HAC deposit is a highly hydrogenated amorphous carbon that has a polymeric alkane, and diamond-like, structure and composition which absorbs strongly in the ultraviolet but hardly at all in the visible. The ambient ultraviolet radiation field will progressively graphitise and dehydrogenate the polymeric material into a material that absorbs strongly in the visible (3). A balance between the accretion of newly accreted HAC and the photoprocessing (photodarkening) of older HAC is achieved in diffuse interstellar clouds such that in any carbon grain or mantle about half the material is in polymeric/diamond-like form and half in a graphitised form. In energetic regions where the grains are heated by an intense ultraviolet radiation field, and where there is a high abundance of hydrogen atoms it is in fact possible to reverse the dehydrogenation of HAC by hydrogen atom reactions in which the covalent bonding of the aromatic ring structures is opened up (4).

2 The carbon cycle in the interstellar medium

Diffuse clouds are, of course, dynamic regions and carbon accretion cannot proceed to completion because of the interaction of interstellar shocks with the cloud matter. Weak shocks (few $km\,s^{-1}$) will rapidly sputter away the softer polymeric carbon from the grains whilst faster shocks (several tens $km\,s^{-1}$) also sputter away the diamond-like component and the underlying graphitic HAC, possibly eroding back to the original silicate core. Thus, the extinction of starlight through shocked regions of the interstellar medium will be a function of the shock type i. e. weak shocks will give rise to decreased ultraviolet extinction whilst strong shocks will produce decreased extinction in the visible as well as in the ultraviolet.

Interstellar carbon is therefore subjected to a dynamic cycle of events (the interstellar carbon cycle), namely, accretion onto grains as a polymeric alkane-type material, photodarkening and dehydrogenation by ultraviolet photons, rehydrogenation by hydrogen atom reactions with hot grains, and sputtering of all or part of the mantle back into the gas phase in interstellar shocks. The dynamics and chemistry of HAC particles in the interstellar medium has been extensively discussed (5).

3 The UIR features and the HAC model

In Table I we show an interesting comparison of the absorption bands of laboratory HAC (a-C:H) (6) and the unidentified infrared emission (UIR) bands (7). Regardless of the assignment of the features there does appear to be a remakable similarity between the laboratory data for solid hydrogenated carbons and the observed UIR bands.

Table I A comparison of the a-C:H absorption bands (6) with the UIR emission bands (7).

a-C:H (absorption) and UIR (emissiion) Bands	
a-C:H cm^{-1} (μm)	UIR cm^{-1} (μm)
3300 (3.03)	—
3045 (3.29)	3040–3050 (3.28–3.29)
3000 (3.33)	2995 (3.34)
2920 (3.42)	2940 (3.40)
—	2890 (3.46)
2850 (3.51)	2845 (3.51)
—	2805 (3.57)
—	1920 (5.21)
—	1905 (5.25)
—	1760–1780 (5.62–5.68)
1620 (6.17)	1610–1615 (6.19–6.21)
1570 (6.37)	—
1509 (6.63)	—
1430 (6.99)	1450–1460 (6.85–6.90)
1367 (7.32)	1350 (7.41)
1300 (7.69)	1250–1315 (7.60–8.00)
1150 (8.70)	1150–1160 (8.62–8.70)
880 (11.36)	885–890 (11.24–11.30)
—	840 (11.90)
780–820 (12.20–12.82)	787 (12.71)

It has been shown (8), contrary to the general view that large gas-phase polycyclic aromatic hydrocarbons (PAHs) are responsible for the UIR bands, that small linear PAH molecules, such as anthracene, naphthalene and tetracene, give an excellent match to the observed features (Table II). Further, the absence of some of the expected anthracene and naphthalene bands (at 10.0, 10.5 and 13.8 μm) provides strong evidence that the emitting species are a component of HAC. The 'missing' bands do not occur in methyl- and ethyl- substituted species where the substituents are on the end rings of the molecules. This is consistent with the structure of HAC (6,9), i.e. polycyclic aromatic hydrocarbons (PAHs) intimately bonded to a polymeric matrix. The size distribution of these PAHs is biased toward the smaller molecules and HAC therefore contains a large fraction of small linear PAH molecules.

Table II Comparison of the major absorption bands in simple polyacenes with the interstellar emission features (8). Laboratory absorption data are for Nujol solutions.

Absorption Bands and Interstellar Emission Features										
Species	Absorption and Emission Features (cm^{-1})									
Interstellar	3040	1905	1754	1615	1460	1315–1250	1150	905	885	787
Naphthalene	3048	—	—	1593	—	1271	1123	—	—	782
Anthracene	3048	1910	1775	1621	1450	1317	1162/1147	—	884	—
Tetracene	3043	—	—	—	—	1297	1122	904	—	—
Pentacene	3043	—	—	—	—	1296	—	908	836	—

4 Conclusion

The structure of HAC is time dependent, initially accreted materials are hydrogen rich and polymeric/diamond-like whilst older HAC has been photodarkened by the ambient ultraviolet radiation to a more graphitic carbon. In the general interstellar medium accretion is continuosly occuring and both types of HAC are present in approximately equal proportions. In energetic regions where the grains are hot rehydrogenation, i. e. PAH conversion to polymeric material can occur.

This model for interstellar carbon can explain the observed variations in interstellar extinction and the observed UIR features in regions adjacent to energetic sources. Small linear PAH molecules, including naphthalene present in interstellar HAC, provide an excellent match to the UIR bands.

Acknowledgement

This research has been supported by SERC and NSERCC. APJ is grateful to SERC and NATO for the award of a postdoctoral fellowship.

References

(1) Duley,W.W. 1987, *M.N.R.A.S.*, **229**, 203.

(2) Duley,W.W., Jones,A.P. and Williams,D.A. 1989, *M.N.R.A.S.*, **236**, 709.

(3) Duley,W.W. and Williams,D.A. 1988, *M.N.R.A.S.*, **230**, 1P.

(4) Rye,R.R. 1977, *Surface Science*, **69**, 653.

(5) Duley,W.W., Jones,A.P. and Williams,D.A. 1990, in preparation.

(6) Robertson,J. 1986, *Advances in Physics*, **35**, 317.

(7) Allamandola,L.J., Tielens,A.G.G.M. and Barker,J.R. 1990, *Ap. J. Suppl.*, in press.

(8) Duley,W.W. and Jones,A.P. 1990, *Ap. J. Letts.*, in press.

(9) Smith,F.W. 1984, *J. Appl. Phys.*, **55**, 764.

Models of the Gas-Grain Interaction in Dense Interstellar Clouds

Paul D. Brown[1], Steven B. Charnley[2,3], and T. J. Millar[4]

[1] *Canadian Institute for Theoretical Astrophysics, University of Toronto, 60 St. George Street, Toronto, Ontario M5S 1A1, Canada*
[2] *Max-Planck-Institut für extraterrestrische Physik, Garching b. München, F. R. G.*
[3] *Physics Department, Rensselaer Polytechnic Institute, Troy, New York 12180, U. S. A.*
[4] *Department of Mathematics, UMIST, PO Box 88, Manchester M60 1QD, U. K.*

We summarise recent results obtained from theoretical models of gas and grain chemistries in cold interstellar clouds and in regions of star formation.

1. Introduction

It has long been recognised that the most abundant interstellar molecule, molecular hydrogen, is formed by reactions between two hydrogen atoms on a grain surface [1]. Also, IR observations have revealed the presence of ice mantles which contain H_2O, CO, CH_3OH and probably NH_3 [2,3]. In the absence of an efficient desorption process, the heavy gas phase component is expected to condense on to the grains on a timescale of *a few* $\times$ $10^9/n$ years, where n is the total hydrogen nucleon number density (cm^{-3}) [4]. Despite these factors, however, most chemical models have ignored the effects of accretion (other than H_2 formation) [5,6,7]. Those exceptions which have included accretion [4,8] and continous desorption (by mantle explosions [9]) have shown that it should be studied using time-dependent models. No region of high obscuration has yet been found which contains an unusually low abundance of heavy gas phase molecules, and there is also uncertainty concerning the general validity of the various desorption processes that have been proposed. We have thus attempted to study the chemistry in an accreting, reacting gas in which some desorption mechanisms operate, to seek possible signatures of grain chemistry. Alternatively, dynamical events which occur within an accretion time, such as shocks and nearby star formation, may be responsible for returning mantle material to the gas phase [10]. In this contribution we summarise the results from our modelling of dense quiescent clouds and describe recent applications of this work to the Orion star-forming region.

2. Quiescent Dense Clouds

The basic gas-grain model has been described by Brown & Charnley [11]. The temperature (10 K) and density (3×10^4 cm^{-3}) are held constant throughout the calculation, that is, the model is pseudo-time-dependent. Accretion occurs in the absence of any desorption process with unit sticking efficiency, and only activationless surface reactions involving H atoms are considered at present. This model has also been extended to include deuterium chemistry, both in the gas and on the grain surface [12,13], and to study the effect of surface-formed methane upon complex molecule formation in the gas [14]. We find that a large degree of deuterium fractionation can occur in surface reactions, supporting the earlier work of Tielens [15].

Figure 1 shows the case of unabated accretion where it may be seen that a total 'freeze-out' of the heavier elements occurs within a few million years. The chemical evolution of gaseous species is, however, non-trivial and accretion does not simply result in a monotonic decrease in the abundance of each species. For example, in the case of OH the abundance initially increases as accretion begins to dominate the chemistry. This is due to a reduction in its gas phase destruction rate following the accretion of the heavy atoms which destroy it. At later times OH also condenses on to the grains and so it eventually does become depleted. However, in the case of the light ions H^+, He^+, H_3^+ and H_2D^+, their abundances remain high because helium atoms and hydrogen molecules do not condense out of the gas in substantial amounts. The formation rates of these ions (driven by cosmic rays) are therefore essentially unchanged by the inclusion of accretion, whereas their destruction rates are drastically reduced by the loss of heavy atoms and molecules with which they react. The fractional ionization is then much enhanced in the highly-depleted gas and this may have interesting consequences for star formation in dense clumps of gas [16]. The condensation of CO molecules on to grains removes important coolant from the gas, this will lead to an increase in temperature and affect pressure support of the cloud [17]. It is interesting to note that if the temperature of the gas is only moderately increased by the initial loss of CO, the abundance of H_2D^+ rises faster than that of H_3^+, due to the barrier in the reverse direction of the reaction:

$$H_3^+ \; + \; HD \; \longrightarrow \; H_2D^+ \; + \; H_2 \; + \; \sim 230 \text{ K}.$$

This leads to a higher level of deuterium fractionation in the molecules remaining in the gas [12], and this could possibly serve as a chemical signpost of a highly-depleted region [18].

We have modelled the dense cloud TMC-1 [14]. By comparing our calculated abundances with observations over a range of model times, we are able to determine an 'optimal time' at which the agreement for all the observed molecules is a maximum, within prescribed limits. We find good agreement for most species, including the cyanopolyynes and other complex hydrocarbons, at a model time which is consistent with estimates for the dynamical age of the cloud [19,20]. These results are at least as good as those obtained at 'early times' and at steady-state in models which neglect accretion. By allowing the ejection of surface-formed methane, we find the agreement is further improved and persists over a longer timescale [14]. We are currently investigating the possible role of grain surface chemistry in resolving outstanding problems with sulphur chemistry in dark clouds [21].

Our modelling has shown that it is unnecessary to invoke an efficient continous desorption mechanism (or a small sticking coefficient) to explain many of the abundances observed in TMC-1. This does not rule out the possibility that such a process is acting and so we have performed a study of the desorption of surface molecules by cosmic rays [22,18]. The passage of a cosmic ray particle may heat the whole grain (for small grains) or part of it, and desorb species with low binding energies (*e.g.* CO, N_2, O_2 and CH_4). From Figure 2 it is clear that this process is not sufficient to prevent substantial condensation on to the grains. Nevertheless, a total 'freeze-out' does not occur and a *pseudo* steady-state is reached. The abundances of the light ions do not increase as much as in the case of unabated accretion as these ions break down the desorbed volatiles, leading to the formation of many other molecules found in dense clouds. The calculated ice mantles are dominated by H_2O, NH_3, CH_4 and CO. IR observations suggest that methanol is a major constituent of the ice [3]. This is not the case in our simple picture and we are presently investigating routes to the formation of more complex molecules.

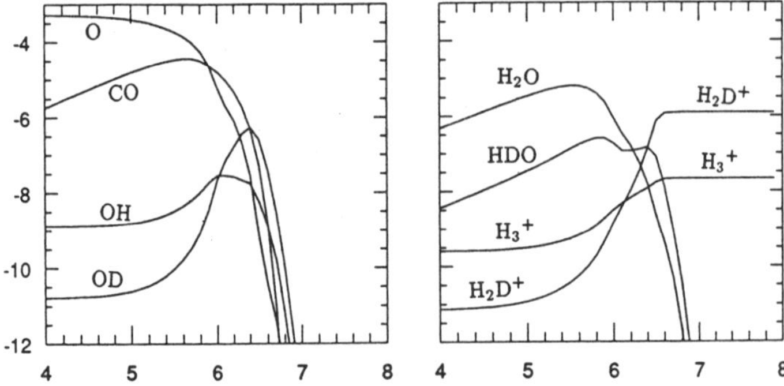

Figure 2. As for Figure 1 except cosmic ray desorption of volatiles is included.

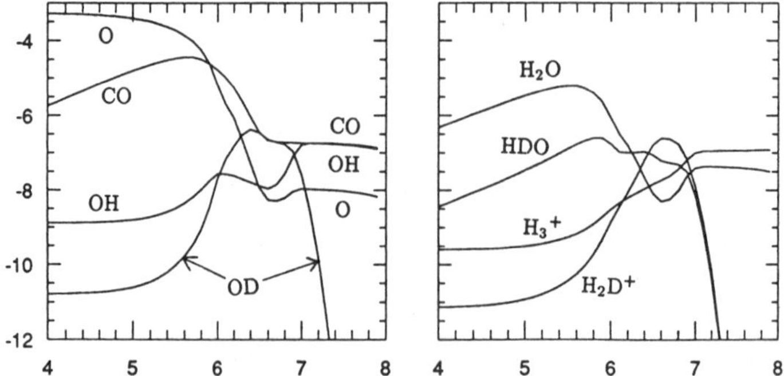

Figure 1. Fractional abundances vs. log(time) for the case of unabated accretion.

3. Orion Star-Forming Region

We have previously applied our model to describe the chemistry of hot molecular cores such as the one in the Orion-KL region [23]. This includes a cold, isothermal free-fall collapse and the assumption that processes associated with the formation of a nearby star halt the collapse and heat the grains. The dust mantles then evaporate. These models provide an explanation for the enhanced abundances of H_2O and NH_3 found in such regions, as well as for the high HCN/HNC ratio, and deuterium fractionation ratios of XD/XH $\sim a\ few \times 10^{-3}$ in gas at temperatures of $\sim 150 - 200$ K. An extension of this work suggests that detectable abundances of multi-deuterated species, such as NHD_2, may exist in the Orion hot core [13]. Recent observations tentatively support this (Turner, *private communication*).

The so-called 'compact ridge' component in Orion-KL is spatially and chemically distinct from the hot core although it has a similarly high density and temperature [24]. It is oxygen-rich and contains many large O-bearing molecules such as $HCOOCH_3$, CH_3OCH_3 and $HCOOH$, as well as large quantities of methanol and formaldehyde. An outstanding theoretical problem for the chemistry in Orion is to account for the abundances and degree of isomerization observed in the compact ridge. Recent calculations [25] have shown that standard pseudo-time-dependent models, based upon radiative association reactions, cannot reproduce the observed abundances. Further calculations have shown that models in which water molecules are injected into the ridge gas, presumably from the O-rich wind of IRc2 [24], also fail in this respect. The main problem with such models is that radiative association of CH_3^+ and H_2O is inefficient in warm gas and the formation of protonated methanol, an essential step for the complex chemistry, is inhibited. Models in which methanol is injected into the evolving gas, however, do produce large abundances of these molecules. This methanol is presumed to be from the sublimation of grain mantles in the warm gas. In this scenario an ion-molecule chemistry pertains in which some of the neutral molecules have been synthesized on grains, and in which inefficient gas phase routes to them have consequently been bypassed. For example, the presence of large amounts of methanol drives the gas chemistry preferentially away from ethanol formation and towards its isomer, dimethyl ether, and, if present, formaldehyde can react with protonated methanol and lead to methyl formate.

If the sublimed methanol was originally formed during mantle photolysis, then the results of ice analogue experiments suggest that copious amounts of CO_2 should also be present when the mantles evaporate. Solid CO_2 has recently been detected in the interstellar medium [26] and it has been postulated that the high observed abundance of protonated carbon dioxide (HCO_2^+) in Sgr B2 [27], a source known to have several hot cores, is due to the release of CO_2 from grains and subsequent reaction with H_3^+. This result cannot be accounted for by current gas phase theories. If surface-formed molecules do play an important role in the compact ridge chemistry, as outlined above, then detectable abundances of HCO_2^+ may also be present in this source.

References

1. Hollenbach, D. & Salpeter, E.E. 1970, *J. Chem. Phys.*, **53**, 79.
2. Whittet, D.C.B., Longmore, A.J. & McFadzean, A.D. 1985, *M.N.R.A.S.*, **216**, 45p.
3. Tielens, A.G.G.M. & Allamandola, L.J. 1987, in *Physical Processes in Interstellar Clouds*,
      eds. G.E. Morfill & M. Scholer, Reidel, p.333
4. Iglesias, E. 1977, *Ap.J.*, **218**, 697.
5. Millar, T.J. & Freeman, A. 1984, *M.N.R.A.S.*, **207**, 405.
6. Herbst, E. & Leung, C.M. 1986, *M.N.R.A.S.*, **222**, 689.
      —————————————— 1989, *Ap.J.Suppl.*, **69**, 271.
7. Langer, W.D. & Graedel, T.E. 1989, *Ap.J.Suppl.*, **69**, 241.
8. Millar, T.J. & Nejad, L.A.M. 1985, *M.N.R.A.S.*, **217**, 507.
9. d'Hendecourt, L.B., Allamandola, L.J. & Greenberg, J.M. 1985, *Astr.Ap.*, **152**, 130.
10. Charnley, S.B., Dyson, J.E., Hartquist, T.W. & Williams, D.A. 1988, *M.N.R.A.S.*, **235**, 1257.
11. Brown, P.D. & Charnley, S.B. 1990, *M.N.R.A.S.* , in press.
12. Brown, P.D. & Millar, T.J. 1989, *M.N.R.A.S.*, **237**, 661.
13. Brown, P.D. & Millar, T.J. 1989, *M.N.R.A.S.*, **240**, 25p.
14. Brown, P.D. & Charnley, S.B. 1990, *M.N.R.A.S.* , in press.
15. Tielens, A.G.G.M. 1983, *Astr.Ap.*, **119**, 177.
16. Hartquist, T.W. & Williams, D.A. 1989, *M.N.R.A.S.*, **241**, 417.
17. Brown, P.D. 1988, *Ph.D. Thesis*, Manchester.
18. Brown, P.D. & Charnley, S.B. 1990, *M.N.R.A.S.* , submitted.
19. Heyer, M.H. 1988, *Ap.J.*, **324**, 311.
20. Olano, C.A., Walmsley, C.M. & Wilson, T.L. 1988, *Astr.Ap.*, **196**, 194.
21. Charnley, S.B., Brown, P.D. and Smith, R.A. 1990, *in preparation*.
22. Léger, A., Jura, M. & Omont, A. 1985, *Astr.Ap.*, **144**, 147.
23. Brown, P.D., Charnley, S.B. & Millar, T.J. 1988, *M.N.R.A.S.*, **231**, 409.
24. Blake, G.A., Sutton, E.C., Masson, C.R. & Phillips, T.G. 1987, *Ap.J.*, **315**, 621.
25. Millar, T.J., Herbst, E. & Charnley, S.B. 1990, *M.N.R.A.S.* , submitted.
26. d'Hendecourt, L.B. & Jourdain de Muizon, M. 1990, *Astr.Ap.*, **223**, L5.
27. Minh, Y.C., Irvine, W.M. & Ziurys, L.M. 1988, *Ap.J.*, **334**, 175.

Acknowledgment. PDB is grateful to SERC & NATO for a postdoctoral fellowship. SBC was supported by Max-Planck-Gesellschaft.

III
CHEMICAL KINETICS AND THEORY

Growing Silicon-Bearing and PAH Molecules with Ion/Molecule Reactions

Diethard K. Bohme

*Department of Chemistry and Centre for Research in Earth and Space Science, York University,
North York, Ontario, M3J 1P3, Canada*

Results of recent gas-phase laboratory studies are highlighted which provide
insight into of the formation of silicon-bearing molecules with chemistry initiated
by ground-state atomic silicon ions, the formation of naphthalene ions from
benzene ions, and the influence of naphthalene on the intrinsic chemistry of
atomic silicon ions.

INTRODUCTION

Interstellar clouds and circumstellar envelopes are partially ionized environments of gas and dust
in which gas-phase and surface chemistry concomitantly can lead to the growth of molecules by
ion/molecule reactions. Figure 1 shows that such growth occurs indirectly along the path of
ionization, ion/molecule reaction and neutralization and that it complements direct growth by neutral
reactions. Higher-order chemistry can lead to still larger ions and molecules which ultimately may
reach particle size much like, for example, the chemistry in flames which can lead to the formation
of soot or silicates. These particles can then promote the occurrence of surface chemistry.

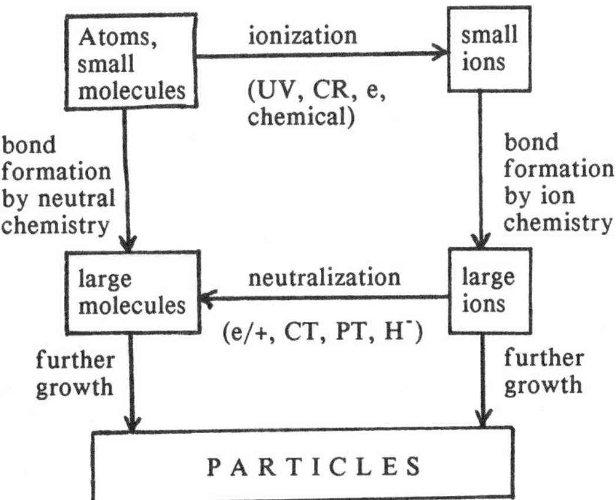

Fig.1. Schematic view of the chemistry for the growth of molecules in partially ionized
environments.

For several years now, research in the Ion Chemistry Laboratory at York University has been
directed to understanding the growth of interstellar and circumstellar molecules and ions by
ion/molecule reactions. The research has been driven by the need to identify chemical pathways for
the formation of known interstellar and circumstellar molecules and ions, and also by the exciting
prospect of identifying new molecules and ions of possible interest to the laboratory spectroscopist,
the theoretical chemist, the chemical modeller and the radioastronomer. Here we highlight recent
results of our laboratory measurements which address the following questions. What kind of silicon-

bearing molecules can be expected to be formed by chemistry initiated by ground-state atomic silicon ions in the gas phase? Can polycyclic aromatic hydrocarbon (PAH) molecules be grown in the gas phase from small acyclic molecular units? Will atomic silicon ions attach to graphitic surfaces such as PAH molecules in the gas phase, and how, if at all, will the presence of such surfaces modify the silicon chemistry?

FORMATION OF SILICON-BEARING MOLECULES INITIATED BY ATOMIC SILICON IONS

The silicon experiments performed in our laboratory so far have been restricted to chemistry initiated by ground-state atomic silicon ions. Although inert toward H_2 and CO, these ions have been found generally to be quite reactive toward other, less abundant interstellar molecules. Often these reactions lead to the formation of chemical bonds with Si. We have now identified ion/molecule reactions of $Si^+(^2P)$ leading to Si-H, Si-C, Si-N, Si-O and Si-S bond formation (1,2,3,4,5). Over 100 silicon-ion/molecule reactions have been measured (a compilation is available upon request). Chemical pathways have been reported which can lead to the following Si-bearing molecules:

SiH, $SiCH$, $SiCH_3$, SiC_2, SiC_2H_2, SiC_3H_2, SiC_4

SiO, SiO_2, SiS, SiS_2

$HSiOOH$, $SiOCH_2$, $SiOC(H)CH_3$, $HSi(OH)_3$, $HSi(O)(OCHO)$, $HSi(O)(OC_2H_5)$, $HSi(OCHO)(OH)_2$, $Si(OCH_3)_3$, $Si(OC_2H_5)_2$, $HSi(OC_2H_5)_3$

$SiNH$, $SiNCH_3$, H_2SiNH

The isomeric structures of the ground states of many of these molecules are uncertain. Little spectroscopic information and few quantum chemical calculations are available to provide insight. Several are likely to be cyclic molecules as indicated below. For example, SiC_2 is known to be cyclic

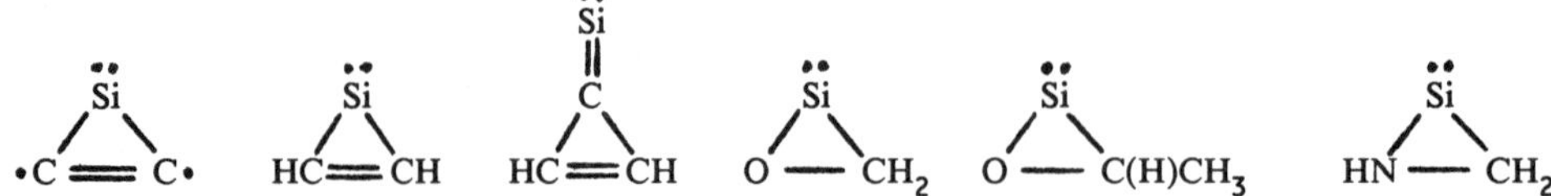

(6) or at least T-shaped (7). In the case of SiC_2H_2, 3-silacyclopropenylidene has been found to be a global minimum in the singlet potential energy hypersurface (8).

Fig. 2 provides a partial summary of the chemistry initiated by Si^+ with interstellar and circumstellar molecules which leads to the formation of molecules containing Si-C bonds. The silicon-carbide molecules SiCH and $SiCH_3$ are formed directly as neutral products of ion/molecule reactions. Formation of the remaining molecules, including SiC_2 and SiC_4, requires neutralization by recombination with electrons or by proton-transfer. Both SiC_2 and SiC_4 have been observed in the circumstellar envelope of IRC+10216 (9,10).

Similar schemes can be developed for hydrides, nitrides and oxides. For example, our experiments have shown in detail that silicon can become increasingly coordinated to oxygen in sequential reactions with many oxygen-containing molecules and so form a number of novel substituted silanes. This can be illustrated most vividly by the rich chemistry which we have seen to be initiated and propagated by hydroxyl-containing molecules (1). Ground-state silicon ions were found to react with H_2O, CH_3OH, C_2H_5OH, HCOOH and CH_3COOH to produce the silene cation $SiOH^+$ which may neutralize to form silicon monoxide:

$$Si^+ \;+\; XOH \;\;-----> \;\; SiOH^+ \;+\; X$$

$$SiOH^+ \;+\; e\,(B) \;\;-----> \;\; SiO \;+\; H\,(BH^+)$$

$SiH_3O_2^+$, which may neutralize to form silanoic acid, is the predominant product of the reaction of $SiOH^+$ with H_2O, C_2H_5OH and HCOOH, while direct formation of silanoic acid is likely in the reaction with CH_3COOH. Further chemistry can be propagated by $SiH_3O_2^+$. It associates with H_2O and C_2H_5OH and produces $SiCH_5O_3^+$ and $SiH_5O_3^+$ with formic acid. Reaction sequences identified with CH_3OH, C_2H_5OH and HCOOH are postulated to lead to the complete saturation of Si^+ forming ions of the type $HSi(OCH_3)_3^+$, $HSi(OC_2H_5)_3^+$ and $HSi(OH)_3H^+$ which may neutralize to generate trimethoxysilane, triethoxysilane and trihydroxysilane, respectively. Analogous reactions can be proposed which lead to the formation of tetrahydroxysilane which is a known building block for condensational synthesis of hydrated silica networks (11,12). It is tempting to speculate about an analogous formation of silica networks in the gas phase.

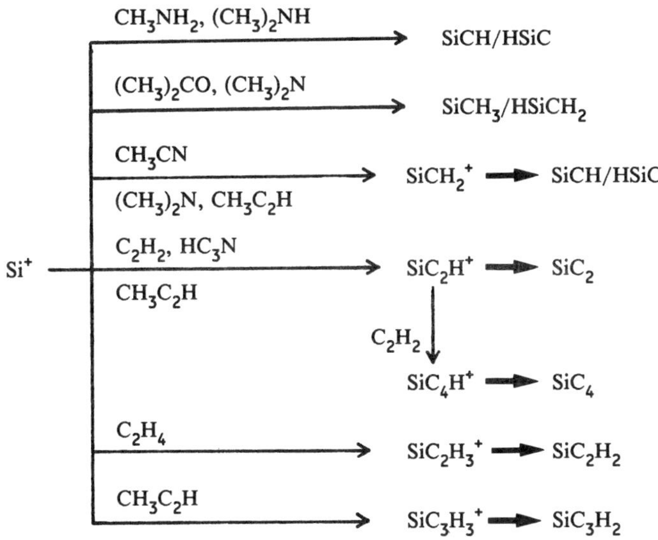

Fig. 2. Limited reaction scheme for the synthesis of silicon carbide molecules initiated by atomic silicon ions. The heavy arrows indicate neutralization reactions such as recombination with electrons and proton transfer.

GROWING PAH MOLECULES

PAH ions and molecules are attracting considerable attention in astrochemistry, particularly in view of their possible contribution to observed IR emission features in nebulae (13,14) and the diffuse interstellar absorption bands (15,16,17). Very little is known about the fundamental gas-phase chemistry which may lead to the growth of PAH ions and molecules in these environments. We have discussed previously how such growth can be envisaged to proceed either directly by cross-bonding in the side-on approach of extended chain-like molecules such as cumulenes or polyacetylenes, or through the successive completion of hexagonal rings by the addition of two two-carbon units or one four-carbon unit (18). A case study has now been completed in our laboratory which demonstrates the gas-phase formation of ionized naphthalene by the addition of diacetylene to ionized benzene in the manner shown below.

Ionized benzene, which itself may be grown from three or two-carbon units such as allene, methylacetylene and perhaps acetylene in the gas phase (18,19), was found to be unreactive toward deuterium and acetylene but reacted rapidly with diacetylene in helium buffer gas at 296 ± 2 K and 0.35 Torr, at nearly the collision-limited rate (20). Chemical reactivity tests were used to confirm the identity of the adduct ion as ionized naphthalene. Ionized naphthalene also was observed to be unreactive toward deuterium and acetylene. Its subsequent reaction with diacetylene was immeasurably slow but a C$_{14}$H$_{10}$$^+$ ion, possibly ionized anthracene or phenanthrene, was observed to be formed.

We can conclude from these observations that the successive grafting of hexagonal rings to the benzene cation appears to be possible in single steps with the four-carbon unit diacetylene, at least at room temperature in helium at moderate pressures, and seems to be preferred to grafting in two steps with the two-carbon unit acetylene. Other more dehydrogenated four-carbon units such as C_4 or C_4H may provide a more effective route, but experiments with these neutral species are currently prohibitive. The single-step mechanism has been proposed for the high-temperature polymerization of PAH's (21) as illustrated below and is the basis for the perceived growth of the large aromatic ions which are believed to be precursors in the ionic formation of soot in hydrocarbon flames (22).

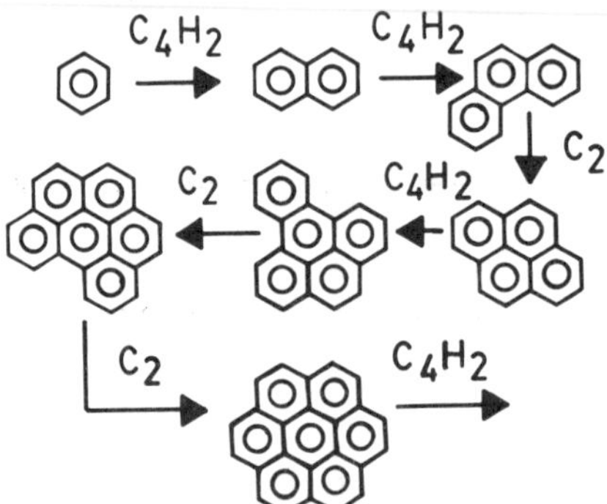

PAH "SURFACE" REACTIONS OF ATOMIC SILICON IONS IN THE GAS PHASE

Other authors previously have speculated about how PAH molecules may attach electrons, ions or atoms and molecules and how ensuing chemistry may provide new routes for neutralization of ions and the synthesis of molecules in the interstellar gas (23,24,25,26). Again however, supporting experimental evidence is severely lacking. Such evidence is difficult to ascertain in the laboratory because of the low vapour pressure of PAH molecules. In the Ion Chemistry Laboratory at York University we have been successful in introducing naphthalene into the flow tube of a SIFT apparatus at room temperature and to investigate the influence of this molecule on the intrinsic chemistry of atomic silicon ions.

In particular, we are interested in identifying novel chemical routes assisted, or perhaps even catalyzed, by the presence of PAH molecules. For example, the following sequence could catalyze the association of Si^+ and M (or their reaction to form bimolecular products):

$$Si^+ \ + \ PAH \quad \longrightarrow \quad (Si^+..PAH)$$

$$(Si^+..PAH) + M \quad \longrightarrow \quad SiM^+ \ + \ PAH$$
$$\overline{}$$
$$Si^+ \ + \ M \quad \longrightarrow \quad SiM^+$$

When the ionization energy of Si^+ is higher than that of the PAH molecule, as is the case with naphthalene, the charge should be transferred in the first step so that the second step amounts to a neutral reaction proceeding in the vicinity, perhaps above the "surface", of a charged PAH molecule:

$$Si^+ \ + \ C_{10}H_8 \quad \longrightarrow \quad (Si..C_{10}H_8^+)$$

$$(Si..C_{10}H_8^+) + M \quad \longrightarrow \quad SiM \ + \ C_{10}H_8^+$$
$$\overline{\phantom{Si^+ + M + C_{10}H_8 \quad \longrightarrow \quad SiM + C_{10}H_8 xxxx}}$$
$$Si^+ + M + C_{10}H_8 \quad \longrightarrow \quad SiM \ + \ C_{10}H_8^+$$

This reaction sequence is not catalytic since the naphthalene is not regenerated as a neutral molecule. But, from the point of view of molecular synthesis, it has the attractive feature that it leads to a neutral product (or neutral products) without requiring a separate neutralization step. The positive charge ends up on the departing naphthalene molecule.

As described elsewhere in these proceedings, the reaction sequence with naphthalene has now been observed with our SIFT apparatus for M = O_2, H_2O, C_2H_2 and C_4H_2 (27). The neutral product (or products) has (have) not been identified. As shown below, SiM may depart as a bound molecule if the reaction exothermicity is sufficiently large. Alternatively, bimolecular products may be formed near the surface and then be detached.

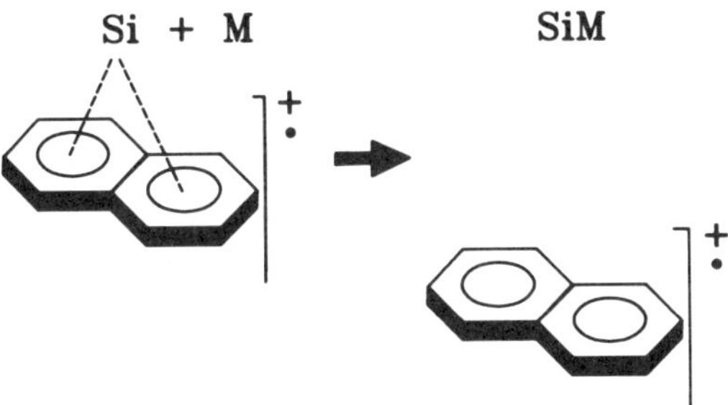

Analogous sequences can be envisaged with other atomic ions, X^+, or atoms, X. In principle, the intermediate $(XPAH)^+$ ion may be formed either by the capture of the atomic ion X^+ by the PAH molecule, or by the capture of the neutral atom X by a positively-charged PAH ion. In the interstellar gas the formation of the intermediate need not be constrained to proceed by radiative association if suitable bimolecular X^+ or X transfer reactions can occur. Nor need the reaction sequence be constrained to occur on the surface of "free" PAH molecules in the interstellar gas if PAH molecules (or graphite, the end member of the PAH series) appended to interstellar grains are available as substrates.

We can expect the gas-phase "surface" mechanism to provide unique opportunities for molecular synthesis. With metal atoms or atomic ions unique pathways should exist to form organometallic molecules. For example, in the case of the reaction of the Si^+.naphthalene adduct ion with acetylene it is reasonable to expect addition with ring insertion to form 3-silacyclopropenylidene in analogy with the neutral gas-phase reaction of Si atoms with acetylene which has been investigated separately by other workers (28). Also, we should note that when a proton is attached to a PAH molecule, an analogous "surface" reaction with a hydrogen atom may provide an efficient route for the recombination of hydrogen atoms to form molecular hydrogen:

$$H^+ \; + \; PAH \qquad -----> \qquad (H..PAH^+)$$

$$(H..PAH^+) + H \quad -----> \qquad H_2 \; + \; PAH^+$$

$$\overline{H^+ + H + PAH \qquad -----> \qquad H_2 \; + \; PAH^+}$$

Other reactions which can lead to the formation of the intermediate $(H..PAH^+)$ include the addition of hydrogen atoms to ionized PAH molecules and exothermic bimolecular proton-transfer reactions in which the proton is delivered by protonated molecules such as H_3^+, for example. The formation of molecular hydrogen in the second step is exothermic, at least for the smaller PAH molecules, and so is expected to occur. Clearly, measurements are now required to confirm these expectations.

ACKNOWLEDGEMENTS

It is a pleasure to thank my co-workers, especially Arnold Fox, Alan Hopkinson and Stanislaw Wlodek, for their contributions to this research and for their stimulating discussions. The research was financially supported by the Natural Sciences and Engineering Research Council of Canada.

REFERENCES

1. Wlodek, S., Fox, A., and Bohme, D.K. 1987, J. Am. Chem. Soc. **109**, 6663.

2. Wlodek, S., Rodriquez, C.F., Lien, M.H., Hopkinson, A.C., and Bohme, D.K. 1988, Chem. Phys. Lett. **143**, 385.

3. Wlodek, S. and Bohme, D.K. 1988, J. Am. Chem. Soc. **110**, 2396.

4. Wlodek, S. and Bohme, D.K. 1989, J. Am. Chem. Soc. **111**, 61.

5. Wlodek, S. and Bohme, D.K. 1989, J. Chem. Soc., Faraday Trans. 2 **85**, 1643.

6. Michalopoulos, D.L., Geusic, M.E., Langridge-Smith, P.R.R., and Smalley, R.E. 1984, J. Chem. Phys. **80**, 3556.

7. Shepherd, R.A. and Graham, W.R.M. 1988, J. Chem. Phys. **88**, 3399.

8. Frenking, G., Remington, R.B., and Schaeffer III, H.F. 1986, J. Am. Chem. Soc. **108**, 2169.

9. Thaddeus, P., Cummins, S.E., and Linke, R.A. 1984, Ap. J. (Letters) **283**, L45.

10. Ohishi, M., Kaifu, N., Kawaguchi, K., Murakami, A., Saito, S., Yamamoto, S., Ishikawa, S., Fujita, Y., Shiratori, Y., and Irvine, W.M. 1989, Ap. J. (Letters) **345**, L83.

11. Abe, Y. and Misono, T. 1984, J. Polym. Sci. Polym. Lett. **22**, 565.

12. Meinhold, R.H., Rothbaum, H.P., and Newman, R.H. 1985, J. Colloid Interface Sci. **108**, 234.

13. Leger, A. and Puget, J.L. 1984, Astron. Astrophys. **137**, L5.

14. Allamandola, A.J., Tielens, A.G.G.M., and Barker, J.R. 1985, Ap. J. (Letters) **290**, L25.

15. Van der Zwet, G.P. and Allamandola, L.J. 1985, Astron. Astrophys. **146**, 76.

16. Leger, A. and d'Hendecourt, L.B. 1985, Astron. Astrophys. **146**, 81.

17. Crawford, M.K., Tielens, A.G.G.M., and Allamandola, L.J. 1985, Ap. J. (Letters) **293**, L45.

18. Bohme, D.K. 1987, in "Structure/Reactivity and Thermochemistry of Ions", P. Ausloos and S.G. Lias (eds.), D. Reidel Publ. Co., p. 219.

19. Lias, S.G. and Ausloos, P. 1985, J. Chem. Phys. **82**, 3613.

20. Bohme, D.K. and Wlodek, S. 1990, J. Am. Chem. Soc., submitted for publication.

21. Stein, S.E. 1978, J. Phys. Chem. **82**, 566.

22. Calcote, H.F. 1981, Combustion and Flame **42**, 215 (1981).

23. Omont, A. 1986, Astron. Astrophys. **164**, 159.

24. Duley, W.W. and Williams, D.A. 1986, Mon. Not. R. astr. Soc. **219**, 859.

25. Lepp, S. and Dalgarno, A. 1988, Ap. J. **324**, 553.

26. Lepp, S., Dalgarno, A., van Dishoeck, E.F., and Black, J.H. 1988, Ap. J. **329**, 418.

27. Bohme, D.K., Wlodek, S., and Wincel, H. 1989, Ap. J. (Letters) **342**, L91.

28. Husain, D. and Norris, P.E. 1978, J. Chem. Soc., Faraday Trans. II **74**, 106.

The Yield of H Atoms through Dissociative Recombination of Polyatomic Ions

B. R. Rowe, A. Canosa, J. C. Gomet, J. L. Queffelec, and M. Morlais

*Département de Physique Atomique et Moléculaire, Université de Rennes I, Campus de Beaulieu,
35042 Rennes, France*

The yield of H atoms through dissociative recombination of several polyatomic ions with electrons has been studied using a plasma flow tube experiment. It was found that CH_5^+, H_3O^+, N_2OH^+, NH_4^+, HCO_2^+ produce about one H atom per dissociative recombination. On the other hand, a low branching ratio was obtained concerning $OCSH^+$ and H_3S^+. Different precursor ions were used and a good agreement was found between all measurements.

1 Introduction

In the early seventies the first gas phase models had already given an insight into the problem of complex molecule formation in interstellar clouds (1,2). However, at that time, there was a serious lack of relevant experimental data and astrochemistry models had to rely on various extrapolations and guesses. Even if these shortenings were necessary and sometimes founded on theoretical basis, laboratory works were clearly needed. Regarding some points, the last ten years have seen remarkable breakthroughs. For example, ion-molecule reaction rate coefficients are now measured down to extremely low temperatures (3,4). It is quite interesting to note that new data often give more questions than it answer (5). Indeed as our knowledge of the physic and chemistry of clouds improves, it appears that no simple models are completely realistic and that molecules are probably formed through a complicated pathway involving gas phase grain surfaces and shocks. Nevertheless, the simple gas phase model remains extremely useful. It has to be noticed that, even in this case, numerous processes have not been studied in laboratory yet. For example, neutral rate coefficients are not measured at temperatures relevant to interstellar clouds (see the paper by M.M. Graff in these proceeding).

A process which plays a key role in the gas phase formation of molecules is dissociative recombination (here in after D.R.) of polyatomic ions with electrons. While there remains some controversy in some cases (see the contribution of T. Amano in this volume concerning H_3^+), D.R. rate coefficients or cross sections have been studied by several groups using a large variety of techniques including the stationary (6) and flowing afterglow (FALP) techniques (7) as well as the merged beam apparatus (8).

Much less data are available concerning the nature of the D.R. neutral products. Previous theoretical and experimental work has mainly concentrated on the state of the products of diatomic ions involved in atmospheric chemistry : production of $N(^2D)$ from NO^+ (9) or N_2^+ (10) and of $O(^1S)$ and $O(^1D)$ from O_2^+ (11). Until recently, the only complete direct measurements of neutral branching ratios for polyatomic ions concerned H_3^+ (14) and H_2O^+ (12). In many of these studies, the ions were vibrationally excited to unknown degrees.

As a part of a collaborative program with the Birmingham group (N.G. Adams and D. Smith), our previous apparatus, which used V.U.V absorption to detect neutral atoms formed in the D.R. of excited ions (11–13) was modified in order to study polyatomic ions in their ground state. The present paper reports on the results obtained concerning the yield of H atoms through D.R. of several polyatomic ions.

2 Experimental

In Rennes laboratory, a plasma flow tube apparatus has been used for several years to study the yield of atoms through dissociative recombination of ions with electrons. Initial investigations (10–13) used highly ionised plasma $(n_e \simeq 10^{11}$ up to 10^{13} cm$^{-3})$ giving vibrationally excited ions. The procedure used for the present measurements remains similar to that used in these previous works. Therefore, some details of the present technique have already been described elsewhere (10,13) . We thus only insist on the new features of the apparatus, the originality of

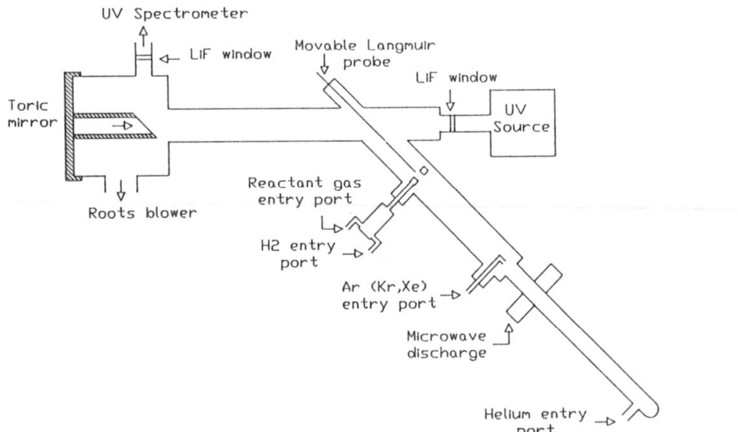

Figure 1: Sketch of the apparatus

which consists in associating our V.U.V. absorption technique with the FALP technique developed by D. Smith and N. Adams in Birmingham (7). Figure 1 gives a sketch of the latest design of our device.

The method is based on the creation of a polyatomic ion via a proton transfer between a reactant gas and H_3^+ or by chemi-ionisation via a charge transfer with atomic ions (such as Xe^+ or Kr^+) followed by a proton transfer with the reactant gas. The terminal ions are then quickly destroyed by D.R. and H atoms are detected by V.U.V. absorption.

A helium buffer gas (pressure: 1.6 Torr) is introduced into one end of a pyrex glass flow tube (5 cm in diameter and 50 cm long) and is constrained to flow down this tube by a roots blower (4000 m^3/h). Upstream of the pyrex tube, helium gas goes through a zeolythe trap cooled at liquid nitrogen temperature. Ionisation, created in a microwave discharge which can move along the pyrex tube, results in the production of an afterglow plasma along the length of the stainless steel tube ($\approx$ 1.4 m long, 8.5 cm in diameter). The upstream region of the afterglow comprises He^+ and He_2^+ ions and neutral metastable atoms He^M, in addition to the ground state helium carrier gas atoms. In this experiment, typical electronic densities are in the range of $n_e \simeq 10^9$ up to 10^{11} cm^{-3}. Such a density enables the quenching of the studied ions prior to D.R.. A movable Langmuir probe allows a precise measurement of this parameter along the axis of the tube.

An axial entry port, placed 20 cm downstream of the microwave discharge, permits to inject a rare gas such as argon, xenon or krypton. An Ar^+ (respectively Xe^+ or Kr^+) dominated afterglow plasma is then generated. Tables 1 and 3 give a detailed description of the chemistry occuring under our experimental conditions when Ar^+, Xe^+ or Kr^+ is taken as the precursor ion.

The terminal molecular ions are formed by the addition of a suitable reactant gas via a second entry port. When Ar^+ is the forerunner ion, a mixing room, situated above the entry port, allows the injection of an homogeneous mixture of the desirable gas with H_2. This second inlet, placed 40 cm downstream of the first one, is a cylindrical stainless steel ring having a 3 cm diameter, positioned symmetrically about the axis of the flow tube. Six small holes (0.5 mm in diameter) drilled in the upstream facing part of the ring direct the effusing gas against the carrier gas flow. This design yields a faster conversion of primary ions into the desired polyatomic ions (15). These ions are then destroyed very efficiently by D.R. before the bend, thus producing H atoms.

A U.V. source yields the Lyman-α line which is used for absorption measurements. This line is monitored using a high resolution three meter spectrometer with a V.U.V. photomultiplier. The U.V source is placed perpendicular to the optical axis of the spectrometer. This provides us with a greater absorption length ($\approx$ 1.1 m) but a toric mirror is needed to reflect the U.V. signal towards the slot of the spectrometer. Preliminary experiments have been carried out with the U.V. source located on the optical axis of the U.V. spectrometer. This led to an absorption length of 8 cm (more details concerning this configuration are available in (12)), but measured U.V. absorptions were found to be very weak ($\simeq 1\%$). The increase of the absorption length gave a greater U.V. absorption. In order to do so, the bend configuration presented in this paper was choosen. Absorptions of the order of 10% were then accessible, thus reducing uncertainties in our experiments. The density of H atoms was then obtained from the measured absorption, as previously described (16). To be reliable, this technique requires a good knowledge of the line profile. The Lyman-α profile was obtained using the 10 m V.U.V. spectrometer of the Meudon Observatory in collaboration with Launay (12).

A very interesting characteristic of our apparatus is its flexibility. Indeed, it was built in several parts put together by means of clutches. In particular, this allows to replace the toric mirror by a quadrupole mass spectrometer. It was also possible to dismantle the upstream bend segment of the device and fix it directly on the pumping chamber. We have therefore been able to check the nature of ions close to the ring entry port.

Table 1: Implied chemistry when H_3^+ is the precursor ion

Characteristic times are calculated for the following experimental conditions:
P=1.6 Torr, T=300 K, $n_{He} = 5.1\ 10^{16}$ cm^{-3}, $n_{H_2} = 1.3\ 10^{14}$ cm^{-3}, $n_{Ar} = 9.7\ 10^{13}$ cm^{-3},
$n_{H_2O} = 1.3\ 10^{12}$ cm^{-3} and $n_X = 1.3\ 10^{13}$ cm^{-3} (where X=CH$_4$, N$_2$, CO$_2$, N$_2$O, OCS, NH$_3$ or H$_2$S).
[a]: assumed, [b]: assumed from (23), *: n=1 for binary reactions, n=2 for ternary reactions

Chemical reaction	Rate coeff.* $k\ cm^{3n}s^{-1}$	Char. time (μs)	Ref.
$He^M + Ar \longrightarrow Ar^+ + He + e$	10^{-10}	100	(17)
$He^+ + 2\,He \longrightarrow He_2^+ + He$	$8.3\ 10^{-32}$	4540	(18)
$He_2^+ + Ar \longrightarrow Ar^+ + 2\,He$	$2\ 10^{-10}$	50	(19)
$Ar^+ + H_2 \longrightarrow ArH^+ + \mathbf{H}$	$9\ 10^{-10}$	9	(19)
$ArH^+ + H_2 \longrightarrow H_3^+(v) + Ar$	$5\ 10^{-10}$	15	(19)
$H_3^+(v) + H_2 \longrightarrow H_3^+ + H_2$	$2.7\ 10^{-10}$	29	(20)
$H_3^+(v) + e \longrightarrow$ products	$\sim 10^{-7}$	400	[a]
$H_3^+ + H_2 + He \longrightarrow H_5^+ + He$	$\leq 2\ 10^{-30}$	≥ 75000	[b]
$H_3^+ + e \longrightarrow$ products	$\leq 10^{-8}$	≥ 4000	(21)
$H_3^+ + H_2O \longrightarrow H_3O^+ + H_2$	$4.3\ 10^{-9}$	180	(19)
$H_3^+ + CH_4 \longrightarrow CH_5^+ + H_2$	$2.4\ 10^{-9}$	32	(19)
$H_3O^+ + e \longrightarrow$ products	$7\ 10^{-7}$	57	(22)
$H_3O^+ + H_2O + He \longrightarrow H_3O^+.H_2O + He$	$6.6\ 10^{-28}$	23000	(19)
$CH_5^+ + e \longrightarrow$ products	$1.1\ 10^{-6}$	36	(21)
$CH_5^+ + CH_4 + He \longrightarrow CH_5^+.CH_4 + He$	$2.5\ 10^{-28}$ (80 K)	> 6000	(21)

3 Results and discussion

3.1 H_3^+ taken as the precursor ion

In this experiment, the helium plasma has been converted into an Ar^+ plasma by injecting a small flow rate of argon. Table 1 specifies the whole chemistry implied in this work. In spite of the slow conversion of He^+ into He_2^+, Ar^+ is the dominant ion at the ring entry port, as it has been shown by mass spectrometric experiments. This has been made possible by keeping a sufficiently large distance between the argon entry port and the ring entry port ($L \approx 40$ cm). The hydrodynamic time was then large enough to allow conversion ($\tau_h = 10800\ \mu$s to compare with $\tau_{He^+ \rightarrow He_2^+} = 4540\ \mu$s).

The reactant gas X (X= N$_2$, CO$_2$, OCS, N$_2$O, CH$_4$, H$_2$O, NH$_3$, H$_2$S), mixed with H$_2$, was added to the plasma via the ring entry port. Before H$_2$ has reached the mixing room, it has gone through a zeolythe trap cooled at -40°C. The H$_2$ flow rate was taken ten times greater than the reactant gas one. This was necessary to make sure that Ar^+ was destroyed only by H$_2$, leading to H_3^+ via reactions:

$$Ar^+ + H_2 \quad \longrightarrow \quad ArH^+ + H \qquad (1)$$
$$ArH^+ + H_2 \quad \longrightarrow \quad H_3^+(v) + Ar \qquad (2)$$

H_3^+ produced in reaction 2 was probably vibrationally excited but in the presence of H$_2$ it was deexcited via:

$$H_3^+(v) + H_2 \quad \longrightarrow \quad H_3^+ + H_2 \qquad (3)$$

This deexcitation occured before the D.R. of $H_3^+(v)$. Then, we had a plasma dominated by ground state H_3^+. A very efficient proton transfer allowed the creation of the terminal ion before H_3^+ has recombined or formed H_5^+ by an association reaction. D.R. followed (before any ternary reaction may occur) and H atoms were produced. Here, we must point out that some H atoms are undesirable: reaction 1 generates extra H atoms. Of course, we have taken into account this problem. The study of N_2H^+ has enabled us to calibrate our experiment and to deduce the additional absorption due to H atoms from reaction 1. Indeed, D.R. of N_2H^+ gives only one H atom :

$$N_2H^+ + e \quad \longrightarrow \quad N_2 + H \qquad (4)$$

Thus, for one destroyed electron, 2 H atoms are formed, with only one proceeding from D.R.. Furthermore, we have observed that the Ar^+ plasma gave a very small background absorption. As a consequence, we had to subtract the background H atom density from the H density obtained for N_2H^+. Then, by dividing this result by two, we have obtained the additional H density and the H density resulting from actual N_2H^+ D.R..

Table 2: Branching ratios obtained for H atoms produced in dissociative recombination of studied ions

[a] N_2H^+ taken as a reference.
[b] CH_5^+ taken as a reference, instead of 1 we have used the value obtained for H_3^+ measurements to have a better comparison between H_3^+ and Kr^+ results.
[c] NH_4^+ taken as a reference, same remarks as for [b].

	PRECURSOR IONS		
	H_3^+	Kr^+	Xe^+
N_2H^+	1^a		
HCO_2^+	0.83		
N_2OH^+	0.98		
$OCSH^+$	0.31		
CH_5^+	1.19	1.19^b	
H_3O^+	1.16	1.04	
NH_4^+	1.03	1.36	1.03^c
H_3S^+	0.38	0.19	0.24

Knowing the extra H atom density, we have then been able to obtain the real H density for other polyatomic ions in D.R.. To this end, the additional and background densities were subtracted from the density deduced from absorption measurements for the studied ion.

This procedure has been applied to the study of seven different polyatomic ions: HCO_2^+, N_2OH^+, $OCSH^+$, CH_5^+, H_3O^+, NH_4^+ and H_3S^+. The experimental protocol has always been the same: first the background absorption was measured, then we determined the absorption proceeding from N_2H^+ D.R., (a good statistic was generally achieved after a forty minutes time) and finally, N_2 was replaced by another gas and absorption measurements were repeated. The deduced H density was then directly compared to the density obtained a few minutes before for the N_2H^+ D.R.. The ratio of these two values gave us the branching ratio for H atoms. This process has allowed a better precision in branching ratios. It has been noticed that the absorption, for given plasma conditions, evolves within a few weeks. Obviously, this was due to a change in the U.V. source. However in a daytime this occurrence is negligeable and branching ratios are not affected. After cleaning the source, absorption usually comes back to its previous value.

Our results are presented in table 2. Uncertainties in absorption measurements are of the order of $\pm 10^{\%}$, so branching ratios are subject to an error of about $\pm 20^{\%}$. We can see that CH_5^+ and H_3O^+ produce slightly more than one H atom via D.R.. N_2OH^+ and NH_4^+ give the same number of hydrogen than N_2H^+. HCO_2^+ generates slightly less than one H atom per D.R.. It is interesting to note the low branching ratio we have obtained concerning $OCSH^+$ and H_3S^+. This clearly indicates that the production of H atoms is not the main channel of D.R. of these two ions.

3.2 Kr^+ (Xe^+) taken as the primary ion

To be more confident in our experiments, we have decided to use other possible precursor ions such as Kr^+ or Xe^+. Here again, mass spectrometric experiments have been carried out . It has been checked that Kr^+ (Xe^+) was the dominant ion in the vicinity of the ring entry port. The chemistry was then different than for H_3^+. The reactant gas MH has been injected through the ring entry port against the Kr^+ (Xe^+) plasma. The following reactions occur:

$$MH + Kr^+ \ (Xe^+) \quad \longrightarrow \quad MH^+ + Kr \ (Xe) \tag{5}$$

$$MH^+ + MH \quad \longrightarrow \quad MH_2^+ + M \tag{6}$$

$$MH_2^+ + e \quad \longrightarrow \quad products \tag{7}$$

Of course, it is only possible to study hydrogenated species (i.e CH_4, NH_3, H_2O, H_2S). Furthermore, Xe^+ doesn't react with CH_4 and H_2O. Table 3 indicates the whole chemistry implied by these experiments. The reactant gas flow rate must be choosen to allow the formation of MH_2^+ prior to MH^+ D.R.. The procedure used to obtain the density of H atoms formed by MH_2^+ D.R. remains strickly identical to that presented in section 3.1. The only difference lies in the absence of additional H atoms in the chemistry.

The hydrogen atom density obtained after recombination of CH_5^+ is shown in figure 2 vs the initial electron density (ring injector position) in the center of the flow tube, displaying a good linearity as expected when the main atom source is the plasma recombination. It should be noted that the slope of the line is approximately 0.5. This is due to the difference between the initial radial electronic profile (close to the ring injector) and the radial H atomic profile in the absorption area. The latter is flat as it has been shown in previous experiments (12). The former is nearly parabolic. This has been checked using a movable Langmuir probe placed 1 cm upstream of the ring injector. The experimental electronic profile we have obtained, was well fitted by a second order polynome: $n_e(0)(1 - X^2)$ where X is the ratio of the radial position of the probe to the radius of the flow tube and $n_e(0)$ is the electronic density in the center of the tube.

Table 3: Implied chemistry when Kr^+ (Xe^+) is the precursor ion

Characteristic times are calculated for the following experimental conditions:
P=1.6 Torr, T=300 K, $n_{He} = 5.1\ 10^{16}$ cm^{-3}, $n_{Xe} = n_{Kr} = 9.7\ 10^{13}$ cm^{-3} and $n_X = 1.3\ 10^{14}$ cm^{-3} (where X=CH$_4$, H$_2$O, NH$_3$ or H$_2$S).
[a]: assumed

Chemical reaction	Rate coeff. $k\ cm^{3n}s^{-1}$	Char. time (μs)	Ref.
$He^M + Kr \longrightarrow Kr^+ + He + e$	$9.94\ 10^{-11}$	104	(24)
$He^M + Xe \longrightarrow Xe^+ + He + e$	$1.2\ 10^{-10}$	83	(24)
$He^+ + Xe \longrightarrow Xe^+ + He$	$7\ 10^{-12}$	1480	(19)
$He^+ + 2\,He \longrightarrow He_2^+ + He$	$8.3\ 10^{-32}$	4540	(18)
$He_2^+ + Kr \longrightarrow Kr^+ + 2\,He$	$1.85\ 10^{-11}$	560	(19)
$Kr^+ + CH_4 \longrightarrow CH_4^+ + Kr$	$1.1\ 10^{-9}$	7	(19)
$CH_4^+ + CH_4 \longrightarrow CH_5^+ + CH_3$	$1.5\ 10^{-9}$	5	(19)
$CH_4^+ + e \longrightarrow$ products	$6\ 10^{-7}$	55	(22)
$CH_5^+ + e \longrightarrow$ products	$1.1\ 10^{-6}$	30	(21)
$Kr^+ + H_2O \longrightarrow H_2O^+ + Kr$	$1.19\ 10^{-9}$	6.5	(19)
$H_2O^+ + H_2O \longrightarrow H_3O^+ + OH$	$1.6\ 10^{-9}$	5	(19)
$H_2O^+ + e \longrightarrow$ products	$7.6\ 10^{-7}$	45	(22)
$H_3O^+ + e \longrightarrow$ products	$7\ 10^{-7}$	49	(22)
$Kr^+ + NH_3 \longrightarrow NH_3^+ + Kr$	$8\ 10^{-10}$	10	(19)
$Xe^+ + NH_3 \longrightarrow NH_3^+ + Xe$	$8.3\ 10^{-10}$	9	(19)
$NH_3^+ + NH_3 \longrightarrow NH_4^+ + NH_2$	$2.2\ 10^{-9}$	3.5	(19)
$NH_3^+ + e \longrightarrow$ products	$3.1\ 10^{-7}$	106	(25)
$NH_4^+ + e \longrightarrow$ products	$1.35\ 10^{-6}$	25	(15)
$Kr^+ + H_2S \longrightarrow H_2S^+ + Kr$	10^{-9}	8	(19)
$Xe^+ + H_2S \longrightarrow H_2S^+ + Xe$	$9.9\ 10^{-10}$	8	(19)
$H_2S^+ + H_2S \longrightarrow H_3S^+ + SH$	10^{-9}	8	(19)
$H_2S^+ + e \longrightarrow$ products	10^{-6}	33	[a]
$H_3S^+ + e \longrightarrow$ products	$3.7\ 10^{-7}$	89	(25)

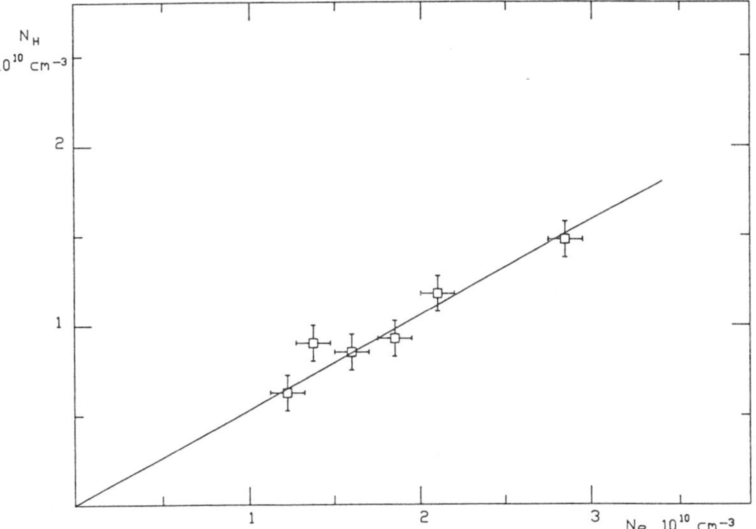

Figure 2: H atom density vs initial electronic density

Results are presented in table 2 and show a good agreement with those obtained when H_3^+ is the precursor ion. More particularly, it has confirmed that H_3S^+ D.R. doesn't give a great amount of H atoms. Likewise, it has born out that CH_5^+, H_3O^+ and NH_4^+ give a similar amount of H atoms (in the error bars of these experiments). Kr^+ and Xe^+ investigations have not been carried out as deeply as for H_3^+ measurements. The latter has been repeated several times and a good statistic was then allowed. The former has only been realised once for each reactant gas. Thus, the error in branching ratios is probably higher for Kr^+ and Xe^+ works than for the H_3^+ study. Particularly, the branching ratio obtained from a Kr^+ primary plasma for NH_4^+ is probably too high.

4   Conclusion

The yield of H atoms through D.R. of several polyatomic ions has been studied using a FALP apparatus with V.U.V. resonance absorption. The results are in good agreement with indirect measurements made in Birmingham using OH laser induced fluorescence (H atoms being converted into OH through chemical reactions) (26). It is clear that more results are needed concerning other atoms and radicals formed in these reactions in order to obtain the complete branching ratio. However, interesting information can already be deduced from the present results, like the inefficiency of H_3S^+ DR as a source of interstellar H_2S. Together with the data obtained in Birmingham, the present measurements will allow a quite complete knowledge of the nature of the product for at least several reactions.

Acknowledgments : The authors thank N.G. Adams and D. Smith for kindly providing drawing and details of their FALP apparatus as well as for stimulating suggestions.

References

[1] P.M. Solomon and W. Klemperer 1972, *Astrophys. J.*, **178**, 389.

[2] E. Herbst and W. Klemperer 1973, *Astrophys. J.*, **185**, 505.

[3] B.R. Rowe 1988, in *Rate Coefficients in Astrochemistry*, eds. T.J. Millar and D.A. Williams, Kluwer Academic Publishers

[4] S.E. Barlow, J.A. Luine and G.H. Dunn 1986, *Int. J. Mass Spectrom. Ion Proc.*, **74**, 97.

[5] J.B. Marquette, C. Rebrion and B.R. Rowe 1988, *J. Chem. Phys.*, **89**, 2041.

[6] F.J. Mehr and M.A. Biondi 1969, *Phys. Rev.*, **181**, 264.

[7] D. Smith, N.G. Adams, A.G. Dean and M.J. Church 1975, *J. Phys. D Appl. Phys.*, **8**, 141.

[8] D. Auerbach, R. Cacak, R. Caudano, T.D. Gaily, C.J. Keyser, J.W. McGowan, J.B.A. Mitchell and S.F.J. Wilk 1977, *J. Phys. B*, **10**, 3797.

[9] D. Kley, G.M. Lawrence and E.J. Stone 1977, *J. Chem. Phys.* **66**, 4157.

[10] J.L. Queffelec, B.R. Rowe, M. Morlais, J.C. Gomet and F. Vallée 1985, *Planet. Space Sci.*, **33**, 263.

[11] J.L. Queffelec, B.R. Rowe, F. Vallée, J.C. Gomet and M. Morlais 1989, *J. Chem. Phys.*, **91**, 5335.

[12] B.R. Rowe, F. Vallée, J.L. Queffelec, J.C. Gomet and M. Morlais 1988, *J. Chem. Phys.*, **88**, 845.

[13] F. Vallée, B.R. Rowe, J.C. Gomet, J.L. Queffelec and M. Morlais 1986, *Chem. Phys. Lett.*, **124**, 317.

[14] J.B.A. Mitchell, J.L. Forand, C.T. Ng, D.P. Levac, R.E. Mitchell, P.M. Mul, W. Claeys, A. Sen and J.W. McGowan 1983, *Phys. Rev. Lett.*, **51**, 885.

[15] E. Alge, N. G. Adams, D. Smith 1983, *J. Phys. B*, **16**, 1433.

[16] M. Dudeck, G. Poissant, B.R. Rowe, J.L. Queffelec and M. Morlais 1983, *J. Phys. D Appl. Phys.*, **16**, 995.

[17] T.P. Parr, D.M. Parr and R.M. Martin 1982, *J. Chem. Phys.*, **76**, 316.

[18] J.D.C. Jones, D.G. Lister, D.P. Wareing and N.D. Twiddy 1980, *J. Phys. B At. Mol. Phys.*, **13**, 3247.

[19] Y. Ikezoe, S. Matsuoka, M. Takebe and A. Viggiano 1987, in *"Gas phase ion-molecule reaction rate constants through 1986"*, Maruzen Company, Tokyo, Japan.

[20] J.K. Kim, L.P. Theard and W.T. Huntress 1974, *Int. J. mass. spectrom. and ion phys.*, **15**, 223.

[21] N.G. Adams, D. Smith and E. Alge 1984, *J. Chem. Phys.*, **81** 1778.

[22] J.W. McGowan, P.M. Mul, V.S. D'Angelo, J.B.A. Mitchell, P. Defrance and H.R. Froelich 1979, *Phys. Rev. Lett.*, **42** 373.

[23] R. Johnsen, C.M. Huang and M.A. Biondi 1976, *J. Chem. Phys.*, **65**, 1539.

[24] W. Lindinger, A.L. Schmeltekopf and F.C. Fehsenfeld 1974, *J. Chem. Phys.*, **61**, 2890.

[25] J.B.A. Mitchell *Phys. Rep. (Rev. Sect. Phys. Lett.)*, (To be published)

[26] D. Smith and al. *J. Chem. Phys.*, (To be submitted)

The CRESUS Apparatus: A Tool for Studying Mass-Selected Ion-Molecule Reactions at Interstellar Temperatures

J. B. Marquette[1], B. R. Rowe[2], and C. Rebrion[2]

[1] *Laboratoire d'Aérothermique du CNRS, 4ter route des Gardes, 92195 Meudon Cédex, France*
[2] *Département de Physique Atomique et Moléculaire, Université Rennes 1, Campus de Beaulieu, 35042 Rennes Cédex, France*

The CRESUS (Cinétique de Réactions en Ecoulement Supersonique Uniforme avec Sélection) experiment allows rate coefficients of reactions of mass-selected ions with molecules to be determined down to 20 K. As in the previous CRESU configuration a uniform supersonic flow of buffer gas (helium in the present case) is used as reactor. Primary ions are injected from a quadrupole mass filter through the wall of the converging-diverging nozzle which generates the buffer flow. Then they are drifted in the nozzle toward the flow axis by a transverse electric field. The reaction of interest is studied downstream the nozzle exit where this electric field is vanishingly small. Among the first processes studied at both 20 and 70 K are the third-body association reactions $C^+ + nH_2 + He \rightarrow CH_2^+ + He$ and $CH_3^+ + nH_2 + He \rightarrow CH_5^+ + He$. Both of them have a ternary rate coefficient k_{3b} showing a "plateau" with decreasing temperature (at 20 K: $k_{3b} < 10^{-28}$ and $k_{3b} = 3.5 \times 10^{-27}$ cm^{6}s^{-1} for the former and latter reaction, respectively). This result is compared to other experimental studies and discussed in the light of association reaction theories including the related interstellar radiative association processes.

1 Introduction

During the two last decades, the need of data about binary ion-molecule processes involved in astrochemistry was one of the main reasons in designing various experimental setups. More particularly, to obtain rate coefficients for a large amount of reactions, the SIFT apparatus (1) has appeared as the ultimate tool at different temperatures, as low as 80 K if the neutral of interest is a non-condensable species. Thus it was natural to keep the structure of this extremely versatile swarm experiment in an experiment specific to the 10 – 80 K temperature range, i.e. (i) production of mass-selected primary ions; (ii) reaction zone; (iii) detection system.

Such low temperatures can be achieved by isentropic expansions through supersonic contoured nozzles. If the nozzle profile is suitably calculated, the expansion yields a uniform flow used as a flow reactor *without walls* (3) in which the study of reactions involving condensable species is possible. It has to be recalled here that a huge pumping facility is necessary to obtain a good compromise between the flow pressure, as low as possible to reduce significantly the influence of ternary processes, and a flow rate as high as possible to optimise the flow size against the viscous boundary layer which grows along the nozzle wall.

The first configuration of the experiment was called CRESU. This french acronym means Kinetics of Reactions in Uniform Supersonic Flow. This apparatus was a sort of "supersonic flowing afterglow" since, as in the pioneering flow tube experiments by Ferguson *et al.* (2), ions were created by adding a parent gas to the buffer. The buffer itself was locally ionised (and partly excited) by electron impact from a 20 keV electron beam whose main purpose was to check the flow pattern (3). Both binary and ternary reactions were studied in the 8 – 163 K temperature range either in monatomic (helium and argon) or diatomic (nitrogen and oxygen) buffer gases and various behaviors with temperature were observed (4,5). However, the yield of primary ions was limited to those released by reactions of buffer gas ions with the parent species. Also important was the fact that Penning reactions of the metastable buffer gas with the ion parent gas could be a problem. In order to overcome these various limitations and to gain in flexibility, an ion injector has been developed which is presented below, together with some results about ternary processes involving C^+ and CH_3^+ at 20 and 70 K.

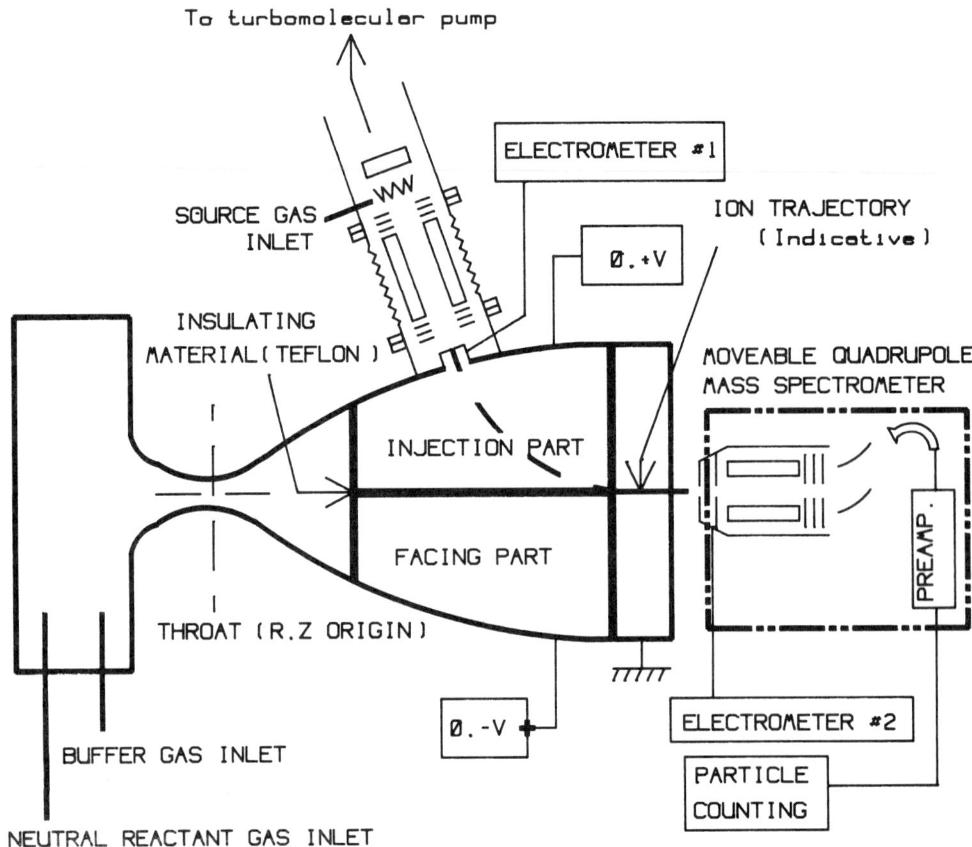

Figure 1: Schematic sketch of the CRESUS apparatus.

2 Experimental setup

The design of the CRESUS experiment is basically the same as the previous CRESU configuration already described in detail elsewhere (3,4). This new apparatus is schematically represented in Figure 1. Some characteristics about the nozzle size and flow conditions are given in Table 1.

The great improvement is the ion injector allowing primary ions to be mass-selected in a way similar to that used in SIFT experiments (1). However, it should be noticed that the CRESUS apparatus has no Venturi inlet at the injection orifice since the pressure at that point is always less than 0.08 Torr while the pressure in the quadrupole chamber is less than 10^{-4} Torr. In addition, such a device would seriously damage the flow uniformity by generating a supersonic and cellular flow (6) perpendicularly to the main stream.

One can see on Figure 1 that the nozzle is divided into three parts where different potentials are applied. Indeed, diffusion phenomena are negligible in supersonic flows. In addition streamlines are parallel (or quasi-parallel in the nozzle where the expansion is nearly achieved) if the flow is uniform. On the other hand, incoming ions will rapidly loose the memory of their origin by collisions with the buffer gas particles, even with a significant initial velocity. As a result, they will remain confined in the vicinity of the nozzle wall in the viscous boudary layer or, in the best case, will penetrate the isentropic core but quite far from the flow axis where measurements are performed (4). Thus, it is necessary to force the ions to transit toward the flow axis by applying a transverse electric field, E, in the nozzle. The potential of the semi-nozzle including the injection orifice has the same sign as the ion charge while the part facing this orifice can be polarised with the opposite sign. The nozzle exit is a thick ring connected to the ground potential in order to prevent

Table 1: Main characteristics of the CRESUS nozzle.

parameter	value
length (from the throat)	38.2 cm
throat diameter	4.15 cm
stagnation temperature	80 K (resp. 300 K)
flow temperature	20 K (resp. 70 K)
stagnation pressure	0.54 Torr (resp. 2.1 Torr)
flow pressure	21 mTorr (resp. 64 mTorr)
Mach number	3
flow velocity	7.87×10^4 cm s^{-1} (resp. 1.51×10^5 cm s^{-1})
flow density	9.64×10^{15} cm^{-3} (resp. 9.20×10^{15} cm^{-3})
Injection orifice	22 cm from the throat
	5.45 cm from the axis
	1 mm diameter

the penetration of the electric field in the reaction zone downstream. In the same way, a Faraday cage is installed around the jet itself to limit the possible influence of the electric fields from the various electric cables connected inside the experimental chamber.

Therefore, the trajectory of a test particle can be calculated if the velocity field is known from the injection orifice to the nozzle exit. The flow velocity is determined during the nozzle calculation (3). The drift velocity is, at first approximation, proportional to the electric field, $viz\ v_d = KE$, where the so-called mobility coefficient K is almost always tabulated for different buffer gases under its reduced form K_0 which is relative to the standard conditions of temperature and pressure (7). The electric field has been calculated in a tridimensional geometry: (i) in the Ecole Centrale de Lyon by A. Nicolas and L. Krähenbühl who have developed a method of surface integral; (ii) in the ENSIEG de Grenoble by A. Meunier who used a finite difference method in the volume. Both calculations are in excellent agreement.

To drift the ions toward the flow axis in the vicinity of the nozzle exit plane it is only necessary to adjust the potential applied to the nozzle wall, taking into account the mobility properties of the species under consideration. The Figure 2 shows the "worst case" which is when the nozzle is operating with room temperature as stagnation temperature and when a low mobility ion, namely He$^+$, is injected in He buffer gas. The theoretical nozzle wall potential was around 10 V while the radial profile of the ion count rate shown in Figure 2 was obtained with a wall potential of 9.25 V. On the other hand, the Figure 3 shows that the presence of the injection orifice does not disturb the flow pattern: the density measured along the flow axis by an induced fluorescence technique (3) remains close to a mean value from about 25 cm downstream the nozzle exit.

3 Association reactions

Among the various processes already examined with the CRESUS apparatus (8,9), particularly important are the ternary associations reactions of C$^+$ and CH$_3^+$ ions with H$_2$ since the corresponding radiative associations are two key routes in the synthesis of interstellar carbon-bearing molecules (10). Such radiative processes are extremely difficult to study experimentally: only the pioneering group of Dunn in Boulder has obtained directly the rate for the formation of CH$_5^+$ in a cooled ion trap at 13 K (11–13). More recently this radiative rate coefficient has been deduced by Gerlich and Kaefer (14) from RF ion trap measurements at 80 K on the ternary process.

In view of this paucity of data a great deal of efforts has been made in the field of theoretical calculations. Among the most recent of them, the possibility of radiative relaxation of the excited complex via an electronic transition has been examined (15–17) to account for the fact that the data of Dunn and coworkers (11–13) imply that the radiative stabilisation must be more greater than can be possible via vibrational emission by one order of magnitude. Conversely, the data of Gerlich and Kaefer (14) show that a vibrational relaxation is most likely. On the other hand, Smith (18) has calculated the effect of quantum mechanical tunneling through the centrifugal barrier of the effective potential. This yields enhancements of the radiative

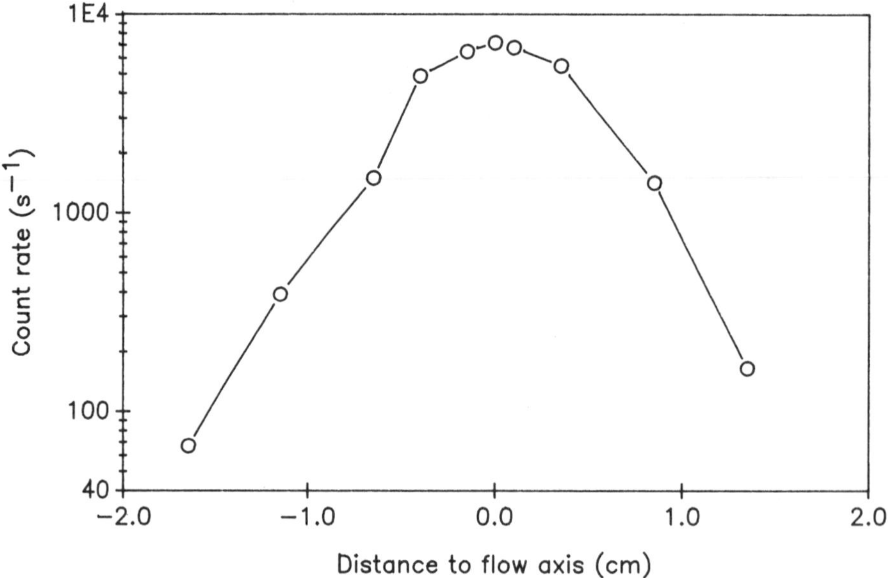

Figure 2: Radial profile of the He^+ count rate.

associations rates at 10 K by a factor of 2.2 and 2.0 for $C^+ + H_2$ and $CH_3^+ + H_2$, respectively. The conclusion of this study is that electronic transitions must play a role in these processes. Thus, it appears that further experimental and theoretical work is still needed.

Figure 4 gives the CRESUS data (open and filled circles) compared to various other experimental values and, for $CH_3^+ + H_2$, theoretical calculations at 80 K. In both cases there is at 20 K a significant departure from the power law (dashed lines) which fits very well the SIFT determinations (open squares) obtained by the Birmingham group (19) and which has theoretical justifications (15–17). Indeed, the *functional* fit to both the CRESUS and SIFT data (dotted lines) has nearly the same shape as the curves reported by Smith (18) in the comparison of the values of the rate of the radiative process $C^+ + H_2(3:1)$, with or without tunneling, with previous calculations by Herbst (20). The same comparison can be made in the case of $CH_3^+ + H_2$. This means that the temperature dependence of these association processes is independent of the stabilisation step.

While the first flowing afterglow measurement (filled inverse triangle) at 90 K (21) on $C^+ + H_2$ was not confirmed by subsequent SIFT measurements, the comparison of the data in the 70 – 100 K region for $CH_3^+ + H_2$ is extremely interesting. For their RF ion trap measurement (open inverse triangle) Gerlich and Kaefer (14) report an uncertainty of +20 K on the temperature which can shift that point toward the SIFT fit. In addition, this RF experiment was performed in a low pressure regime, 3 – 4 orders of magnitude below the SIFT regime. On the other hand, the CRESUS experiment is operating in a median pressure regime, 3 – 20 times lower than the SIFT one. Thus, the fact that the three experiments are in good agreement makes a pressure effect as predicted by statistical phase space theory (22) very unlikely: the corresponding point at 80 K (open diamond) is indeed much higher than observed. In the same way, Bates (23) has corrected the SIFT value (open diamond) by arguing that the ambient helium pressure was too high for the third order kinetics. In such a case the RF datum should be *above* both CRESUS and SIFT data. Another key point is that this agreement between different experiments gives further credit to the RF determination of a radiative rate characteristic of a vibrational relaxation and obtained, as stressed by the authors (14), in the same conditions than the ternary rate.

4 Perspectives

With the ability to inject mass-selected ions in a uniform supersonic flow, the CRESUS experiment has gained in versatility. In fact, the potentialities are nearly the same than in a SIFT apparatus. For example, it is now possible to investigate branching ratios at very low temperatures or to study isotope exchange processes, two fields in which astrochemistry has an urgent need for data.

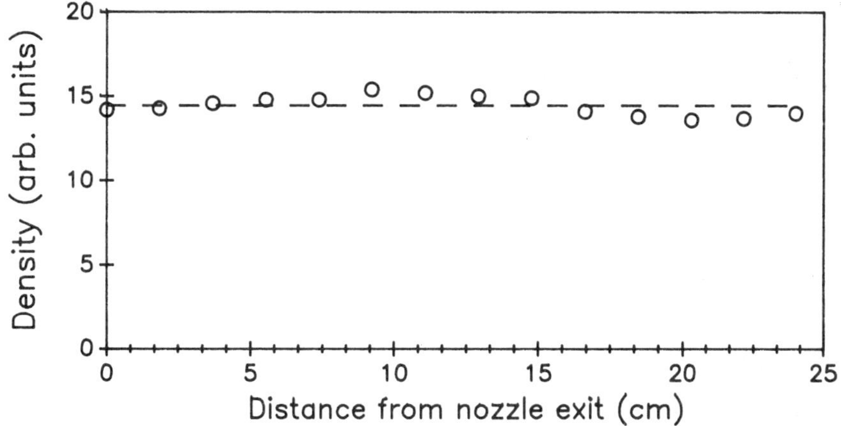

Figure 3: Buffer density profile along the flow axis downstream the nozzle exit.

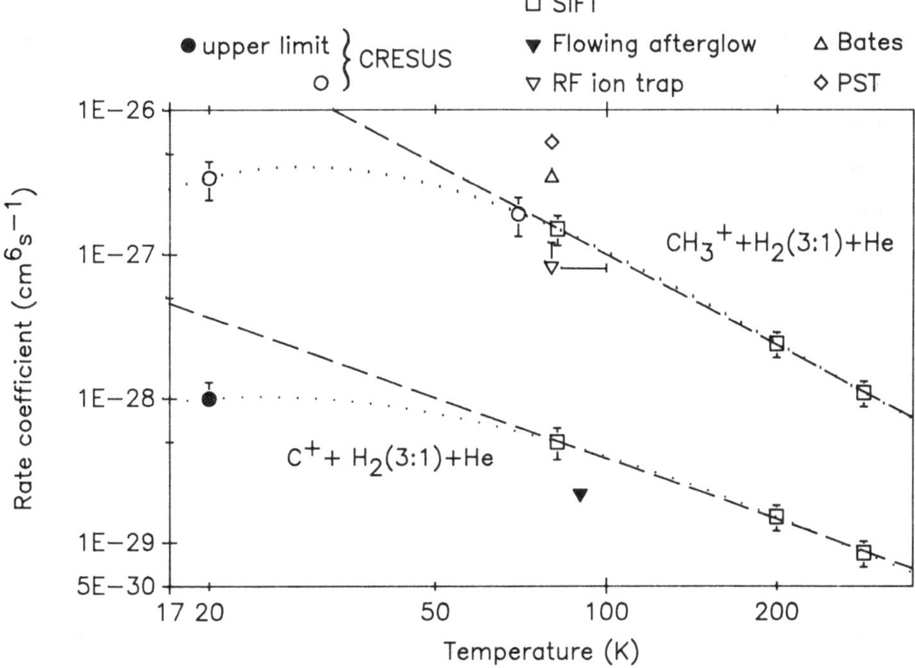

Figure 4: Ternary association reactions of C^+ and CH_3^+ with normal H_2. The meaning of the symbols is given in the text.

Acknowledgments

The CRESUS group is supported by the CNRS through grants from the GDRs "Physico-Chimie des Molécules Interstellaires" and "Dynamique des Réactions Moléculaires".

References

(1) Smith, D. and Adams, N.G. 1987, *Adv. Atom. Mol. Phys.*, **24**, 1.

(2) Ferguson, E.E., Fehsenfeld, M.C., and Schmeltekopf, A.L. 1969, *Adv. Atom. Mol. Phys.*, **5**, 1.

(3) Dupeyrat, G., Marquette J.B., and Rowe, B.R. 1985, *Phys. Fluids* **28**, 1273.

(4) Rowe, B.R., and Marquette, J.B. 1987, *Int. J. Mass Spectrom. Ion Proc.* **80**, 239.

(5) Rebrion, C., Marquette, J.B., and Rowe, B.R. 1990, this volume.

(6) Dupeyrat, G., Rowe, B.R., Fahey, D.W., and Albritton, D.L. 1982, *Int. J. Mass Spectrom. Ion Phys.* **44**, 1.

(7) McDaniel, E.W., and Mason, E.A. 1973, *The Mobility and Diffusion of Ions in Gases* (New York:Wiley).

(8) Rebrion, C., Rowe, B.R., and Marquette, J.B. 1989 *J. Chem. Phys.* **91**, 6142.

(9) Rowe, B.R., Marquette, J.B., and Rebrion, C. 1989, *J. Chem. Soc. Faraday Trans. 2* **85**, 1631.

(10) Smith, D. and Adams, N.G. 1989, *J. Chem. Soc. Faraday Trans. 2* **85**, 1613.

(11) Barlow, S.E., Dunn, G.H., and Schauer, M. 1984, *Phys. Rev. Letters* **52**, 902.

(12) Barlow, S.E., Dunn, G.H., and Schauer, M. 1984, *Phys. Rev. Letters* **53**, 1610.

(13) Luine, J.A., and Dunn, G.H. 1985, *Ap. J.* **299**, L67.

(14) Gerlich, D., and Kaefer, G. 1989, *Ap. J.* **347**, 849.

(15) Bates, D.R. 1986, *Phys. Rev. A* **34**, 1878.

(16) Bates, D.R. 1987, *Ap. J.* **312**, 363.

(17) Herbst, E., and Bates, D.R. 1988, *Ap. J.* **329**, 410.

(18) Smith, I.W.M. 1989, *Ap. J.* **347**, 282.

(19) Adams, N.G., and Smith, D. 1981, *Chem. Phys. Lett.* **79**, 563.

(20) Herbst, E. 1982, *Ap. J.* **252**, 810.

(21) Fehsenfeld, F.C., Dunkin, D.B., and Ferguson, E.E. 1974, *Ap. J.* **188**, 43.

(22) Bass, L.M., and Jennings, K.R. 1984, *Int. J. Mass Spectrom. Ion Proc.* **58**, 307.

(23) Bates, D.R. 1986, *J. Chem. Phys.* **85**, 2624.

CRESU Measurements of Ion-Molecules Reactions Down to 20 K

C. Rebrion[1], J. B. Marquette[2], and B. R. Rowe[1]

[1] *Département de Physique Atomique et Moléculaire, Université de Rennes I, Campus de Beaulieu,
35042 Rennes, France*
[2] *Laboratoire d'Aérothermique du C. N. R. S., 4ter route des Gardes, 92190 Meudon, France*

The few experimental techniques allowing measurements of ion-molecule reaction rate coefficients at very low temperatures are reviewed. For numerous reactions, capture theories which consider only the long-range part of the intermolecular potential are in satisfying agreement with the experimental data. Two of them, having an analytical expression, can be used for straightforward calculations. Special attention has been paid to the influence of rotational energy.

1 Introduction

As our understanding of circumstellar shells and of diffuse and dense interstellar clouds improves, it appears that the simple gas phase models are certainly not completely realistic. It is clear that the formation of interstellar molecules could involve a variety of processes including grain surface and shock wave chemistry (1). However, the lack of relevant laboratory data often impedes to obtain improved models.

Given that they explain the formation of many interstellar molecules, gas phase models remain extremely useful. A large advantage of this approach is that many of the gas processes can be evaluated from the current theoretical and experimental knowledge. However, there is always a lack of laboratory data, either because some processes have not been studied at all (for various reasons: very small cross section, unstable species, ...) or because they have not been studied under the relevant conditions. Consequently, uncertainties remain important and a large effort has to be made in this field.

The present paper deals with the improvements which have been obtained concerning our knowledge of ion-molecule chemistry. The last decade has seen a major breakthrough since four techniques have allowed ion-molecule reaction rate constants to be measured at extremely low temperatures. Here, some figures give an idea of the current state of the problem. Many of the chemical models of the clouds include several hundred species and as many as two thousand reactions (2). One of the most recent data compilation (3) has accumulated over 9,300 rate coefficient measurements from over 1,100 references. This shows how active research in ion-molecule chemistry is. But most of these data have been obtained with stable reactant molecules (for example, there is a lack of results for reactions with carbon atoms) and at room temperature.

Concerning results below 80K, the most productive technique (CRESU(S), see the paper by J.B. Marquette in these proceedings) has studied more than 60 reactions between 8 and 160K. It is clear that the models will have to rely on theoretical extrapolations of room temperature data for quite a long time. The aim of this paper is to provide a guidance for these extrapolations based on low temperature data. A special emphasis will be put on the role of rotational states in the reactions.

2 Experimental devices

If a great variety of techniques allow ion-molecule rate constants measurements above 80K (flowing afterglow, SIFT and other swarm techniques, ICR, beam techniques ...), only four devices in the world make it possible under 80K. This fact is related to the difficulty to perform experiments at so low temperatures. These apparatus can be separated into two groups, according to their cooling strategy.

2.1   Cryogenic helium-cooling

2.1.1   The ion trap of G. Dunn et al. (T=11K)

Ions are confined in a helium cooled experiment cell at very low pressure using an adequate electromagnetic field. According to the very long storage time (10^7s), all ions are radiatively deexcited and are in their ground state. The neutral gas is then added and the ion density is checked versus time using radiofrequence impulsions. As for fast reactions sampling time would be of the same order of magnitude as reaction time, this method is restricted to slow reactions ($k=10^{-11}$ to 10^{-15} cm^{3}s^{-1}), yet, taking advantage of the extremely low pressure, it allows measurements of radiative association rate coefficients (4).

2.1.2   The static drift tube of H. Bohringer et al. (T=45K)

As it could be operated at relatively high pressure, this device allowed fast and slow reaction rates ($k{\geq}10^{-13}$cm^{3}s^{-1}), as well as third body rates to be measured (5). Yet no more results have appeared in the literature for a long time, since this apparatus is no longer used.

The helium-cooling generates a great disadvantage for both the ion trap and static drift tube: most of neutrals condense on the wall of the apparatus, limiting their possibilities to hydrogen. To make measurements with condensable species possible, the cooling strategy has to be changed.

2.2   Aerodynamical cooling

2.2.1   The CRESUS apparatus

It has been described in detail in the accompanying paper by J.B. Marquette. Let us remember that it allows fast reaction rates to be measured in the 8-163K range, under true thermal conditions for most neutrals: except for the rotational levels of H_2 and the vibrational levels of heavy molecules like $C_2H_2Cl_2$ which are very close to each other and do not relax during the expansion of the buffer, all energy levels of molecules are populated according to the thermodynamical equilibrium distribution.

2.2.2   The free jet reactor of M. Smith et al.

This supersonic flow kinetic technique has been recently developed in Tucson (USA) and allows reaction rate studies with extremely low collision energies (6).

The premixed gases are formed in a free jet by expansion through a fraction of millimeter orifice in a vacuum chamber. Downstream, the ions are selectively created by a pulsed dye laser via resonant multiphoton ionization, and monitoring of the chemical composition via time of flight mass spectrometry allows to determine the absolute rate coefficient.

If the basic principle is very close to that of the CRESU apparatus, the main difference is that in this kind of free jet, local thermodynamical equilibrium cannot be carried out. Because of the very important gradients of density, of the large velocity and of the low pressure, at a small distance from the orifice there are not enough collisions to ensure an efficient relaxation toward a real Maxwell-Boltzmann distribution function for the velocity.

As a consequence, the translational distribution is different if one considers the parallel and perpendicular degrees of freedom in the jet, and heavy calculations are needed to deconvolute the results in order to obtain a rate coefficient versus a unique kinetic temperature. Another consequence is that the rotational energy is higher than the translational one and that the real rotational population is unknown. Here again, only a mean rotational energy and therefore a mean effective "temperature" can be estimated. This phenomenon is particularly important in the case of H_2, the resulting rotational temperature of which ranges between 150-200K, as the translational distribution can be represented by a temperature of about 2K (6).

A rapid survey of the literature shows that supersonic expansion techniques have been far more productive than cryogenic cooling methods in terms of measured reactions. Only the CRESUS method allows measurements in true thermal conditions. However the results obtained with the free jet reactor can be very instructive when compared with the CRESUS ones since this comparison can highlight the effects of rotational energy.

3   Overview of some ion-molecule theories

For a very large number of ion-molecule reactions, rate coefficients are required to satisfy the increasingly complex ion-chemical models. As it is not possible to experimentally study every ion-molecule reaction supposed to be involved in a model, the help of a theoretical prediction would be welcome.

Many hundreds of studied ion-neutral reactions indicate that a large majority of reactions has no activation barrier and proceed with unit collision efficiency when they are exothermic. These characteristics lead us to compare the experimental results with predictions of some capture theories.

3.1 The capture approximation

Capture theories are based on the assumption that the potential energy surface corresponding to the reaction can be divided into two parts:

- the long-range part, corresponding to the long-range intermolecular forces driving the collision partners close together. From this part the capture rate constant k_C can be extracted.

- the short-range part, where chemical bonds are created or broken. The effect of short range forces is represented by the reaction probability P.

In the capture approximation, the short-range part is neglected and P is set to unity. The capture rate constant is then derived from the long-range part only and $k=k_C$.

Let us consider a system formed by an ion and a molecule approaching each other. The notations are:

ion	q:	charge	neutral	α:	polarisability	system	μ:	reduced mass
				D:	dipole moment		v:	relative velocity
				B:	rotational constant		r:	ion-molecule separation

The kinetic energy is the translational energy (E_c) plus the energy of relative rotation (E_{rot}). If $\mathbf{V}$ is the classical electrostatic potential, the relative energy E_{rel} of the system is :

$$E_{rel} = E_c + E_{rot} + \mathbf{V} = \tfrac{1}{2}\mu\, v^2$$

Since the rotational energy can be seen as a centrifugal potential energy, the effective potential of the ion-molecule system is :

$$V_{eff} = E_{rot} + \mathbf{V}$$

If the impact parameter of the collision is different from zero, E_{rot} creates a "centrifugal barrier" in $\mathbf{V}_{eff}$. In most of capture theories, the height $(\mathbf{V}_{eff})_{max}$ of this barrier has to be determined. Then, the capture occurs only if $E_{rel} \geq (\mathbf{V}_{eff})_{max}$. In the other cases, the centrifugal barrier prevents capture and the reaction does not occur.

It is clear that the capture rate constant calculation depends on the nature of the ion-molecule system: the expression of $\mathbf{V}$ strongly depends on the electrostatic nature of the neutral.

3.2 Ion-induced dipole theory

Also called Langevin theory, it is valid for molecules having no dipole and no quadrupole moments. The ion-induced dipole potential leads to the well known Langevin rate constant $k_L = 2\pi q\,(\alpha/\mu)^{0.5}$ which is typically 10^{-9} cm^{3}s^{-1} and temperature independent.

3.3 Ion-dipole theories

In this case no exact mathematical treatment can be applied. As a consequence, various approximations have been made, and among the theoretical treatments applied to the problem there are the ADO and AADO theories (7), variational transition state theory (8), average free energy calculations (9), PRS method (10–13), adiabatic invariance method (14–16), ACCSA method (17,19), classical trajectory calculations (16) and the statistical adiabatic channel model (SACM) (20,21).

All theories are in good agreement above 100K but the calculated rate constants diverge below 100K (22). However, they have a common point: they predict a sharp increase in the rate constant as the temperature decreases.

The ACCSA method and the SACM allow the calculation of a rotational state selected rate coefficient k_j and predict that it increases when the rotational quantum number decreases. This means that k_j is the greatest when $j=0$, and this prediction is very important for rate constant calculations in dark molecular clouds where most molecules are radiatively relaxed to $j=0$ (23).

In the following part, we will limit the comparison of theory and experiment to the theoretical predictions which have an analytical expression depending only on the characteristics of the ion-molecule system. According to these criteria, two of them - in good agreement with each other - have been selected:

the parametrization of classical trajectory calculations (24) and a simplification of the SACM formula (21). It should be noted that the SACM formula has a finite limit when the temperature tends toward zero.

Authors	formula	remarks
Su and Chesnavich (24)	$k = k_L(\dfrac{0.3371 D}{\sqrt{\alpha kT}} + 0.62)$	$kT \leq \dfrac{D^2}{8\alpha}$
Troe (21)	$k = \dfrac{k_a k_b}{k_a + k_b}$ $k_a = k_L \sqrt{1 + \dfrac{D^2}{3\alpha B}}$ $k_b = k_L + 0.4 \, q \, D \, \sqrt{\dfrac{8\pi}{\mu kT}}$	$\lim_{T \to 0} = k_a$

The numerical values correspond to the c.g.s. unit system.

4   Results and recommendations

One of the most widely used rate coefficient in astrochemistry models is the Langevin rate coefficient. As discussed above, it assumes an ion-induced dipole potential and is temperature independent. Using the CRESU(S) apparatus, we have studied numerous reactions of several ions with non polar molecules. Usually is was found that if the reaction is fast at room temperature (i.e. close to Langevin), the rate coefficient is nearly unchanged at low temperature (25,26).

However we have recently measured reactions of $Ar^+\ ^2P_{3/2}$ and N_2^+ ions with H_2. In both cases, the rate coefficient is about two times less at 20K than at 300K. There is not yet a satisfying theoretical explanation for this behavior which clearly shows the need of experimental data at low temperature.

An other case of reaction with H_2 which deserves careful thought is that with N^+. While quite fast at room temperature (27), the reaction becomes extremely slow at the lowest temperatures. The analysis of the results obtained for normal as well as for para-H_2 is detailed in two papers (28,29). It is based on a capture model developed by Levine and Bernstein (30) for endothermic reactions and shows that the experimental data can be explained on the basis of a very small endothermicity. The main conclusion of this study was that this reaction is not an efficient pathway for interstellar ammonia formation.

The effect of quadrupolar terms in the intermolecular potential has also been investigated using molecules of large quadrupole moments (31). The theory predicts a variation of about 30% around the Langevin rate which was found experimentally to remain a good approximation.

The theoretical predictions have also been confirmed in the case of dipolar molecules: we observed a large increase in the rate constant at low temperature. Table 1 summarizes some results obtained in this field with the CRESU(S) method (32) with values deduced from the selected theories. A first glance shows that a rather good agreement is often obtained.

When there is no agreement, specific calculations are required. Recently Clary (33) has made a complete theoretical study of the reaction between $C^+(^2P)$ and $HCl(X^1\Sigma^+)$ indicating the necessity of incorporating the multi-surface nature of open-shells systems for some ion-molecule reactions. This means that the triple degeneracy of the ^2P state of C^+ is lost in the presence of HCl, and for comparison with experimental results, the splitting of the potential energy surface has to be taken into account.

M. Smith has measured a rate coefficient of $4.7 \ 10^{-9} \ cm^3 s^{-1}$ for the reaction $NH_3^+ + NH_3$ at a temperature of 2.5K. According to the parametrization of Su et al., the rate constant at this temperature should be about $2.5 \ 10^{-8} \ cm^3 s^{-1}$ and the measured value corresponds to a temperature of 75K. This clearly shows that the rate constants of M. Smith are underestimated (6). This can be understood by the fact that the rotational mean energy is much higher than the translational one in this experiment. The calculated corresponding temperature is in good agreement with the authors estimation of its rotational temperature. This shows that a knowledge of the internal states of the reactant is necessary (when not in thermal equilibrium) in order to use laboratory data for interstellar chemistry models.

In this short review we do not detail the case of reactions that cannot be described by capture theories, since they obviously need specific theoretical treatments. However several studies including $O_2^+ + CH_4$ (34), $N_2^+ + O_2$ (35) and reactions of Ar^+ ions (36) with several molecules have shown that if the rate coefficient is slow at room temperature, it can greatly increase at the lowest temperatures, sometimes approaching the capture rate coefficients.

		27-30K				68K		
		H_3^+	He^+	C^+	N^+	He^+	C^+	N^+
H_2O	exp		4.3	12.0	9.9	1.8	5.2	6.0
	Su et al.		14.2	9.6	9.1	9.5	6.4	6.1
	Troe		14.6	9.9	9.4	11.0	7.4	7.1
NH_3	exp	9.1	4.5	4.6	5.2	3.0	3.2	3.2
	Su et al.	13.2	11.7	7.9	7.6	11.7	7.9	5.2
	Troe	14.4	12.7	8.6	8.3	8.4	5.7	5.5
HCl	exp		11.0	3.8		4.6	1.9	
	Su et al.		8.6	5.4		5.9	3.7	
	Troe		8.6	5.4		5.6	3.5	
SO_2	exp	11.0	8.2	5.7		6.5	4.1	
	Su et al.	14.2	12.4	7.6		8.3	5.1	
	Troe	14.3	12.5	7.6		8.0	4.9	
H_2S	exp	6.5	4.6	4.8		4.6	1.9	
	Su et al.	9.6	8.2	5.2		5.7	3.6	
	Troe	7.8	6.7	4.3		4.4	2.8	

All values are given in 10^{-9} cm^{3}s^{-1}.

5 Conclusion

The present review demonstrates that a large majority of ion-molecule reaction rates can be estimated using simple analytical formulae. However, it has to be kept in mind that for polar molecules, rotational energy is a crucial parameter and this has to be taken into account if, for some reasons, the rotational states are not populated according to thermodynamical equilibrium.

Despite the success of these theoretical predictions, numerous results have also been presented (N_2^+ + H_2, C^+ + HCl for example) which clearly show that much more experimental data are needed. Even if most of the next results can be foreseen, some surprises are also presumable which will lead to more realistic models. Of course, improved models will have to rely on breakthroughs in other fields than in the very active ion-molecule research field. Some encouraging results are already obtained on branching ratios for dissociative recombination of polyatomic ions with electrons (see the paper of Rowe et al. in these proceedings) but other challenges have to be overcome, like neutral or surface reactions at low temperature.

References

[1] D.A. Williams, 1988, in *Rate Coefficients in Astrochemistry*, eds. T.J. Millar and D.A. Williams, Kluwer Academic Publishers.

[2] E. Herbst, C.M. Leung, 1986 Astrophys. J.,**310**, 378.

[3] Y. Ikezoe, S. Matsuoka, M. Takebe, A. Viggiano, 1987, *Gas Phase Ion-Molecule Reaction Rate Constants Trough 1986*, Maruzen Company, Tokyo, Japan.

[4] S.E. Barlow, G.H. Dunn, M. Schauer, 1984, Phys. Rev. Lett. **52**, 902.

[5] H. Bohringer, F. Arnold, 1982, J. Chem. Phys. **77**, 5534.

[6] M. Hawley, T.L. Mazely, L.K. Randeniya, R.S. Smith, X.K. Zeng, M.A. Smith, 1990, Int. J. Mass Spectr. Ion Proc. in press.

[7] T. Su, M.T. Bowers, 1979, in *Gas Phase Ion Chemistry*, Academic, New York..

[8] W.J. Chesnavich, T. Su, M.T. Bowers, 1980, J. Chem. Phys. **72**, 2641.

[9] F. Celli, G. Weddle, D.P. Ridge, 1980, J. Chem. Phys. **73**, 801.

[10] K. Takayanagi, 1978, J. Phys. Soc. Jpn. **45**, 976.

[11] K. Sakimoto, 1982, Chem. Phys. **68**, 155.

[12] K. Sakimoto, 1984, Chem. Phys. **85**, 273.

[13] K. Sakimoto, 1985, Chem. Phys. Lett. **116**, 86.

[14] D.R. Bates, 1982, Proc. R. Soc. London Ser. A **384**, 289.

[15] D.R. Bates, I. Mendas, 1985, Proc. R. Soc. London Ser. A **402**, 245.

[16] W.L. Morgan, D.R. Bates, Astrophys. J. **314**, 817.

[17] D.C. Clary, 1984, Mol. Phys. **53**, 3.

[18] D.C. Clary, 1985, Mol. Phys. **54**, 605.

[19] D.C. Clary, 1987, J. Chem. Soc. Faraday Trans. 2 **83**, 139.

[20] J. Troe, 1985, Chem. Phys. Lett. **122**, 425.

[21] J. Troe, 1987, J. Chem. Phys. **87**, 2773.

[22] T. Su, 1988, J. Chem. Phys. **88**, 4102.

[23] N.G. Adams, D. Smith, D.C. Clary, 1985, Astrophys. J. **296**, L31.

[24] T. Su, W. J. Chesnavich, 1982, J. Chem. Phys. **76**, 5183.

[25] B.R. Rowe,J.B. Marquette, G. Dupeyrat, E.E. Ferguson, 1985, Chem. Phys. Lett. **113**, 403.

[26] J.B. Marquette, C. Rebrion, B.R. Rowe, 1989, Astron. Astrophys. **213**, L29.

[27] N.G. Adams, D. Smith, 1985, Chem. Phys. Lett. **117**, 67.

[28] J.B. Marquette, B.R. Rowe, G. Dupeyrat, E. Roueff, 1985, Astron. Astrophys. **147**, 115.

[29] J.B. Marquette, C. Rebrion, B.R. Rowe, 1988, J. Chem. Phys. **89**, 2041.

[30] R.D. Levine, R.B. Bernstein, 1972, J. Chem. Phys. **56**, 2281.

[31] C. Rebrion, J.B. Marquette, B.R. Rowe, N.G. Adams, D. Smith, 1987, Chem. Phys. Lett. **136**, 495.

[32] B.R. Rowe, J.B. Marquette, C. Rebrion, 1989, J. Chem. Soc. Faraday Trans. 2 **85**, 1631.

[33] C.E. Dateo, D.C. Clary, 1990 J. Chem. Phys. in press.

[34] S.E. Barlow, J.M. VanDoren, C.H. DePuy, V.M. Bierbaum, I. Dotan, E.E. Ferguson, N.G. Adams, D. Smith, B.R. Rowe, J.B. Marquette, G. Dupeyrat, M. Durup-Ferguson, 1986, J. Chem. Phys. **85**, 3851.

[35] P. Gaucherel, J.B. Marquette, C. Rebrion, G. Poissant, G. Dupeyrat, B.R. Rowe, 1986, Chem. Phys. Lett. **132**, 63.

[36] C. Rebrion, B.R. Rowe and J.B. Marquette, 1989, J. Chem. Phys. **91**,6142.

Ab initio Study of the Potential Energy Surfaces for the Radiative Association Reaction $C^+ + H_2 \rightarrow CH_2^+ + h\nu$

Agee OZEKI and Suehiro IWATA*

Department of Chemistry, Faculty of Science and Technology, Keio University, Hiyoshi, Kohoku-ku, Yokohama 223, Japan

Abstract : Potential energy surfaces of $C^+ + H_2$ system are calculated with the *ab initio* CI method to study the possible radiative association reaction pathway. It is demonstrated that the non-adiabatic transition near the conical crossing of the $^2A_1$ and $^2B_2$ states should play a role in the process.

1   INTRODUCTION

One of the crucial initial steps in the proposed intersteller carbon chemistry is the radiative association: $C^+ + H_2 \rightarrow CH_2^+ + h\nu$. Because of its astronomical interest, the reaction rate has been theoretically studied by Herbst and his coworkers [1]. In the present work we critically examine the potential energy surfaces and the transition moment functions of the reaction. To study the radiative association it is important to examine whether the system can reach the molecular (product) region on the excited state potential energy surface without any extra kinetic energy.

2   METHOD

The multi-reference POLCI with a complete active space as the reference configurations are carried out for the $^2A_1, ^2B_1$, and $^2B_2$ states of C_{2v}, and for $^2A'$ and $^2A''$ states for Cs. The active orbitals comprise $2a_1, 3a_1, 4a_1, 1b_2, 2b_2$ and $1b_1$. The total number of geometries studied is more than 200. The *ab initio* program package used in the study is our MOLYX system, which runs both on the main frame computers and on the UNIX based engineer workstations(EWS).

3   RESULTS AND DISCUSSION

Figures 1, 2, and 3 are the potential energy surfaces (PES) of the lowest state of the $^2A_1, ^2B_1$, and $^2B_2$ symmetry for the C_{2v} approach. The energy of the dissociation limit, D_{reac}, in our calculation is -38.4502 Hartree. The calculated structures and relative energy of the characteristic points on the surfaces are in good agreement with those of Sakai et al [2] when available. The crossing seam between $^2A_1$ and $^2B_2$ is shown in Figures 1 and 2. Because the two states become $^2A'$ in the C_s conformation, the vertex of the

---

* To whom correspondence should be sent.

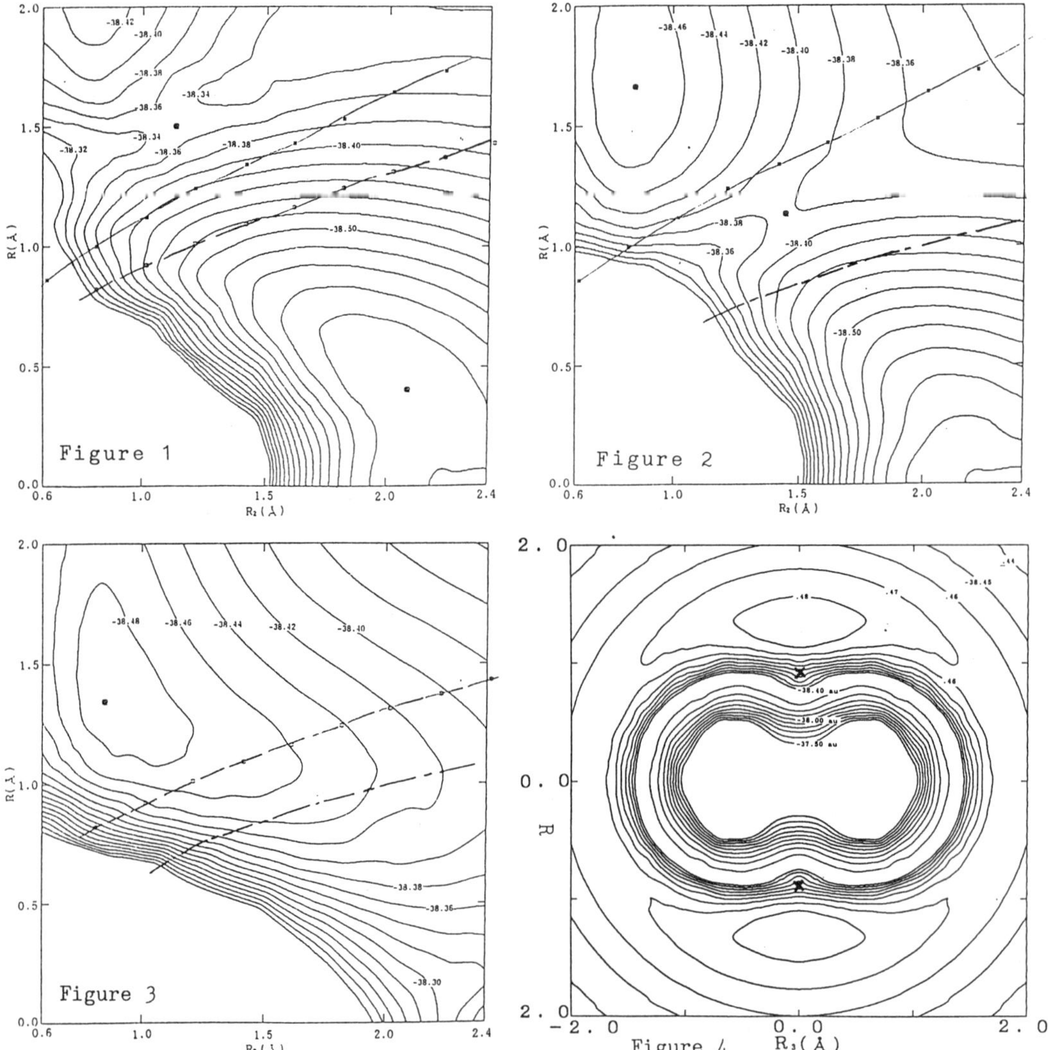

Figure 1. The PES of the $^2A_1$ state for the C_{2v} approach.
The crossing seam between $^2A_1$ and $^2B_1$: ———— , and the crossing seam between $^2A_1$ and $^2B_1$: — — — .
The energy unit of the contour is the Hartree. R: the distance between C^+ and the center of H_2. R_2: the H–H distance.

Figure 2. The PES of the $^2B_1$ state. The crossing seam between $^2A_1$ and $^2B_1$: ———— , and the crossing seam between $^2B_1$ and $^2B_2$: — ·· — · .

Figure 3. The PES of the $^2B_2$ state. The crossing seam between $^2A_1$ and $^2B_2$: — — — , and the crossing seam between $^2B_1$ and $^2B_2$: —— · — ·· .

Figure 4. The PES of the $1^2A'$ state at a fixed H–H distance ($R_2 = 1.0$Å). The C^+ ion is at $(R_3, R) = (x, y)$, and H_2 lies on the x axis.

conical intersection between the two $^2A'$ surfaces runs on this seam. The crossing seam between the $^2A_1$ and $^2B_1$ states is also shown in Figures 1 and 2. In this crossing region the non-adiabatic interaction through the rotational coupling term becomes important. Although the dominant term of the long-range interaction between a positive charged atom and a neutral molecule is the polarization interaction and it is expressed in terms of polarizability α of the neutral system as $-\alpha/R^4$, the figures demonstrate the state-dependence of the long-range attractive term. The $^2B_2$ state is most attractive, and the $^2B_1$ state is also attractive , but the $^2A_1$ state repulsive except at a very long-range. Because of this difference, a carbon ion C^+ slowly approaching H_2 must be on the surfaces whose main character is the $^2B_2$ or $^2B_1$ state in the C_{2v} configuration. In Herbst's mechanism the system must reach the region where the $^2A'$ and $^2A''$ states are close enough for the non-adiabatic transition to be significant; the two states are degenerate on the crossing seams between the $^2B_1$ and $^2A_1$ states or between the $^2B_1$ and $^2B_2$ states. These seams in our calculations lie well above D_{reac}; that is, if the initial kinetic energy is small, the non-adiabatic transition between the $^2A'$ and $^2A''$ states is unfavorable. On the other hand, because the crossing seam between the $^2B_2$ and $^2A_1$ states lies below the energy, D_{reac}, at the dissociation limit, the system can cross it with rather large relative kinetic energy, even when the initial kinetic energy is zero, and it climbs up the upper state surface. But this crossing is only possible for the C_{2v} approach. In the C_s symmetry this crossing is avoided because both states become $^2A'$. Thus, for the system to go to the excited surface, the non-adiabatic transition through radial coupling has to take place near the seam line of the C_{2v} approach. Figure 4 is the PES of the $1^2A'$ state for a fixed H–H distance (R_2); the bond distance $R_2 = 1.0$Åis an intermediate one between those of the reactant H_2 and of the product CH_2^+. A large region of the attractive well is seen in Figure 4. Inside of the conical crossing (denoted by X) at this particular R_2, PES is repulsive. For larger R_2 the crossing point falls below D_{reac}. The system might approach the crossing region with a relative kinetic energy large enough to hop to the upper surface. Our surfaces suggest that $1^2A' - 2^2A'$ non-adiabatic transition near the conical crossing is the most possible pathway to produce the excited state at very low temperature.

4   REFERENCE

[1] Herbst,E., Schubert,J.G., and Certain,P.R., 1977, Ap.J., <u>213</u>, 696; Herbst,E. 1982, Ap.J., <u>252</u>, 810. [2] Sakai,S., Kato,S., and Morokuma,K. 1981, J.Phys.Chem., <u>75</u>, 5398.

Time-of-Flight Analysis of $C_3H_3^+$ Ions Produced by the Reactions $C_3H_4^+ + C_3H_4$

Teruhiko OGATA[1], Shinzo SUZUKI[2], and Inosuke KOYANO[3]

[1] *Faculty of Liberal Arts, Shizuoka University, Ohya, Shizuoka 422, Japan*
[2] *Department of Chemistry, Tokyo Metropolitan University, Minamiōsawa, Hachiōji, Tokyo 193-02, Japan*
[3] *Department of Material Science, Himeji Institute of Technology, 2167 Shosha, Himeji 671-22, Japan*

The time-of-flight (TOF) coincidence spectra of $C_3H_3^+$ ions produced by the reactions $C_3H_4^+ + C_3H_4$ have been investigated using the Threshold Electron – Secondary Ion Coincidence (TESICO) technique. The reactions studied include all combinations of reactant ions and neutrals from allene, cyclopropene, and allene-d_4. The TOF spectra of $C_3H_3^+$ were found to be exceptionally broad and composed of several successive peaks. The complexity of the peaks has been interpreted as due to multiple production mechanisms and structures for the $C_3H_3^+$ ions. The branching ratios of various $C_3X_3^+$ (X = H or D) products from the reaction $C_3H_4^+ + C_3D_4$ indicate that H and D atoms are almost completely randomized in the reaction, suggesting that a complex mechanism is dominant.

There has recently been considerable interest in the ion–molecule reactions $C_3H_4^+ + C_3H_4$ (1 – 5). This reaction is suited for the study of reaction mechanisms because C_3H_4 has three isomers, cyclopropene (CP), allene (AL), and methylacetylene (MA), and thus the structural dependence of the products and cross sections of the reaction are expected to be studied. The reactions have also drawn attention recently as a possible source of various forms of the $C_3H_2^+$ ions, whose structures, reactivities, and roles in interstellar molecule formation are current issues. Most of the studies performed so far have been concerned only with the allene and methyl acetylene system and very little has been performed on the cyclopropene system. We, therefore, initiated a systematic study of the $C_3H_4^+ + C_3H_4$ reactions including all three isomers for both reactant ions and neutrals. In the course of this study, which utilizes our TESICO (Threshold Electron – Secondary Ion Coincidence) technique, we found that the time-of-flight (TOF) spectra of the $C_3H_3^+$ product ions (mass-selected) were abnormally broad, spreading over the range covering the TOFs of $C_3H_4^+$ and $C_6H_7^+$, regardless of the nature of the C_3H_4 neutrals. We thus investigated the TOF spectra of $C_3H_3^+$ in some detail. The present paper reports the results from the CP^+ and AL^+ reactions.

Experimental

The ion–molecule reactions were studied using the TESICO technique (6), which essentially measures the TOF distribution of the primary and secondary ions of the reaction in a beam – chamber mode. The wavelengths used for the ionization of cyclopropene and allene were 126.1 nm and 123.8 nm, respectively, corresponding to an internal energy content of the reactant ions of 0.22 eV and 0.33 eV, respectively. The impact energy of the ions was 5 eV (c.m.). The pressure in the reaction chamber was kept at about 6 mTorr throughout the experiment. Cyclopropene was prepared by the method of Closs and Krantz (7) and purified by trap-to-trap distillation. Allene and deuterated allene-d_4 were purchased from Takachiho Shoji Ltd.

Results and Discussion

Time-of-Flight Coincidence Spectra. The TOF spectra of the $C_3H_3^+$ product ions from reactions of CP^+ and AL^+ with CP, AL, and AL-d_4 are shown in Figure 1. The spectra of the $C_3H_3^+$ ions were found to be much broader than those of other product ions ($C_3H_5^+$ and $C_6H_7^+$) and reactant ion $C_3H_4^+$, spreading over the whole range between the peaks of $C_3H_4^+$ (43.8 μs) and $C_6H_7^+$ (62.7 μs), regardless of the reaction system. The spectra from the CP^+ + CP reaction consist of at least four components, giving peaks at (1) 43.8 μs, (2) 47.1 μs, (3) 55.7 μs, and (4) 62.7 μs. The TOF spectra of the $C_3H_3^+$ ions produced from other combinations of reactant ions and neutrals are found to be composed of all or several of these peaks, as summarized in Table I. Peak (1), which is characteristic of the reaction systems involving the CP^+ ion, almost disappears when the contribution from the $C_3H_3^+$ ions produced in the ionization chamber is subtracted from the signal. Peak (2) is also characteristic of the CP^+ reactant and can be ascribed to its reaction with C_3H_4 in the reaction chamber. Peak (3) is observed in all reaction systems studied, even in the reaction with allene-d_4. On the contrary, peak (4) which appears in almost all reaction systems is never observed when the neutral reactant is allene-d_4.

From these comparisons among reaction systems, the following conclusions may be drawn. Peak (1) is due to $C_3H_3^+$ ions produced in the ionization chamber by the reaction CP^+ + CP. The $C_3H_3^+$ ions corresponding to peaks (2) and (3) originate from $C_3H_4^+$ ions by transferring an H atom to a neutral reactant. Ions giving peak (4) apparently originate from C_3H_4 neutrals by an H^- abstraction. The $C_3H_3^+$ ions corresponding to peak (2) cannot be considered to be produced by a complex mechanism,

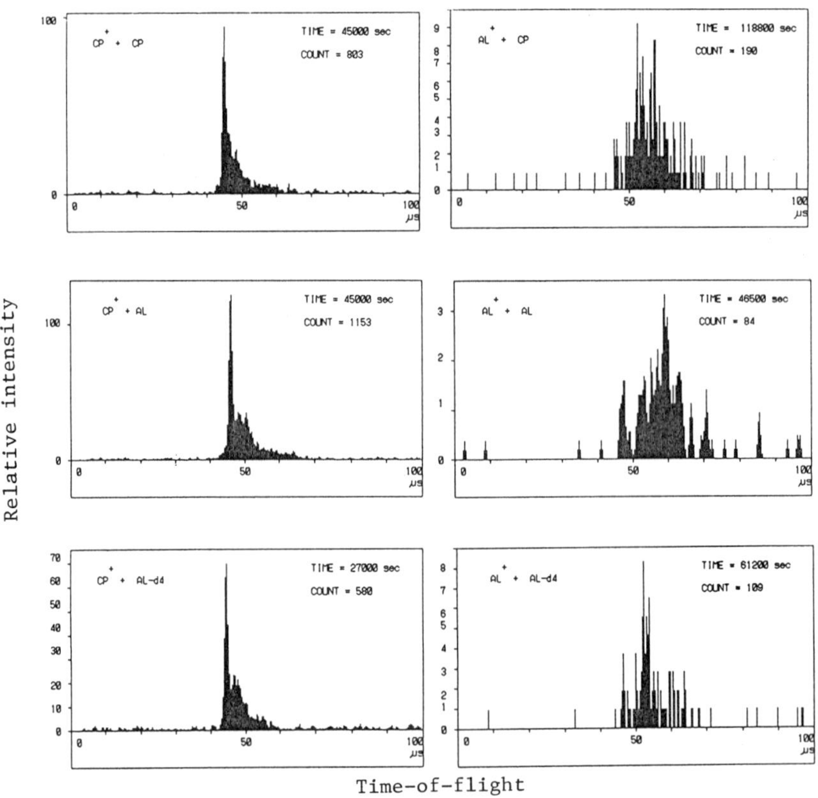

Figure 1. Time-of-flight coincidence spectra of $C_3H_3^+$ from six reaction systems.

Table I. Existence (Y) and nonexistence (N) of the four characteristic $C_3H_3^+$ peaks in six reaction systems.

Reaction System	Peak (1)	Peak (2)	Peak (3)	Peak (4)
CP^+ + CP	Y	Y	Y	Y
+ AL	Y	Y	Y	Y
+ AL-d$_4$	Y	Y	Y	N
AL^+ + CP	N	N	Y	Y
+ AL	N	N	Y	Y
+ AL-d$_4$	N	N	Y	N

since the peak is rather sharp and its TOF is not far apart from that of the parent ion $C_3H_4^+$. So, these $C_3H_3^+$ ions are probably produced by a direct mechanism. Smith and Adams (8) have presented evidence for the occurrence of two stable structures for the $C_3H_3^+$ ions produced in many systems. It is not clear at present, however, whether the two peaks observed here, peaks (2) and (3), correspond to these two structures, namely, cyclic and linear structures. The TOF difference between these two peaks seems to be too large to be ascribed to the difference in the kinetic energy release between the reactions forming these two structural isomers.

Branching Ratios in the $C_3X_3^+$ (X = H or D) Ions. As pointed out above, the TOF of the $C_3H_3^+$ ions produced in the ionization chamber is sharp and thus these ions are readily distinguished from those produced in the reaction chamber. The isotopic distribution in the $C_3X_3^+$ (X = H or D) product ions formed in the reactions of CP^+ and AL^+ with AL-d$_4$ was examined after correcting for the $C_3H_3^+$ ions produced in the ionization chamber. The results are summarized in Table II. If a complete randomization of the H and D atoms occurred during the reaction, the intensity ratios $C_3H_3^+$: $C_3H_2D^+$: $C_3HD_2^+$: $C_3D_3^+$ should be 1 : 6 : 6 : 1. The results in Table II show that this is evidently not the case. However, the occurrence of a fair amount of $C_3H_2D^+$ and $C_3HD_2^+$ in both reactions suggests that at least a complex mechanism is operative in these reactions as a dominant one.

Structure- and Isotope-Dependence of the Reaction Cross Section. Table III shows the number of the $C_3X_3^+$ ions produced per 10^5 reactant ions in various reaction systems. This number may be taken as representing the relative reaction cross sections. It can be clearly seen that the reactivity lowers drastically in the order of CP^+, AL^+, and AL-d$_4^+$ for each fixed target. The extremely low reactivity of the AL-d$_4^+$ ions (as compared with that of AL^+) is particularly noteworthy. The reason for this is not clear at present, however. We have also examined the reactions using Ar as target. The results, also shown in Table III, show that essentially no $C_3X_3^+$ products are produced, suggesting that the origin of the $C_3H_3^+$ ions in the CP^+ and AL^+ reactions is not a simple collision induced dissociation of the reactant ions.

Table II. Isotopic distribution in the $C_3X_3^+$ product ions from the AL^+ + AL-d$_4$ and CP^+ + AL-d$_4$ reactions.

	AL^+ + AL-d$_4$	CP^+ + AL-d$_4$	if complete-ly randomized
$C_3H_3^+$	94	909 (2945)[a]	6
$C_3H_2D^+$	1534	1670	24
$C_3HD_2^+$	44	488	24
$C_3D_3^+$	31	489	4

a) The number in the parenthesis includes the ions in peak (2).

T. OGATA et al.

Table III. Number of the $C_3H_3^+$ ions produced per 10^5 reactant ions in various $C_3H_4^+$ + C_3H_4 reaction systems.

Reactant Ion	Neutral			
	CP	AL	AL–d_4	Ar
CP^+	1438	1146	1009	-16[a]
AL^+	167	140	126	5
$AL–d_4^+$	–	9[b]	–	–

a) Minus sign means that the number of background ions exceeds that of the ions produced in the reaction chamber.
b) The number of $C_3D_3^+$ ions.

References

(1) Myher, J. J., and Harrison, A. G. 1965, J. Phys. Chem., 72, 1905.
(2) Nato, A., Niwa, M., Honma, K., Tanaka, I., and Koyano, I. 1980, Internat. J. Mass Spectrom. Ion Phys., 34, 287.
(3) Lifshitz, C., and Gleitman,Y. 1982, J. Chem. Phys., 77, 2383.
(4) Lifshitz, C., and Gleitman, Y. 1981, Internat. J. Mass Spectrom. Ion Phys., 40, 17.
(5) van Pijkeren, D., van Eck, J., and Niehaus, A. 1986, Chem. Phys., 103, 383.
(6) Koyano, I., and Tanaka, K. 1980, J. Chem. Phys., 72, 4858.
(7) Closs, G. L., and Krantz, K. D. 1966, J. Org. Chem., 31, 638.
(8) Smith, D., and Adams, N. G. 1987, Internat. J. Mass Spectrom. Ion Process., 76, 307.

Laboratory Studies of Possible Interstellar Isomeric Ions

Colin G. Freeman and Murray J. McEwan

Department of Chemistry, University of Canterbury, Christchurch, New Zealand

Many simple ions and molecules containing as few as 3 atoms can exist in stable isomeric forms. We have used the selected ion flow tube technique to examine the behaviour of four isomeric pairs of ions that are relevant to chemical processes occurring in interstellar clouds: CCN^+ and CNC^+; HOC^+ and HCO^+; HCN^+ and HNC^+; $CCCN^+$ and $c\text{-}C_3N^+$. Although the primary ionization mechanism will usually result in mixtures of the isomeric forms, it is the ensuing chemistry that controls the interstellar abundance, and which often leads to a marked imbalance in the densities of each isomer.

Introduction

In recent years there has been a growing awareness of the importance of structural isomers of simple ions and molecules and of their role in interstellar cloud chemistry. Although structural isomerism does not present any major problem to the techniques of observational astronomy where spectroscopic detection is used, the same cannot be said for laboratory investigations where the observational technique most commonly used is mass spectrometry. For example, the structural isomers HCN and HNC have both been identified in a variety of interstellar objects including TMC 1, Ori A and Sgr B2 and their respective J = 1-0 transitions are very different in energy. In the laboratory, however, when mass spectrometric detection is used, each of these species appears at the same mass and therefore different or additional identification techniques are required. A further distinction between the identification of isomers in intersteller clouds and in the laboratory is in the relationship between the isomeric forms. In the interstellar medium the widely separate spectroscopic lines of isomers leads observers to emphasise the unique character of each isomer, whereas in the laboratory the techniques required to produce one isomer invariably produce mixtures of isomers. Thus, in the laboratory the fact that isomers usually have common production sources is emphasised. For example, if dicyanogen, C_2N_2, is subject to electron impact and the resulting C_2N^+ fragment ions at $m/z = 38$, after mass selection in the selected ion flow tube (SIFT) (1), are injected into the flow tube, then it can be shown that the C_2N^+ contains species of two different types. The distinction is made on the basis of reactivity of the isomers. By choosing an appropriate neutral reactant gas, such as CH_4 for C_2N^+, it was observed that one of the C_2N^+ species was reactive with CH_4 and the other unreactive (2). Laboratory techniques that rely on mass spectrometric detection can therefore be modified to yield information on isomeric species.

We summarize next, the results of laboratory investigations into the isomeric pairs of ions CCN^+, CNC^+; HOC^+, HCO^+; HCN^+, HNC^+ and $CCCN^+$, $c\text{-}C_3N^+$.

Results

CCN^+/CNC^+

Haese and Woods (3) first suggested the possible different role of the two C_2N^+ isomers in interstellar clouds. They also suggested that the product of the reaction (1) was CNC^+ rather than CCN^+.

$$C^+ + HCN \rightarrow C_2N^+ + H \qquad\qquad (1)$$

Two recent laboratory studies, Bohme et al (3) and Knight et al (2), have revealed the different ion chemistry of these isomers.

Electron impact (typically 25eV) on C_2N_2, CH_3CN and HC_3N generated mixtures of the isomers CCN^+ and CNC^+ with the lower energy isomer (CNC^+) being the predominant species (the ratio being typically $\sim$ 4:1). A summary of their reaction chemistry is given in Table I.

Table I

Rate coefficients $k(10^{-9}\ cm^3\ s^{-1})$ and product distributions of C_2N^+ isomers with the given neutral reactant[a].

reactant	products[b]	branching ratio	$k(CCN^+)$	$k(CNC^+)$	k_{coll}[c]	$-\Delta H,\ kJ\ mol^{-1}$[d]	
						CCN^+	CNC^+
H_2	$H_2CN^+ + C$ [e]	0.9				62	
			0.9	<0.0001	1.53		
	$H_2C_2N^+$ [e]	0.1				$\sim$512	
CH_4	$C_2H_3^+ + HCN$	0.6				399	293
	$H_2CN^+ + C_2H_2$	0.1	0.7	0.004	1.12	477	371
	$H_2C_3N^+ + H_2$	0.3				512	406
H_2O	$HCO^+ + HCN$	0.92				525	419
			1.63	0.07	2.55		
	$H_2CN^+ + CO$	0.08				648	542
HCN	$HC_3N_2^+$	1.0	0.42[f]	0.42[f]	3.48		
O_2	$C_2NO^+ + O$ [e]	>0.97					
			$\sim$0.4	<0.0001	0.70		
	$O_2^+ + C_2N$ [e]	<0.03					
CO_2	$C_2NO^+ + CO$	1.0	1.1	<0.0001	0.83		

[a] Condensed from a more extensive table in ref 2. [b] The product distributions shown were obtained from a mixture of CCN^+ and CNC^+ isomers in the ratio $CCN^+/CNC^+ \sim 0.25$. [c] Calculated collision rate. [d] Enthalpies are based on values for ΔH_f of $CCN^+ = 1726\ kJ\ mol^{-1}$ and $CNC^+ = 1620\ kJ\ mol^{-1}$. [e] Products refer to reaction with CCN^+ only. [f] Pseudobimolecular rate coefficient at a reaction tube pressure of 0.30 Torr.

We note that reaction (1) does in fact yield CNC^+ exclusively and that the two isomers have a very different reactivity with H_2. CCN^+ is reactive and CNC^+ unreactive. Thus CNC^+ should be present at very much larger densities than CCN^+ in dense interstellar clouds.

HOC^+/HCO^+

HCO^+ was the first polyatomic ion identified in dense interstellar clouds and is also the most abundant ion so far detected (5). Ever since the first tentative observation by Woods et al (6) of its isomer HOC^+ towards Sgr B2, there has been considerable interest in the laboratory in elucidating the chemistry of the formyl, HCO^+, and isoformyl, HOC^+, ions (Illies et al (7); Wagner-Redeker et al (8); Jarrold et al (9)). Much of the difficulty in achieving appreciable densities of the higher energy HOC^+ species in the earlier studies was resolved when HOC^+ was identified as the major product of the reaction between C^+ and H_2O (10). We also made a study of a number of reactions of HOC^+

including the identification of two new production sources in addition to the reaction of $H_3^+ + CO$ (11).

$$CO^+ + H_2 \begin{cases} \xrightarrow{0.48} HOC^+ + H \\ \xrightarrow{0.52} HCO^+ + H \end{cases} \qquad k = 1.5 \times 10^{-9} \text{ cm}^3 \text{ s}^{-1} \qquad (2)$$

$$C^+ + H_2O \begin{cases} \xrightarrow{0.84} HOC^+ + H \\ \xrightarrow{0.16} HCO^+ + H \end{cases} \qquad k = 2.3 \times 10^{-9} \text{ cm}^3 \text{ s}^{-1} \qquad (3)$$

The most important reaction of HOC^+ in governing its interstellar abundance is the reaction with H_2 (reaction (4)) in which we found isomerization competing with proton transfer (11).

$$HOC^+ + H_2 \begin{cases} \xrightarrow{0.57} H_3^+ + CO \\ \xrightarrow{0.43} HCO^+ + H_2 \end{cases} \qquad k = 4.7 \times 10^{-10} \text{ cm}^3 \text{ s}^{-1} \qquad (4)$$

This result is of course a room temperature result and although it is very difficult to predict what will happen at low temperatures, it does appear likely that in contrast to HCO^+, relatively large densities of HOC^+ cannot be achieved in interstellar clouds. Proton transfer from HOC^+ to H_2 is about 4kJ mol^{-1} endothermic (11, 12) and although this channel may decrease at low temperatures, the isomerization channel should still proceed efficiently. The mechanism for this reaction has been discussed in more detail in reference 12 and has been treated theoretically in reference 9. We note also that HOC^+ is isomerised efficiently by CO, reaction (5), (11):

$$HOC^+ + CO \rightarrow HCO^+ + CO \qquad k = 6 \times 10^{-10} \text{ cm}^3 \text{ s}^{-1} \qquad (5)$$

Amano and Nakanaga (13) have also detected HOC^+ in the laboratory from its IR absorption and although their rate coefficient for reaction (4) was similar to ours, they did not identify the H_3^+ channel.

HCN^+/HNC^+

Two isomers both at m/z = 27 were identified by their reactions with CF_4: HCN^+ was reactive and HNC^+ unreactive (12). We found that HCN, when subjected to ~30 eV electrons, produced a mixture of HCN^+ and HNC^+, with HNC^+ accounting for ~ 25% of the total m/z = 27 signal. Some of the reactions of relevance to astrochemistry that we have studied (14) are shown in Table II.

We also observed that the products at mass 27 in reaction (6) are equally distributed between HCN^+ and HNC^+.

$$CN^+ + H_2 \begin{cases} \xrightarrow{0.5} HCN^+ + H \\ \xrightarrow{0.5} HNC^+ + H \end{cases} \qquad k = 1.0 \times 10^{-9} \text{ cm}^3 \text{ s}^{-1} \qquad (6)$$

The most notable feature of the reactions of HCN^+ and HNC^+ are that each isomer reacts rapidly with H_2 to produce $HCNH^+$, an ion identified only comparatively recently in interstellar clouds (15).

Table II

Rate coefficients k (10^{-9} cm^3 s^{-1}) and product distributions of HCN^+ and HNC^+ with the given neutral reactant.

						$-\Delta H$, kJ mol^{-1} [b]	
reactant	products	branching ratio	k(HCN^+)	k(HNC^+)	k_{coll} [a]	HCN^+	HNC^+
H_2	$HCNH^+ + H$ [c]	1.0	0.86	0.70	1.54		282
C_2H_2	$C_2H_2^+ + HCN$ [d]	1.0	1.5		1.20	212	
C_2H_2	$C_2H_2^+ + HCN$ [e]	≥0.3		1.5	1.20		138
	$H_2C_3N^+ + H$	≥0.5					255
CO	$HNC^+ + CO$ [d]	1.0	0.46	≤0.001	0.91	74	
CO_2	$HNC^+ + CO_2$ [d]	>0.99	0.50		0.92	74	
CO_2	products [e]			≤0.0012	0.92		
O_2	$O_2^+ + HCN$ [d]	1.0	0.5		0.77	147	
O_2	$HNCO^+ + O$ [e]	0.75		0.36	0.77		109
	$NO^+ + HCO$	0.25					344

[a] Calculated collision rate coefficient (10^{-9} cm^3 s^{-1}). [b] Estimated on the basis that ΔH_f [HNC^+] = 1373 kJ mol^{-1} and ΔH_f[HCN^+] = 1447 kJ mol^{-1}. [c] Identical products for HCN^+ and HNC^+.
[d] products for reaction with HCN^+. [e] products for reaction with HNC^+.

$CCCN^+$ and c-C_3N^+ (or $CCNC^+$)

Two isomeric forms of ions at m/z = 50 generated by electron impact on HC_3N were identified (16). One of these was reactive with H_2 (10% of the m/z = 50 signal) and one was not. Ab initio calculations of the various structures of the C_3N^+ ion suggest that the lowest energy form is the $^3\Sigma^-$ $CCCN^+$ isomer with possible candidates for the other isomer being $CCNC^+$ or cyclic C_3N^+ (17). If the isomer unreactive with H_2 (k ≤ 5 x 10^{-12} cm^3 s^{-1}) is indeed $CCCN^+$, then its large dipole moment (2.27 D) together with the fact that HC_3N has been observed in a wide variety of interstellar objects would indicate that $CCCN^+$ is a potential candidate for radio observation.

Conclusions

We find that where an ion can exist in more than one low energy isomeric form, then electron impact on the source molecule invariably produces mixtures of the two (or more) isomers. In the interstellar environment, the same processes should occur and therefore it is the ensuing chemistry that determines the relative abundance of the isomeric forms. Most chemical sources also generated mixtures of isomers when the reaction was sufficiently exothermic to form the higher energy isomer.

Acknowledgements

We thank the New Zealand Universities Grants Committee for financial support.

References

1. Knight, J.S., Freeman, C.G., McEwan, M.J., Adams, N.G., and Smith, D. 1985, Int. J. Mass Spectrom. Ion Proc., 67, 317.

2. Knight, J.S., Petrie, S.A.H., Freeman, C.G., McEwan, M.J., McLean, A.D., and DeFrees, D.J. 1988, J. Am. Chem. Soc., 110, 5286.

3. Haese, N.N., and Woods, R.C. 1981, Astrophys. J., 246, L51.

4. Bohme, D.K., Wlodek, S., Raksit, A.B., Schiff, H.I., Mackay, G.I., and Keskinen, K.J. 1987, Int. J. Mass Spectrom. Ion Proc., 81, 123.

5. Buhl, D., and Snyder, L.E. 1970, Nature, 228, 267; Snyder, L.E., Hollis, J.M., Lovas, F.J., and Ulrich, B.L. 1976, Astrophys. J. 209, 67.

6. Woods, R.C., Gudeman, C.S., Dickman, R.L., Goldsmith, C.F., Huguenin, G.R., Irvine, W.M., Hjalmasson, A., Nymein, L.A., and Olofsson, H. 1983, Astrophys. J. 207, 10.

7. Illies, A.J., Jarrold, M.F., and Bowers, M.T. 1983, J. Amer. Chem. Soc., 105, 2562.

8. Wagner-Redeker, W., Kemper, P.R., Jarrold, M.F., and Bowers, M.T. 1985, J. Chem. Phys., 83, 1121.

9. Jarrold, M.F., Bowers, M.T., De Frees, D.J., McLean, A.D., and Herbst, E. 1986, Astrophys. J., 303, 392.

10. Freeman, C.G., and McEwan, M.J. 1987, Int. J. Mass Spectrom. Ion. Proc., 75, 127.

11. Freeman, C.G., Knight, J.S., Love, J.G., and McEwan, M.J. 1987, Int. J. Mass Spectrom. Ion Proc., 80, 255.

12. Petrie, S., Freeman, C.G., Mautner, M., McEwan, M.J., and Fergusson, E.E. 1990, J. Amer. Chem. Soc., (In press).

13. Amano, T., and Nakanaga, T. 1988, Astrophys. J., 328, 373.

14. Petrie, S., Freeman, C.G., McEwan, M.J. and Ferguson, E.E. To be submitted.

15. Ziurys, L.M., and Turner, B.E. 1986, Astrophys. J., 302, L31.

16. Petrie, S., Freeman, C.G., and McEwan, M.J. To be submitted.

17. Harland, P.W., and Maclagan, R.G.A.R. 1987, J. Chem. Soc., Faraday Trans., 83, 2133.

Fast Neutral Reactions in Cold Interstellar Clouds

M. M. Graff

School of Physics, Georgia Institute of Technology, Atlanta, GA 30332-0430, U. S. A.

The dynamics of exothermic neutral reactions between radical species are examined, with particular attention to reactivity at the very low energies characteristic of cold interstellar clouds. Long-range interactions (electrostatic and spin-orbit) are considered within the adiabatic capture-infinite order sudden approximation. Analytic expressions are provided for cross sections and rate constants of exothermic reactions between atoms and dipolar radicals at low temperatures. A method for approximating the adiabatic potential surface for the reactive state is presented. The reaction systems O+OH and O+CH are both predicted to be fast at low temperatures. The systems C+CH and C+OH are expected to be nonreactive from the lowest fine-structure levels, and upper limits of rate constants for these reactions have been estimated. General predictions are made for other reaction systems. Implications for interstellar chemistry are discussed.

In cold interstellar clouds, reaction is expected to be an important process only for exothermic systems with attractive long-range interactions. For neutral systems, reactivity at low temperature is generally limited to radical-radical collisions. Here we examine the interactions of four such systems:

$$O+OH \rightarrow O_2+H, \tag{R1}$$
$$O+CH \rightarrow CO+H, \tag{R2}$$
$$C+CH \rightarrow C_2+H, \text{ and} \tag{R3}$$
$$C+OH \rightarrow CO+H. \tag{R4}$$

Reactions (R1)-(R4) have attractive long-range potential surfaces for at least some of the reactant fine-structure levels. Because the reactions are highly exothermic, it is probable that there exists, for geometries matching those of reaction intermediates (HO$_2$, HCO, C$_2$H, and HOC), a downhill path to products for each reaction. Reaction is probable if a collision populates a reactive state at small separations. These basic considerations allow reasonable estimates of the reaction rate constants to be made for low temperatures.

The adiabatic-capture, infinite-order sudden approximation (ACIOSA) offers a straightforward method for estimating cross sections and rate constants for reactions whose dynamics are dominated by long-range forces (Clary 1984). Further details about the method and its application to specific reaction systems may be found in Graff (1989) and Graff and Wagner (1990). The basic adiabatic-capture approximation assumes that reaction occurs in all collisions occurring on a reactive surface for which the total energy E exceeds the maximum in the effective long-range potential energy surface. The reaction cross section is of the form

$$\sigma(E,\theta) = \frac{\pi}{k_E^2} \left(J_{max}(E,\theta) + 1 \right)^2 ,$$

(1)

where $J_{max}(E,\theta)$ is the highest reactive partial wave and k_E is the wave number for collision energy E.

Low-energy collisions are dominated by the dipole-quadrupole interaction ($V{\sim}R^{-4}$). The ACIOSA reaction cross section for the dipole-quadrupole interaction has been calculated (Graff 1989) by angular average over θ to be

$$\sigma^{AC} = 2.47\,\pi \left[\frac{|\mu_1 Q_2|}{E} \right]^{1/2} ,$$

(2)

where μ_1 is the molecular dipole moment, Q_2 is the quadrupole moment of the most attractive sublevel of the lowest atomic fine-stucture level, and all quantities expressed in atomic units. The resulting rate constant, which depends on temperature only through the fine-structure distribution of reactants:

$$k^{AC}(T_{fs}) = 5.00{\times}10^{-10}\,\pi \left[\frac{|\mu_1 Q_2|}{\mu} \right]^{1/2} p(T_{fs}) ,$$

(3)

where μ is the collisional reduced mass and the rate constant is in units of $cm^3\,s^{-1}$. The function $p(T_{fs})$, the probability of initiating a collision on the reactive surface for fine-structure distributions described temperature T_{fs}, is given by the electronic partition function.

The above expression indicates that rate constants for reactions with long-range dipole-quadrupole attractive interactions are finite at T=0. This general result differs from the functional form $k{\sim}T^{1/2}$ generally assumed for neutral reactions at low temperatures. Higher-order terms (e.g. the quadrupole-quadrupole interaction, for which $V{\sim}R^{-5}$) cause the rate constant to increase slightly with temperature. These effects are minor, however, in cold clouds: for the prototypic reaction (R1), Eq. (7) agrees with the more complete calculation (including terms through R^{-5}) within 10% for the temperature range 0-100 K.

The method outlined here applies directly only for systems whose lowest potential surface varies as R^{-4}. Examples include (R1), (R2), and other reactions of atomic oxygen or sulfur with dipolar radicals. For reactions with more complicated long-range adiabatic potential surfaces (e.g. reactions (R3) and (R4), which involve $C(^3P_0)$), careful consideration of the collision process is required. Individual cases are discussed below.

a)   Reactions involving atomic oxygen

The lowest fine-structure level of atomic oxygen, $O(^3P_2)$, has several distinct quadrupole moments corresponding to the five m_j sublevels: $Q_m=+0.8$, -0.4, and $-0.8\,ea_0^2$ for $m_j=\pm2$, ±1, and 0, respectively (Gentry and Giese 1977). The quantum mechanical state with most attractive potential is a mixture of these sublevels. For the purpose of estimating the rate constant, the form of the lowest surface may be approximated classically using $|Q_0|=0.8\,ea_0^2$; the next two states may be approximated by $|Q_0|=0.4\,ea_0^2$.

Reaction (R1) is likely to proceed via the single ground-state surface of HO_2, which correlates adiabatically to the lowest long-range surface. The rate constant may then be calculated, using $\mu_{OH}=0.656\,ea_0$ (Meerts and Dymanus 1975), to be $7.9{\times}10^{-11}\,cm^3s^{-1}$ (Table I). Analysis of the energetics of reaction (R2) indicates that three surfaces may proceed to reaction. Approximating the reaction as three independent single-state processes, the total rate constant may be calculated as the algebraic sum of rate constants for individual states. The value of the rate constant for reaction (R2) shown in the table is derived using the value $\mu_{CH}=0.574\,ea_0$ (Follmeg, Rosmus, and Werner 1987) and effective quadrupole moments for the lowest three states. Consideration of energetics and reaction geometry indicates that reaction (R2) should lead entirely to products CO+H.

b)   Reactions involving atomic carbon

Collisions involving atomic carbon illustrate the importance of considering the long-range behavior for all states correlating to separated reactants. The quadrupole moment of the lowest fine structure level of atomic carbon $C(^3P_0)$ is identically zero, because the total electronic angular momentum for this level is zero (Gentry and Giese 1977). The lowest potential surfaces for collision systems involving atomic carbon are therefore flat (through terms of order R^{-5}) at large distances. At relatively large internuclear separations, surfaces that correlate to excited fine-structure levels intersect with the flat surface in weakly avoided crossings. The quadrupole-quadrupole interaction that couples these states is weak at large separations, so collision is expected to be diabatic through the crossing. Reactive flux is therefore likely to originate from the excited fine-structure levels, and collisions originating from the lowest fine structure level are expected to be nonreactive. The upper limit listed in Table I for reactions (R3) and (R4) was obtained by estimating a maximum transition probability of $\sim 10^{-2}$ at the curve crossings.

Table I. Low-temperature rate constants for simple neutral reactions.

Reaction	$k(T=0)$ (cm^3s^{-1})
$O+OH \rightarrow O_2+H$	7.9×10^{-11}
$O+CH \rightarrow CO+H$	1.9×10^{-10}
$C+CH \rightarrow C_2+H$	$<1.0 \times 10^{-12}$
$C+OH \rightarrow CO+H$	$<1.0 \times 10^{-12}$

Reactivity for reactions (R1)-(R4) is thus determined largely by the quadrupole moment(s) of the lowest spin-orbit state. Atoms for which $j \neq 0$ in the lowest fine-structure level (e.g. oxygen, sulfur) have nonzero quadrupole moments and therefore attractive long-range potential surfaces from the lowest fine-structure level. Reactions of these atoms with radicals are generally expected to be relatively fast ($k \sim 10^{-11}$-10^{-10} cm^3 s^{-1}) at low temperatures: the reactive surface(s) may be accessed directly from the lowest fine-structure level of the collision system. By contrast, reactions of atoms for which $j=0$ (and therefore $Q_m=0$) in the lowest fine-structure level (e.g. carbon, silicon) are expected to be slow or nonreactive ($k<10^{-12}$ cm^3 s^{-1}) at low temperatures: collisions originating from the lowest fine-structure level of separated reactants are not expected to reach the reactive surface. The implications for silicon chemistry are discussed by Langer and Glassgold in this volume.

The chemical balance of interstellar clouds is very sensitive to those few reactions that are fast at low temperature. Reactions (R1) and (R2) represent a major loss mechanism for radicals OH and CH; reactions of atomic oxygen with other polar species may have similar effects on the radical pool. Modifications of interstellar chemical models may be indicated. Further work on other radical reactions is necessary for assessing the importance of neutral reactions in the chemical kinetics.

REFERENCES

Graff, M.M., *Ap. J.*, **339**, 239 (1989).
Graff, M.M., and Wagner, A.F. 1990, *J. Chem. Phys.* , **92**, 2423.
Clary, D.C. 1984, *Molecular Physics*, **53**, 3.
Gentry, W.R., and Giese, C.F. 1977, *J. Chem. Phys.*, **67**, 2355.
Follmeg, B., Rosmus, P., and Werner, H.-J. (1987), *Chem. Phys. Letters*, **136**, 562.
Meerts, W.L., and Dymanus, A. 1973, *Chem. Phys. Letters*, **23**, 45.

Rate Constants of Ion-Molecule Reactions: Importance of Long-Range Forces

Kazuhiro Sakimoto

The Institute of Space and Astronautical Science, Yoshinodai, Sagamihara 229, Japan

I will summarize some recent theoretical work on low-temperature ion-molecule reactions. Particular emphasis is laid on the importance of long-range interaction. The low temperature limit of the rate constant and quantum mechanical effects are also discussed.

1. Introduction

Ion-molecule reactions are a very important process for molecular formation in interstellar clouds. In the last five years, many experimental studies have been done on ion-molecule reactions at temperatures much lower than 100K. Although the main purpose of these studies is the application to astrochemistry, the understanding of the mechanism of very-low temperature reactions is of fundamental interest.

In this report, I will summarize some theoretical work on ion-molecule reactions at low temperatures. I will discuss mainly the role of long-range interactions.

2. Capture (Langevin) Model

In order to calculate the rate constant of ion-molecule reactions, a capture model or so-called "Langevin" model is sometimes introduced. The idea of the Langevin model is as follows.

We assume that the reaction is exothermic and a polarization interaction $-a/(2R^4)$ is dominant between the ion and the molecule. Here, a is the polarizability of the molecule and R is the intermolecular distance. By using the polarization force, we calculate an orbiting impact parameter, which we label b_{orb}. Then, we assume that the reaction probability is unity if the impact parameter b is less than b_{orb}, and otherwise zero. In that case, the reaction rate constant is simply given by the Langevin expression,

$$k_L = 2\pi(a/m)^{1/2}. \tag{1}$$

Here, m is the reduced mass of the colliding system. (Atomic units are used unless otherwise stated.)

The Langevin rate constant has two important features. It is independent of temperature T and has a typical value of about $10^{-9} cm^3/s$. Although the Langevin model is based on a very simple idea, it explains many ion-molecule reaction processes.

3. Capture model for anisotropic interaction

The polarization force is not necessarily the most important long-range interaction. When the molecule possesses a dipole or quadrupole moment, of much more importance are ion-dipole or ion-quadrupole interactions. These interactions have a range longer than that of the polarization force. Hence, we expect that they may have a significant influence on the reaction. However, it is not easy to estimate the capture rate for these interactions because they are anisotropic. The strength of the interaction depends on the molecular orientation. That causes rotational excitation of the molecule. Thereby we must solve the collision dynamics including the molecular rotation to study the capture process.

3.1. Theoretical method

There are plenty of works that treat the capture model for anisotropic interactions. (See the references cited in [1,2].) Among early works, classical trajectory calculations of Dugan and his colleagues [3,4] gave an important contribution. In an average dipole orientation (ADO) model, an average potential approach was proposed [5]. A variational transition state theory (TST) was employed to estimate directly the capture rate [6,7]. An adiabatic method was introduced by Takayanagi [8]. Later, many other workers have started to use the adiabatic approach.

In this section, I will only briefly comment on the idea of the adiabatic method, which is especially powerful for low-energy collisions. The idea is just the same as that of the Born-Oppenheimer approximation employed in quantum chemistry. In the adiabatic method, we treat the rapid and the slow motions separately. In the present case, we have two kinds of time scale: the rotational period of the molecule, t_{rot}, which is roughly given by $1/B$ (B being the rotational constant), and the collision time $t_{coll} = r/v$, where r represents the interaction range and v is the collision velocity. For a dipole interaction, the collision time is roughly estimated by $t_{coll} = (D/B)^{1/2}/v$, where D is the dipole moment [8]. Here we use the so-called Massey criterion. We define a parameter s by the ratio t_{rot}/t_{coll}. The criterion says that when the parameter s is much smaller than unity, the excitation proceeds adiabatically. In that case, the probability of rotational excitation is very small [8-12].

Table I Collision energy corresponding to $s = 1$.

	CO	HCl	NH_3	H_2O
$E = mv^2/2$ * (for $s = 1$)	0.047eV	2.6	3.3	7.5

* $m = 5$ amu.

In the adiabatic case, we can introduce an adiabatic potential curve as in the case of quantum chemistry. The adiabatic potential curve is given as a function of the rotational state of the molecule. By regarding it as an effective isotropic potential, we can easily calculate the capture rate. For the calculation of the adiabatic potential curve, we can use a WKB method to save the labor of calculation [12-14]. Table I shows the collision energies for which the parameter s is equal to unity. Since the temperature in interstellar molecular clouds is much less than these values, the collisions there are almost adiabatic.

3.2. Scaling formula for thermal rate constant

We consider a thermal capture rate constant $\langle k \rangle$, which is given by assuming that the population of both the translational and the rotational states is in thermodynamic equilibrium.

For polar molecules, by using the variational TST, Chesnavich et al. [6] found that the thermal rate divided by the Langevin value is a function of only the parameter $x = D/(aT)^{1/2}$. Later, Su and Chesnavich [15] made a classical trajectory calculation, and obtained a universal curve for the relation $\langle k \rangle/k_L$ vs. x. This is a very useful finding because all the thermal capture rate can be deduced from the universal curve. Also for quadrupolar molecules, Bates and Mendas [11] obtained the same kind of scaling formula. Very recently, Turulski and Niedzielski [7] introduced a variational TST in which conservation of angular momentum is accurately taken into account. They found no difference in the results obtained by their method and by the classical trajectory calculation.

The above scaling formula was obtained by solving the dynamics using classical mechanics. Hence it is not evident whether the scaling formula is valid at very low temperature. The classical description of the molecular rotation is particularly questionable. The scaling formula was tested against the result of the adiabatic method and found satisfactory unless the temperature is extremely low [12]. The universal curve obtained by Su and Chesnavich overestimates the capture rate in the low temperature limit [2]. For symmetric-top molecules, the thermal capture rate somewhat differs from the universal curve [16] though the difference is not so serious.

The capture rate should provide an upper limit to the reaction rate since the reaction probability is usually less than unity. In this sense, the scaling formula gives a reasonable result as compared with low-temperature experiments made by Rowe's group for polar molecules [17]. In some cases, we have even a good agreement between the capture rate and experimental results [17,18]. (Dubernet and McCarroll [19] pointed out that there are some numerical problems with very low temperature results in ref.[18].) In the case of quadrupolar molecules, the capture rate differs depending on the sign of the quadrupole moment. The theory well explains the difference in the measured rate constants due to the sign of the quadrupole moment [13].

3.3. State-resolved capture rate

In interstellar clouds, the rotational levels may not be populated as in thermodynamic equilibrium. It depends on the density of the cloud. A critical density n_c is given by

$$n_c\, k_{de}(1 \to 0) = A(1 \to 0), \tag{2}$$

where $k_{de}(1 \to 0)$ is the rotational de-excitation rate, and $A(1 \to 0)$ is the transition probability for emitting a photon. Only if the density is much larger than the critical value can we assume thermodynamic equilibrium. Table II shows two critical densities. Use is made of the collision data [20].

Table II Critical density.

	CO	NH_3
n_c	$2.6 \times 10^3 cm^{-3}$	3.4×10^3

Since the cloud density is at most comparable to these critical values, we cannot necessarily assume thermodynamic equilibrium. Then, we need rate

constants for lower rotational states. The ground state is usually the most
important. For symmetric-top molecules, the state with the same value of J
and K is also important because this state is metastable (J and K being the
total rotational quantum number and its projection on the symmetry axis of the
molecule). Recently, Troe has provided an approximate analytical expression
for the state-resolved capture rate [2].

At very low temperatures, the thermal rate coincides with the rate for
the rotational ground state because only the ground state is populated (table
III). However, as the temperature increases, the difference between the two
rates becomes large. For example, the rate for the ground rotational state is
two times larger than the thermal rate at T = 10 K for HCN.

Table III k/k_L for NH_3.

(J,K)	T=0.5 K	5	10	50
(0,0)	12.6	11.9	11.2	7.5
(1,0)	4.7	4.5	4.6	4.8
(1,1)	17.9	8.4	6.9	5.4
thermal	12.6	11.8	9.5	4.3

For symmetric-top molecules, the rate for the state with the same value
of J and K behaves anomalously (table III). It becomes even larger than the
rate for the ground state at very low temperatures. For symmetric-top
molecule with a finite K, a long-range part of the adiabatic potential has a
form of $-1/R^2$. For this reason, the capture rate for a finite K takes a very
large value, and even diverges in the zero temperature limit [2,10]. Hence,
it is very important to know whether the state with the same value of J and K
(particularly J=K=1) is actually populated or not at very low temperatures.

4. Low temperature limit

We have reaction processes which cannot be explained in terms of the
capture model. However, an experiment for the reaction $O_2^+ + CH_4$, made by
Rowe et al. [21], shows that even if the reaction rate is largely different
from the Langevin value, it approaches the Langevin rate in the limit of low
temperatures. As the temperature goes to zero, the life time of the collision
complex should become infinite. Since we are considering exothermic reactions, the products have a finite kinetic energy while the reactants have no
kinetic energy. This means that the collision complex has only a branch
leading to the product channels. Hence the reaction probability should become
unity in the limit of zero temperature. It is highly probable that for any
exothermic reaction the total reaction rate should become the capture rate in
the limit of low temperatures.

Here, I will show the low temperature limit of the capture rate. For
the Langevin rate, the limit is trivial because the rate has no temperature
dependence. For anisotropic interactions, we can easily obtain the low temperature limit by using the adiabatic method [2,3,12,22].

For linear polar molecules, the limit is given by

$$k \longrightarrow k_L \times [1 + D^2/(3aB)]^{1/2}. \tag{3}$$

The rate has a finite value in the limit of zero temperature. Three examples
are shown in table IV. Even for a weakly polar molecule, carbon mono oxide,
the rate becomes more than two times larger than the Langevin value.

As I mentioned already, the rate diverges for symmetric-top molecule
with the same value of J and K. For example,

$$k(J=1,K=1) \longrightarrow k_L \times D/[3(2\pi aT)^{1/2}]. \tag{4}$$

The rate diverges as $T^{-1/2}$.

Table IV $k/k_L (T \longrightarrow 0)$.

	CO	HCl	HCN
k/k_L	2.52	8.78	61.6

For quadrupolar molecules, the rate is larger than the Langevin value at finite temperatures. However in the limit of low temperatures, the rate becomes just equal to the Langevin value.

Quantum mechanical effects are usually important in low energy collisions. The capture model based on classical mechanics, namely orbiting, may break down in the low energy limit. To see if the collision is classical or quantum mechanical, it is necessary to examine how many partial waves contribute to the orbiting. In the low energy limit, we can easily calculate the orbiting angular momentum as follows

$$l_L = (8m^2 aE)^{1/4} \qquad \text{for polarization interaction,} \tag{5}$$

$$l_D = [8m^2 D^2 E/(3B)]^{1/4} \quad \text{for dipole interaction,} \tag{6}$$

where E is the collision energy. Table V shows some examples.

Table V Orbiting angular momentum.

E	10 K	1	0.1	0.01
l_L	9	5	2.5	2
l_D	27	16	8	5

In calculating equations (5) and (6), I chose m = 2000au, a = 10au, D = 0.5au, B = 10^{-4}au. As we see from table V, if the energy considered in our study is as low as 1 K, we have a sufficiently large angular momentum. Hence, it is not unreasonable to evaluate the low temperature limit by using the capture model. However, this result does not mean that the quantum mechanical effects are always negligible. Of particular importance is a resonance effect. In the next section, I will show such an example.

5. $H^- + H \longrightarrow H_2 + e$

Recently, I have made a quantum mechanical calculation on associative detachment of $H^- + H$ by using a complex potential method [23]. In this method, associative detachment is expressed as an absorption due to the imaginary part of the complex potential. In the case of $H^- + H \longrightarrow H_2 + e$, it has been found that the reaction can be well explained in terms of the capture model. We do not have any resonances and other quantum mechanical effects.

I have also made a model calculation in which the magnitude of the imaginary part is artificially changed. The results are as follows. As long as the imaginary part is large, the reaction can be well explained in terms of the capture model. However, when the imaginary part is small, the reaction cross section is smaller than the capture one, and a potential-resonance effect becomes significant. As the imaginary part is reduced further, the resonances cause a drastic enhancement of the cross section. The importance of the resonances can be seen also in the radiative process in the collision

of H^+ + He [24]. The conclusion of this calculation is that when the reaction
rate is comparable to the capture rate, the resonances are negligible. However, when the reaction rate is much smaller than the capture rate, the resonance contribution can be significant.

Although it can be explained in terms of the capture model, the reaction
H^- + H --> H_2 + e has a cross section larger than the Langevin value in some
energy range [23]. In this system, there is no static dipole or quadrupole
interaction. One might expect no long-range forces other than the polarization interaction. However, in this case, the wavefunction of H^- is very
diffuse. Hence, electron exchange interaction becomes very strong. Actually,
an accurate ab initio calculation shows that a distant part of the interaction
is much more attractive than the polarization interaction [25].

References

[1]   K. Takayanagi, in "Physics of Electronic and Atomic Collisions" ed. by
       S. Datz (North-Holland Publishing Company, 1982) p.343.
[2]   J. Troe, J. Chem. Phys. **87** (1987) 2773.
[3]   J.V. Dugan and J.L. Magee, J. Chem. Phys. **47** (1967) 3103.
[4]   J.V. Dugan, Chem. Phys. Letters **21** (1973) 476, and references therein.
[5]   T. Su and M.T. Bowers, J. Chem. Phys. **58** (1973) 3027.
[6]   W.J. Chesnavich, T. Su, and M.T. Bowers, J. Chem. Phys. **72** (1980) 2641.
[7]   J. Turulski and J. Niedzielski, Mol. Phys. **67** (1989) 181.
[8]   K. Takayanagi, J. Phys. Soc. Japan **45** (1978) 976.
[9]   K. Sakimoto, J. Phys. Soc. Japan **50** (1981) 1668.
[10]   K. Sakimoto, Chem. Phys. **63** (1981) 419.
[11]   K. Sakimoto, J. Phys. B: At. Mol. Phys. **15** (1982) 1073.
[12]   K. Sakimoto, Chem. Phys. **85** (1984) 273.
[13]   D.R. Bates and I. Mendas, Proc. R. Soc. London Ser.A402 (1985) 245.
[14]   W.L. Morgan and D.R. Bates, Astrophys. J. **314** (1987) 817.
[15]   T. Su and W.J. Chesnavich, J. Chem. Phys. **76** (1982) 5183.
[16]   K. Sakimoto, Chem. Phys. Letters **116** (1985) 86.
[17]   B.R. Rowe, in "Rate Coefficients in Astrochemistry" ed. by T.J. Millar
       and D.A. Williams (Kluwer Academic Press, 1988) p.135.
[18]   D.C. Clary, D. Smith, and N.G. Adams, Chem. Phys. Letters **119** (1985)
       320; Astrophys. J. **296** (1985) L31.
[19]   M.L. Dubernet and R. McCarroll, Z. Phys. D13 (1989) 255.
[20]   D.R. Flower, Phys. Rep. **174** (1989) 1.
[21]   B.R. Rowe, G. Dupeyrat, and J.B. Marquette, J. Chem. Phys. **80** (1984)
       241.
[22]   D. Hyatt and L. Stanton, Proc. Roy. Soc. A318 (1970) 107.
[23]   K. Sakimoto, Chem. Phys. Letters **164** (1989) 294.
[23]   B. Zygelman, A. Dalgarno, M. Kimura, and N.F. Lane, Phys. Rev. A40
       (1989) 2340.
[25]   J. Senekowitsch, P. Rosmus, W. Domcke, and H.J. Werner, Chem. Phys.
       Letters **111** (1984) 211.

Ab Initio Study of the $NH_3^+ + H_2$ Reaction

D. J. DeFrees[1,2], D. Talbi[1], F. Pauzat[1], W. Koch[2], and A. D. McLean[2]

[1] *Molecular Research Institute, 845 Page Mill Road, Palo Alto, CA 94304, U. S. A.*
[2] *IBM Almaden Research Center, 650 Harry Road, San Jose, CA 95120-6099, U. S. A.*

Abstract The reaction of ammonia cations, NH_3^+, with molecular hydrogen, H_2, is a key step in the gas-phase synthesis of ammonia in dense interstellar clouds. The unusual dependence of the rate of this reaction on temperature (the rate is at a minimum around 100 K and it increases as the temperature is either raised or lowered) has been explained *via* a two-step reaction mechanism. Key to the explanation are formation of an ion-molecule complex in the entrance channel and a subsequent barrier to reaction. *Ab initio* molecular orbital theory has been used to confirm the qualitative features of this hypothesis. The calculated structures, vibrational frequencies, and energies will be used to determine if the surface explains the temperature dependence.

Several reaction schemes have been proposed for the gas-phase synthesis of interstellar ammonia (for example, Herbst and Klemperer 1973; Dalgarno 1974; Herbst, DeFrees, and McLean 1987). In many of these the final two steps are the production of NH_4^+ and its destruction *via* dissociative recombination yielding ammonia:

$$NH_3^+ + H_2 \rightarrow NH_4^+ + H \tag{1}$$

$$NH_4^+ + e^- \rightarrow NH_3 + H \tag{2}$$

Reaction [1] has been observed to be slow at room temperature even though it is exothermic (Fehsenfeld, Lindinger, Schmeltekopf, Albritton, and Ferguson 1975). This observation has inspired many studies of the temperature dependence of the rate of [1] including examinations of the effects of substituting D for H (Smith and Adams 1981; Adams and Smith 1984; Luine and Dunn 1985; Böhringer 1985; Barlow and Dunn 1987). These studies showed an unusual temperature dependence: the rate of the reaction decreases as the temperature is lowered from 800 K to approximately 100 K; this is consistent with an exothermic reaction with a potential energy barrier. However, further decreases in temperature (measurements have been made down to 10 K) result in an increase in the reaction rate. (Extrapolating from the early high temperature data, models of gas phase chemistry in dense interstellar clouds used reaction rates that turned out to be too low by several orders of magnitude.) Similar temperature dependence has been observed for the exothermic ion-molecule reaction of O_2^+ with CH_4 (Rowe, Dupeyrat, Marquette, Smith, Adams, and Ferguson 1984).

These observations have been explained with a two-step mechanism and a qualitative reaction coordinate picture (Barlow and Dunn 1987):

$$NH_3^+ + H_2 \quad \underset{k_b}{\overset{k_c}{\underset{\leftarrow}{\rightarrow}}} \quad [NH_3^+ \cdot H_2] \tag{3}$$

$$[NH_3^+ \cdot H_2] \quad \overset{k_f}{\underset{[NH_3^+ \cdot H \cdot H]^\dagger}{\rightarrow}} \quad NH_4^+ + H \qquad\qquad [4]$$

In simplified terms, the temperature dependance of the reaction rate is postulated to be due to quantum mechanical tunnelling through the potential energy barrier, $[NH_3^+ \cdot H \cdot H]$, separating the complex, $[NH_3^+ \cdot H_2]$, and products, $NH_4^+ + H$. The tunneling rate increases as the temperature is decreased below 100K because the lifetime of the complex increases. Above 100K the rate increases because the number of states of the complex which exceed the height of the barrier increases.

To test this hypothesis for the mechanism of [1] we have performed *ab initio* molecular orbital calculations. The result of these calculations is a quantitative reaction coordinate picture of the reaction, Figure 1. We have determined the geometric structures of the reactants, products, intermediates, and transition structure for the reaction along with their energies and vibrational frequencies. These will be used ultimately to compute the reaction rate as a function of temperature for comparison with observed rates (Herbst, DeFrees, Talbi, Pauzat, Koch, and McLean 1990). However, even without this, the molecular orbital study of the potential energy surface provides a crucial test of the proposed mechanism.

For a general treatise on *ab initio* molecular orbital theory, and a description of the molecular orbital models used in this study, consult Hehre, Radom, Schleyer, and Pople (1986). We have used the GAUSSIAN 86 (Frisch *et al.* 1986) and GAUSSIAN 88 (Frisch *et al.* 1988) computer programs to perform the calculations. The isoelectronic potential energy surface, CH_5, has been extensively studied and the results of these investigations provided a useful starting point for the study of NH_5^+ (Weston and Ehrenson 1971; Morokuma and Davis 1972; Walch 1980; Gordon, Bano, and Boatz 1983).

The potential energy surface was initially explored at the UHF/6-31G(d,p) level of theory. The geometry of each of the reactants was optimized (Table 1) and an extensive mapping of the surface, obtained by exploring many approaches of the reactant molecules toward each other and of the product molecules toward each other, was performed. In the entrance channel is an ion-molecule complex, $NH_3^+ \cdot H_2$, where the hydrogen molecule is perpendicular to the planar NH_3^+ moiety. This is followed on the reaction coordinate by an early transition structure, $NH_3^+ \cdot H \cdot H$, which was found to have C_{3v} symmetry with an H_2 lined up with the three-fold symmetry axis. In the exit channel is an ion-molecule complex $NH_4^+ \cdot H$ with the hydrogen atom along one of the NH bonds. Extensive studies of the potential energy surface using brute-force, manual, and automated techniques (Baker 1986; Baker 1987; Cerjan and Miller 1981; Simons, Jørgenson, Taylor, and Ozment 1983; Banerjee, Adams, Simon, and Shepard 1985) fail to locate any additional, low-energy paths from reactants to products. None of the wavefunctions for these molecules were significantly spin contaminated, the highest value for $\hat{S}^2$ is 0.78, compared to 0.75 for a pure doublet. The character of each structure (minimum-energy structure or transition structure) was confirmed by computing the harmonic vibrational frequencies. These results are qualitatively confirmed at the MP2/6-31G(d,p) level of theory which includes electron correlation effects on the wavefunction. Geometries, energies, and vibrational frequencies computed with this theoretical model are given in Tables 1 and 2.

The relative energies of the five stationary points under discussion on the NH_5^+ potential energy surface were then computed using higher levels of theory. Basis sets were improved from split valence polarized 6-31G(d,p) to triple split valence 6-311 + + G(3df,3pd) (Krishnan, Binkley, Seeger, and Pople 1980; Clark, Chandrasekhar, Spitznagel, and Schlyer 1984; Frisch, Pople, and Binkley 1984) with SCF, MP2, and MP4 wavefunctions. Relative to a reactant $NH_3^+ + H_2$ energy of 0.0 kcal mol^{-1}, the loose $NH_3^+ \cdot H_2$ complex is at -3.2 kcal mol^{-1} and the transition state at 2.2 kcal mol^{-1} in our best MP4 calculations with the largest basis set; these "E" values are reported in Table 3. The comparison with MP2 results indicates satisfactory convergence relative to inclusion of electron correlation; comparison with the SCF results shows the necessity for including electron correlation — the SCF wavefunction is totally inadequate. Our basis set studies show that these results are converged with respect to basis set.

These relative energies are small; because they are key to confirming Barlow and Dunn's proposed mechanism it is necessary to adjust them for basis set superposition error (BSSE) (Boys and Bernardi 1970, Liu and McLean 1973, 1989). For example, the energy difference between transition state and reactants is computed in two steps; first from transition state to reactants frozen at the transition state geometry in calcu-

Table 1. Geometries and energies computed with ab initio molec- ular orbital theory[a]		
Molecule	**UHF/6-31G(d,p)**	**MP2/6-31G(d,p)**
H	E = -0.49823	E = -0.49823
H_2 $D_{\infty h}$	r(HH) = 0.733 E = -1.13133	r(HH) = 0.734 E = -1.15766
NH_3^+ D_{3h}	r(NH) = 1.012 E = -55.88489	r(NH) = 1.020 E = -56.03232
NH_4^+ T_d	r(NH) = 1.012 E = -56.54553	r(NH) = 1.022 E = -56.73765
$NH_3^+ \cdot H_2$[b] C_{2v}	r(NH1) = 1.014 r(NH) = 1.011 r(H1X) = 2.150 r(XH2) = 0.368 ∠(HNH1) = 120.1 E = -57.01828	r(NH1) = 1.023 r(NH) = 1.020 r(H1X) = 1.988 r(XH2) = 0.369 ∠(HNH1) = 120.2 E = -57.19296
$NH_3^+ \cdot H \cdot H$[c] C_{3v}	r(NH1) = 1.422 r(NH) = 1.010 r(HH1) = 0.843 ∠(HNH1) = 102.8 E = -56.99334	r(NH1) = 1.584 r(NH) = 1.019 r(HH1) = 0.784 ∠(HNH1) = 99.3 E = -57.18283
$NH_4^+ \cdot H$[d] C_{3v}	r(NH1) = 1.012 r(NH) = 1.012 r(HH1) = 2.423 ∠(HNH1) = 109.5 E = -57.04449	r(NH1) = 1.023 r(NH) = 1.022 r(HH1) = 2.274 ∠(HNH1) = 109.5 E = -57.23691

[a]Distances are given in Ångstroms, angles in degrees, and energies in Hartrees.

[b]H_2 is perpindicular to the plane of NH_3^+ with its midpoint, X, aligned with one of the NH bonds, N-H1. H2 is one of the hydrogens of the H_2.

[c]H_2 is on the three-fold symmetry axis; H1 is the hydrogen of H_2 closer to the nitrogen.

[d]The hydrogen is aligned with the N-H1 bond of NH_4^+.

lations where both transition state and reactants are computed in the same "supermolecule" transition state basis, second from reactants at the transition state geometry to reactants at their relaxed geometries in calculations in the reactant bases alone. The energy of the $NH_3^+ \cdot H_2$ loose complex will be adjusted in a parallel manner. The adjustments change the relative energies to -3.0 kcal mol^{-1} for the $NH_3^+ \cdot H_2$ complex and to 2.8 kcal mol^{-1} for the transition state, the E_0^c values of Table 3.

Our calculations are of high enough quality that we can assign error bars of ± 0.2 kcal mol^{-1} and ± 0.4 kcal mol^{-1} respectively for these energies. The energy of the zero point nuclear vibrations can now be added to give the energies E_0^c of Table 3, and shown in Figure 1. The zero-point energies are computed from the harmonic frequencies of Table 2, scaled by the the empirical factor 0.945 shown by DeFrees and McLean (1985) to give much improved agreement with experiment, although still requiring a small increase in estimated error bars. Our final, and best, estimate for E_0^c is -1.4 $\pm$ 0.3 kcal mol^{-1} for the $NH_3^+ \cdot H_2$ complex and 4.8 $\pm$ 0.5 kcal mol^{-1} for the transition state. We estimate that hν for vibrations along the reaction coordinate (the lowest a_1 frequency of $NH_3^+ \cdot H_2$ in Table 2) is 0.7 kcal mol^{-1}, indicating that certainly one and possibly two quanta of dissociative vibrational energy can be added before the loose complex breaks up into reactants.

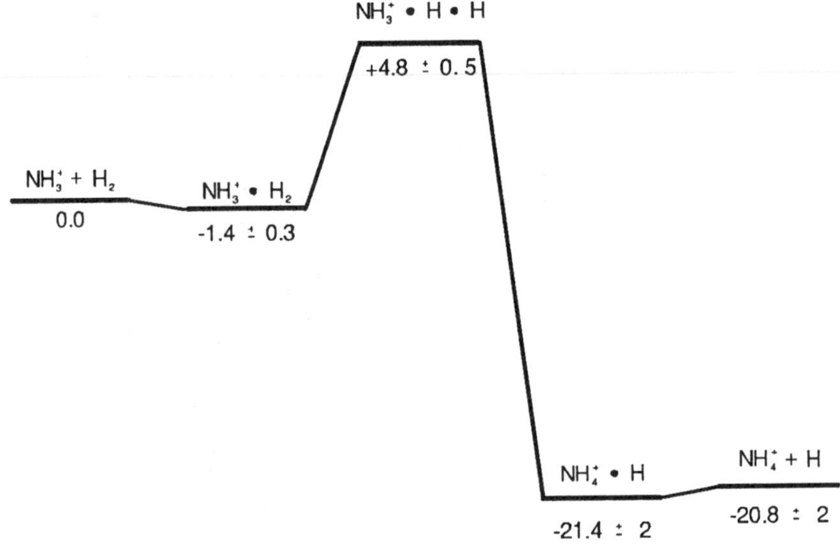

Figure 1. Reaction coordinate diagram.. The energies in kcal mol^{-1} were computed with the MP4/6-311 + + G(3df,3pd)//MP2/6-31G(d,p) theoretical model and inlclude zero-point vibrational energy and a correction for basis set superposition error.

Table 2. MP2/6-31G(d,p) harmonic vibrational frequencies	
Molecule	**MP2/6-31G(d,p) frequencies, cm^{-1}**
H$_2$, D$_{\infty h}$	σ_g: 4610
NH$_3^+$, D$_{3h}$	a'_1: 3462 a''_2: 881 e': 1598, 3672
NH$_4^+$, T$_d$	a_1: 3440 e: 1769 t_2: 1530, 3598
NH$_3^+ \cdot$H$_2$, C$_{2v}$	a_1: 264, 1596, 3440, 3632, 4528 a_2: 68 b_1: 257, 495, 913 b_2: 194, 1612, 3676
NH$_3^+ \cdot$H$\cdot$H, C$_{2v}$	a_1: 979i, 1052, 3253, 3480 e: 486, 822, 1614, 3680
NH$_4^+ \cdot$H, C$_{3v}$	a_1: 209, 1528, 3434, 3578 e: 131, 1534, 1770, 3600

Table 3. NH_5^+ energies (kcal mol⁻¹) relative to $NH_3^+ + H_2$. E is our computed energy with the 6-311++G(3df,3pd) basis at the optimized MP2/6-31G(d,p) geometries. E^c includes basis set superposition error (BSSE) adjustment where appropriate. E_0 includes scaled MP2/6-31G(d,p) zero-point vibrational energies.

Molecule		SCF	MP2	MP4
$NH_3^+{\cdot}H_2$	E	-2.2	-3.1	-3.2
	E^c	-2.1	-2.9	-3.0
	E_0^c	-0.5	-1.3	-1.4
	Best estimate	-1.4 ± 0.3		
$NH_3^+{\cdot}H{\cdot}H$	E	11.8	2.6	2.2
	E^c	12.1	3.2	2.8
	E_0^c	14.1	5.2	4.8
	Best estimate	4.8 ± 0.5		
$NH_4^+{\cdot}H$	E	-18.1	-30.6	-26.0
	E^c	-18.0	-30.5	-25.9
	E_0^c	-13.5	-25.5	-21.4
	Best estimate	-21.4 ± 2		
$NH_4^+ + H$	E	-17.3	-29.4	-24.7
	E_0	-13.4	-25.5	-20.8
	Experiment	-19 ± 3		
	Best estimate	-20.8 ± 2		

In the second half of the reaction, from transition state to the $NH_4^+ + H$ products, the H_2 dissociates and transfers a H to NH_3^+; the geometry and electronic changes are much greater than in going from reactants to transitions state and the energies in Table 3 for the $NH_4^+{\cdot}H$ products, relative to reactants, may have a significant error bar. Based on comparisons of proton affinities (DeFrees and McLean 1987) and heats of formation (Pople, Frisch, Luke, and Binkley 1983; Binkley and Frisch 1983; Pople, Luke, Frisch, and Binkley 1985) computed at this level of theory, we estimate the error to be ± 2 kcal mol⁻¹. This is supported by the experimental value for ΔH^0 (0 K) for reaction [1] of -19 ± 3 kcal mol⁻¹. (Heats of Formation for H and H_2 were taken from Chase, Davies, Downey, Frurip, McDonald, and Syverud 1985; Heats of Formation for NH_3^+ and NH_4^+ were taken from Lias, Bartmess, Liebman, Holmes, Levine, and Mallard 1988 and Bartmess 1990 and were corrected from 298 K to 0 K by assuming the difference is the same as for the isoelectronic species CH_3 and CH_4 given in Chase et al.) The BSSE adjustment to the energy separation between $NH_4^+{\cdot}H$ and products involves product molecule calculations in the $NH_4^+{\cdot}H$ supermolecule basis, followed by relaxation energy calculations, as discussed previously. The energies shown in Figure 1 are our best estimates of relative zero-point energy levels at the five stationary points on the surface.

Calculations in progress (Herbst, DeFrees, Talbi, Pauzat, Koch, and McLean 1990) will use this surface to determine the temperature dependence of the reaction rate to give a quantitative comparison with experiment. At this point, we conclude by saying that the features of the surface confirm the qualitative aspects of the Barlow and Dunn mechanism.

REFERENCES

Adams, N. G., and Smith, D. 1984, *Int. J. Mass Spectrom. Ion Processes*, **61**, 133.

Baker, J. 1986, *J. Comp. Chem.*, **7**, 385.

Baker, J. 1987, *J. Comp. Chem.*, **8**, 563.

Banerjee, A., Adams, N., Simons, J., and Shepard, R. 1985, *J. Phys. Chem.*, **89**, 52.

Barlow, S. E., and Dunn, G. H. 1987, *Int. J. Mass Spectrom. Ion Processes*, **80**, 227.

Bartmess, J. E. 1990, private communication.

Binkley, J. S., and Frisch, M. J. 1983, *Int. J. Quant. Chem. Symp.*, **17**, 331.

Böhringer, H. 1985, *Chem. Phys. Letters*, **122**, 185.

Boys, S. F. and Bernardi, F. 1970, *Mol. Phys.* **19**, 553.

Chase, M. W. Jr., Davies, C. A., Downey, J. R. Jr., Frurip, D. J., McDonald, R. A., and Syverud, A. N. 1985, *JANAF Thermochemical Tables, Third Edition*, J. Phys. Chem. Ref. Data 14(S1).

Clark, T., Chandrasekhar, J., Spitznagel, G. W., Schlyer, P. v. R. 1983, *J. Comp. Chem.*, **4**, 294.

Cerjan, C. J., and Miller, W. H. 1981, *J. Chem. Phys.*, **75**, 2800.

Dalgarno, A. 1974, in Proceedings of the First NATO Conference on Ion-Molecule Interactions, Biarritz, France, June 24-July 6.

DeFrees, D. J., and McLean, A. D. 1985, *J. Chem. Phys.*, **82**, 333.

DeFrees, D. J., and McLean, A. D. 1987, *J. Comp. Chem.*, **7**, 321.

Fehensfield, F. C., Lindinger, W., Schmeltekopf, A. L., Albritton, D. L., and Ferguson, E. E. 1975, *J. Chem. Phys.*, **62**, 2001.

Frisch, M. J., Binkley, J. S., Schlegel, H. B., Raghavachari, K., Melius, C. F., Martin, R. L., Stewart, J. J. P., Bobrowicz, F. W., Rohlfing, C. M., Kahn, L. R., DeFrees, D. J., Seeger, R., Whiteside, R. A., Fox, D. J., Fluder, E. M., Pople, J. A. 1986, *Gaussian 86*, Carnegie-Mellon Quantum Chemistry Publishing Unit, Carnegie-Mellon University, Pittsburgh PA 15213

Frisch, M. J., Head-Gordon, M. Schlegel, H. B., Raghavachari, K., Binkley, J. S., Gonzalez, C., DeFrees, D. J., Fox, D. J., Whiteside, R. A., Seeger, R., Melius, C. F., Baker, J., Kahn, L. R., Stewart, J. J. P., Fluder, E. M., Topiol, S., and Pople, J. A. 1988, *Gaussian 88* Gaussian, Inc., Pittsburgh PA.

Frisch, M. J., Pople, J. A., and Binkley, J. S 1984, *J. Chem. Phys.*, **80**, 3265.

Gordon, M. S., Gano, D. R., and Boatz, J. A. 1983, *J. Amer. Chem. Soc*, **105**, 5771.

Hehre, W. J., Radom, L., Schleyer, P. v. R., and Pople, J. A. 1986, *Ab Initio Molecular Orbital Theory*, (New York: Wiley).

Herbst, E., DeFrees, D. J., and McLean, A. D. 1987, *Ap. J.*, **321**, 898.

Herbst, E., DeFrees, D. J., Talbi, D., Pauzat, F., Koch, W., and McLean, A. D. 1990, in preparation.

Herbst, E., and Klemperer, W. 1973, *Ap. J.*, **185**, 505.

Krishnan, R., Binkley, J. S, Seeger, R., and Pople, J. A. 1980, *J. Chem. Phys.*, **72**, 650.

Lias, S. G., Bartmess, J. E., Liebman, J. F., Holmes, J. L., Levin, R. D., and Mallard, W. G. 1988, *Gas-Phase Ion and Neutral Thermochemistry*, J. Phys. Chem. Ref. Data 17(S1).

Liu, B., and McLean, A. D. 1973, *J. Chem. Phys.* **59**, 4557.

Liu, B., and McLean, A. D. 1989, *J. Chem. Phys.* **91**, 2348.

Luine, J. A., and Dunn, G. H. 1985, *Ap. J.*, **299**, L67.

Morokuma, K., and Davis, R. E. 1972, *J. Amer. Chem. Soc.*, **94**, 1060.

Pople, J. A., Frisch, M. J., Luke, B. T., and Binkley, J. S. 1983, *Int. J. Quant. Chem. Symp.*, **17**, 307.

Pople, J. A., Luke, B. T., Frisch, M. J., and Binkley, J. S. 1985, *J. Phys. Chem.*, **89**, 2198.

Rowe, B., Dupeyrat, G., Marquette, J. B., Smith, D., Adams, N. G., and Ferguson, E. E. 1984, *J. Chem. Phys.*, **80**, 241.

Simons, J., Jørgenson, P., Taylor, H., and Ozment, J. 1983, *J. Phys. Chem.*, **87**, 2745.

Smith, D., and Adams, N. G. 1981, *M. N. R. A. S.*, **197**, 377.

Walch, S. P. 1980, *J. Chem. Phys.*, **72**, 4932.

Weston, Jr., R. E., and Ehrenson, S. 1971, *Chem. Phys. Letters*, **9**, 351.

Ion-Molecule Radiative Association

A. D. Sen[1], V. G. Anicich[1], and M. J. McEwan[2]

[1] *Jet Propulsion Laboratory, CIT, Pasadena, CA, U. S. A.*
[2] *University of Canterbury, Christchurch, New Zealand*

Results of a number of ion–molecule association reactions are presented. Bimolecular and termolecular association rate coefficients have been measured for a number of ion–molecule system of relevance in interstellar cloud chemistry by ion cyclotron resonance spectrometry. Rapid radiative association rates ($> 1 \times 10^{-10}$ cm^3 . s^{-1}) and collisional association rates ($> 1 \times 10^{-22}$ cm^6 . s^{-1}) were observed. Lower limits to the average unimolecular dissociation and radiative decay lifetimes of complexes have been determined.

Ion–molecule radiative association is an important process in the gas–phase ion chemistry of interstellar clouds. The process has been utilized to explain the occurrence and abundances of many complex molecules observed in the clouds. Laboratory measurements of radiative association rate coefficients are hard to measure and only few systems have been studies in any detail (1). Radiative association reactions proceed with the formation of an excited complex. The complex can dissociate back to the reactants at a rate k_d, or can emit a photon and radiatively stabilize at a rate A_r. There is a competing collisional stabilization channel where a third body collides with the complex and removes its excess energy. The stabilization rate is k_s and the stabilization efficiency is β. The details of the treatment can be found elsewhere (2,3).

Bimolecular and termolecular rate coefficients ($k_{rad.}$ and k_3) were measured with ion cyclotron resonance (ICR) spectrometers (2,4,5). Upper limits to k_d and A_r were determined. Unimolecular dissociation (τ_d) and radiative decay (τ_r) lifetimes are the reciprocals of k_d and A_r respectively.

Over the last decade a number of ion–molecule systems exhibiting radiative association reactions of interest in interstellar chemistry have been investigated in detail in our laboratory. The measured bimolecular and termolecular association rates are shown in Table I. Stabilization efficiencies of Helium are also presented. These measurements allow $k_{rad.}$ to be determined indirectly from the appearance of a bimolecular productions for AB^+ at low pressures. It is assumed that the bimolecular process is radiative stabilization.

TABLE I
RADIATIVE AND COLLISIONAL ASSOCIATION RATES

Association Reactions	$k_{rad.}$ (x 10^{-10} cm^3.s^{-1})	$k_3(\beta)$ (x 10^{-23} cm^6.s^{-1})	$\beta_{rel.}$(He)
$CH_3^+ + HCN$	2.0	1.1	0.16
$C_4H_2^+ + C_2H_2$	2.7	5.7	0.38
$C_4H_3^+ + C_2H_2$	2.0	1.3	0.37
$HC_3N^+ + HC_3N$	0.65		
$HC_5N^+ + HC_3N$	5.0	11.6	0.42
$CH_3^+ + CH_3CN$	0.9	19.0	0.53

Association rates for all the reactions shown above are large, especially the termolecular rates. This result is not surprising since these reactions are saturated in drift tube experiments and proceed at almost their capture limit. The reactions are also highly exothermic (- $\Delta H_f \geq 3$ eV).

CH_3^+ + HCN was the first association reaction to be studied in our laboratory (6).

$$CH_3^+ + HCN \longrightarrow CH_3CNH^+ + h\nu \tag{1}$$

Later experiments (7) found the radiative association rate to be much smaller ($< 5 \times 10^{-12}$ cm^3.s^{-1}). On repeating the experiment recently, the loss of CH_3^+ in reaction 1 can only be explained by a rapid bimolecular rate coefficient. The discrepancy is believed to be resolved, but few more confirming results are needed.

$C_2H_2^+/C_2H_2$ system exhibited fast association reactions by the secondary ions $C_4H_2^+$ and $C_4H_3^+$ (8).

$$C_4H_2^+ + C_2H_2 \longrightarrow C_6H_4^+ + h\nu \tag{2}$$

$$C_4H_3^+ + C_2H_2 \longrightarrow C_6H_5^+ + h\nu \tag{3}$$

where $C_4H_2^+$ and $C_4H_3^+$ are products of the reaction $C_2H_2^+ + C_2H_2$ (9). $C_4H_2^+$ and $C_4H_3^+$ are formed with excess internal energy. At low pressures, there is no bimolecular association until the ions cool by collisions with C_2H_2. This causes a delay in the appearance of the adduct ions in reactions 2 and 3. $C_4H_2^+$ requires 4 quenching collisions, while $C_4H_3^+$ requires 5 quenching collisions before they can react by association.

A detailed study is in progress on the reactions of HC_3N^+ with molecules found in interstellar clouds. The dimer ion formation via radiative association (9)

$$HC_3N^+ + HC_3N \longrightarrow H_2C_6N_2^+ + h\nu \tag{4}$$

is a minor channel in the reaction of HC_3N^+ and HC_3N. Work is currently in progress to measure k_3 for the channel.

The prominent association reaction in the system is that of HC_5N^+ (3). Reaction 5 shows the reaction of HC_5N^+ and HC_3N.

$$HC_5N^+ + NC_3N \longrightarrow H_2C_8N_2^+ + h\nu \tag{5}$$

Both the two and three body rates are very fast. HC_5N^+ is the major product channel of the reaction $HC_3N^+ + HC_3N$. It too is formed with excess internal energy and requires 4–5 collisions to relax prior to the onset of bimolecular association channel. The characteristic delay in the onset of $H_2C_8N_2^+$ production is also observed.

The association reaction of methyl ion and acetonitrile (2) seems to follow the pattern exhibited by the other system, competition between bimolecular and termolecular association.

$$CH_3^+ + CH_3CN \longrightarrow CH_3CNCH_3^+ + h\nu \tag{6}$$

TABLE II

AVERAGE LIFETIMES AND RADIATIVE RATES

Collision Complex	τ_d (μs)	τ_r (ms)	A_r (s^{-1})
$(CH_3CNH^+)^*$	0.76	0.013	7.7×10^4
$(C_6H_4^+)^*$	47.1	0.22	4.6×10^3
$(C_6H_5^+)^*$	17.0	0.08	1.3×10^4
$(H_2C_8N_2^+)^*$	15.0	0.08	1.3×10^4
$(CH_3NCCH_3^+)^*$	8.6	0.7	1.4×10^3

The unimolecular dissociation lifetimes of all the systems studied are very large > 1 μs. The long lifetime is responsible for large value of k_3 observed in these reactions. Since the complexes are able to survive for such long times, collisional stabilization is efficient.

Fast bimolecular association rate coefficients are the result of long unimolecular decay lifetimes. The decay rates in the systems that we have studied are consistant with those measured by the PIPECO technique (10). Long unimolecular dissociation lifetimes result in fast termolecular association rate coefficients. The long lifetimes are a consequence of deep potential wells for the stabilized complexes with respect to the reactant energies and many modes in which energy may be stored. The combination of long τ_d and comparable τ_r enables radiative stabilization to be competitive with collisional stabilization.

This paper represents the results of one phase of research carried out in the Jet Propulsion Laboratory, California Institute of Technology, sponsored by the National Aeronautics and Space Administration.

References

1. Bates, D.R. and Herbst, E. 1988, in Rate Coefficients in Astrochemistry, eds. T.J. Millar and D.A. Williams, (Dordrecht:Kluwer Academic Press).

2. McEwan, M.J., Denison, A.B., Huntress, W.T., Jr., Anicich, V.G., Snodgrass, J., and Bowers, M.T. 1989, J. Phys. Chem., **93**, 4064.

3. Sen, A.D. 1989, Ph.D. Thesis, University of Wyoming.

4. McEwan, M.J., Denison, A.B., Anicich, V.G., and Huntress, W.T., Jr. 1987, Int. J. Mass Spectrom. Ion Processes, **81**, 247.

5. Anicich, V.G., Huntress, W.T., Jr., and McEwan, M.J. 1986, J. Phys. Chem., **90**, 2446.

6. McEwan, M.J., Anicich, V.G., Huntress, W.T., Kemper, P.R., and Bowers, M.T. 1980, Chem. Phys. Letters, **75**, 278.

7. Kemper, P.R., Bass, L.M., and Bowers, M.T. 1985, J. Phys. Chem., **89**, 1105.

8. Anicich, V.G., Sen, A.D., Huntress, W.T., Jr., and McEwan, M.J. 1989, submitted to J. Chem. Phys.

9. Sen, A.D., Anicich, V.G., and McEwan, M.J. 1990, to be published.

10. Baer, T. 1986, in Adv. Chem. Phys., eds. I. Prigogine and S.A. Rice (New York: Wiley Interscience), **64**, 111.

Gas-Phase Reactions of $SiC_6H_6^+$ and $SiC_{10}H_8^+$ with Simple Molecules: A Possible Role for Polycyclic Aromatic Hydrocarbons in the Synthesis of Interstellar Molecules

Stanislaw WLODEK, Henryk WINCEL, and Diethard K. BOHME

Department of Chemistry and Centre for Research in Earth and Space Science, York University, North York, Ontario, Canada M3J 1P3, and the Institute of Physical Chemistry, Polish Academy of Sciences, ul. Kasprzaka 44/52, 01-224 Warszawa, Poland

Results of laboratory measurements are presented for the capture of Si^+ by benzene and naphthalene in the gas phase and for the kinetics of the reactions of Si^+.benzene and Si^+.naphthalene adduct ions with the molecules D_2, CO, N_2, O_2, H_2O, NO_2, NH_3, C_2H_2 and C_4H_2 at 297 K. Significant differences are seen for the reactions with C_2H_2 and C_4H_2. The production of ionized naphthalene observed for the reactions with the Si^+.naphthalene adduct ion is suggestive of a novel chemical role for polycyclic aromatic hydrocarbons (PAHs) in the synthesis of interstellar molecules. The reactions can be generalized as follows (where X may be Si or some other atom):

$$XPAH^+ + M \longrightarrow PAH^+ + XM$$

It is proposed that these reactions proceed as "surface" reactions in which the neutral atom X interacts with the incoming molecule M above the plane of the charged PAH molecule.

INTRODUCTION

The growing evidence for the presence of PAH molecules as abundant species in interstellar environments (1,2) has caused considerable interest in the physical and chemical consequences of that presence. For example, Omont has proposed that PAH molecules are the major sink for atomic metal and silicon cations (3). There is no previous experimental evidence for additions of this kind and the subsequent chemistry of the resulting adduct ions. Here we report laboratory data for the capture of atomic silicon ions by naphthalene, the simplest PAH molecule, and benzene, a possible precursor of naphthalene, and present gas-phase kinetics for reactions of the resulting Si^+.naphthalene and Si^+.benzene adduct ions with selected interstellar molecules.

EXPERIMENTAL

All experiments were performed with a SIFT apparatus described previously (4,5). Ground-state Si^+ ions were generated from $Si(CH_3)_4$ by electron impact, mass selected, and introduced into a flowing helium buffer at 0.35 Torr and 297 K. Benzene or naphthalene vapour was introduced into the flow tube upstream of the reaction region. Rapid association reactions were observed to convert the atomic silicon ions into adduct ions according to reactions [1] and [2]. Some charge

$$Si^+ + C_6H_6 \longrightarrow SiC_6H_6^+ \tag{1}$$

$$Si^+ + C_{10}H_8 \longrightarrow SiC_{10}H_8^+ \tag{2a}$$

$$\longrightarrow C_{10}H_8^+ + Si \tag{2b}$$

transfer was observed with naphthalene which has an ionization energy (8.14 eV) slightly lower than that of atomic silicon (8.152 eV). Additions of neutral reactant molecules downstream into the reaction region led to reactions with the adduct ions for which the kinetics was explored in the usual manner (4,5).

RESULTS

Table 1. summarizes the rate constants, products and product distributions determined for the reactions of the adduct ions $SiC_6H_6^+$ and $SiC_{10}H_8^+$ with the molecules D_2, CO, N_2, O_2, H_2O, NH_3, C_2H_2 and C_4H_2. Some of the reactions with $SiC_{10}H_8^+$ have been reported previously (6). Inspection of the results in Table 1. leads to the following observations:

1. Both $SiC_6H_6^+$ and $SiC_{10}H_8^+$ are unreactive with D_2, CO and N_2. This means that these adduct ions will persist in interstellar regions rich in hydrogen and CO with other molecules contributing to their chemical destruction.

2. Molecular oxygen slowly eliminates SiO_2 from $SiC_6H_6^+$ and $SiC_{10}H_8^+$. About 10% of the reaction with the benzene adduct leads to the formation of silicon monoxide and what is likely to be the phenol cation, $C_6H_5OH^+$.

3. Water eliminates $SiOH_2$ from both adduct ions as well as SiOH from the benzene adduct.

4. Ammonia efficiently bonds to both adduct ions.

5. Acetylene and diacetylene both release ionized naphthalene from $SiC_{10}H_8^+$ as do O_2 and H_2O, but show a distinctly different mode of reactivity with $SiC_6H_6^+$. Acetylene forms an adduct ion with $SiC_6H_6^+$ in competition with H-atom elimination while diacetylene displaces the benzene molecule from $SiC_6H_6^+$.

Table 1. Kinetic data obtained for reactions of $SiC_6H_6^+$ and $SiC_{10}H_8^+$ at 297 K in helium buffer gas at 0.35 Torr. B.R. is the branching ratio with an estimated accuracy of ±30%. The rate constant, k, is in units of 10^{-10} cm^3 molecule^{-1} s^{-1} and also has an estimated accuracy of ±30%.

Molecule	$SiC_6H_6^+$			$SiC_{10}H_8^+$		
	Reaction	B.R.	k	Reaction	B.R.	k
D_2	no reaction		<0.003	no reaction		<0.0035
CO	no reaction		<0.0009	no reaction		<0.0031
N_2	no reaction		<0.002	no reaction		<0.004
O_2	$C_6H_6^+ + SiO_2$	0.9	0.03	$C_{10}H_8^+ + SiO_2$	1.0	0.0037
	$C_6H_6O^+ + SiO$	0.1				
H_2O	$SiC_6H_6^+.H_2O$	0.4	2.0	$C_{10}H_8^+ + SiOH_2$	1.0	0.055
	$C_6H_6^+ + SiOH_2$	0.35				
	$C_6H_7^+ + SiOH$	0.25				
NH_3	$SiC_6H_6^+.NH_3$	1.0	3.9	$SiC_{10}H_8^+.NH_3$	1.0	4.1
C_2H_2	$SiC_8H_8^+$	0.6	0.6	$C_{10}H_8^+ + SiC_2H_2$	0.9	0.63
	$SiC_8H_7^+ + H$	0.4		$SiC_{12}H_{10}^+$	0.1	
C_4H_2	$SiC_4H_2^+ + C_6H_6$	>0.3	7.0	$C_{10}H_8^+ + SiC_4H_2$	1.0	10.
	$SiC_6H_6^+.C_4H_2$	<0.7				

DISCUSSION

The most striking feature of these results is that the chemistry of the $SiC_{10}H_8^+$ adduct ion (with the exception of the reaction with ammonia) is dominated by the elimination of ionized naphthalene in reactions of the following type:

$$SiC_{10}H_8^+ + M \ ----> \ C_{10}H_8^+ + (SiM)$$

With the assumption that $SiC_{10}H_8^+$ is a pi-complex with the positive charge distributed over the planar naphthalene molecule (charge transfer to naphthalene is exothermic since IE(Si) > IE(naphthalene)), this reaction may be viewed as a "surface" reaction in which a neutral Si atom interacts with the incoming molecule M above the plane of the charged naphthalene molecule (6). In the analogous benzene complex the positive charge is expected to remain on the silicon atom since the charge transfer is endothermic. Indeed, the observed reaction of the benzene complex with acetylene mimics that observed with the free Si^+ ion which leads to H-atom elimination and formation of the adduct ion $SiC_2H_2^+$ in a branching ratio of 7/3 (7).

The identity of the neutral products of the surface" reactions of the naphthalene complex is intriguing, particularly if SiM leaves as a bound molecule. SiO_2 is well known as a stable molecule in the gas phase. A matrix-isolation study of the reaction of silicon atoms with water has shown that the silicon atom inserts into the O-H bond to from the HSiOH molecule (8). The gas-phase reaction of Si with acetylene has been attributed to efficient addition with ring insertion to form 3-silacyclopropenylidene (9) which is then also a plausible product of the "surface" reaction. Ab initio calculations have shown that 3-silacyclopropenylidene is the most stable isomer of SiC_2H_2 (10). In analogy to the reaction with acetylene, the reaction with diacetylene can be expected to lead to the cyclic attachment of Si to one of the triple bonds. The higher PAH molecules have a lower ionization energy so that "surface" reactions similar to that proposed for naphthalene also can be expected to efficient, at least with molecules containing acetylene units:

$$SiPAH^+ + M \ \text{----}> \ PAH^+ + (SiM)$$

In summary, we postulate on the basis of our experimental results with naphthalene, that PAH molecules may capture atomic silicon ions in interstellar environments (where the capture would need to proceed by radiative stabilization) and promote the synthesis of novel silicon-containing molecules, SiM, where M may be O_2, H_2O, acetylene, polyacetylenes and possibly cyanopolyacetylenes. Furthermore, other atomic cations, X^+, may lead to similar capture and chemistry.

$$X^+ + PAH \qquad \text{----}> \ XPAH^+$$

$$XPAH^+ + M \qquad \text{----}> \ PAH^+ + (XM)$$

We note here that the ionization energies (given in parentheses in eV) of the interstellar atoms of S(10.360), Zn(9.394), Fe(7.870) and Mg(7.646) are all larger than that typical for the larger PAH molecules (7.0). The capture and chemistry of the corresponding atomic ions will be the subject of future studies in our laboratory.

ACKNOWLEDGEMENT

This research was supported by the Natural Sciences and Engineering Research Council of Canada.

REFERENCES

1. Leger, A. and Puget, J.L. 1984, Astron. Astrophys., **137**, L5; Allamandola, L.J., Tielens, A.G.G.M., and Barker, J.R. 1985, Astrophys. J., **290**, L25.
2. Van der Zwet, G.P. and Allamandola, L.J. 1985, Astron. Astrophys., **146**, 76; Leger, A. and d'Hendecourt, L.B. 1985, Astron. Astrophys., **146**, 81; Crawford, M.K., Tielens, A.G.G.M. and Allamandola, L.J. 1985, Astrophys. J., **293**, L45.
3. Omont, A. 1986, Astron. Astrophys., **164**, 159.
4. Mackay, G.I., Vlachos, G.D., Bohme, D.K. and Schiff, H.I. 1980, Int. J. Mass Spectrom. Ion. Phys., **36**, 259.
5. Wlodek, S., Fox, A. and Bohme, D.K. 1987, J. Am. Chem. Soc., **109**, 6663.
6. Bohme, D.K., Wlodek, S. and Wincel, H. 1989, Astrophys. J., **342**, L91.
7. Bohme, D.K., Wlodek, S. and Fox, A. 1988, in "Rate Coefficients in Astrochemistry". eds. T.J. Millar and D.A. Williams, Dordrecht, Kluwer, p. 193.
8. Ismail, Z.K., Hauge, R.H., Fredin, L., Kauffman, J.W. and Margrave, J.L. 1982, J. Chem. Phys., **77**, 1617.
9. Husain, D. and Norris, P.E. 1978, J. Chem. Soc. Faraday Trans. II, **74**, 106.
10. Frenking, G., Remington, R.B. and Schaefer, H. F. III. 1986, J. Am. Chem. Soc., **108**, 2169.

IV
MODELLING

Molecules in the Ejecta of Supernova 1987A

A. Dalgarno, Weihong Liu, and S. Lepp

Harvard-Smithsonian Center for Astrophysics, Cambridge, Massachusetts, U. S. A.

Molecular formation and destruction processes occurring in the
expanding ejecta of Supernova 1987A are discussed. Formation
occurs by radiative association and negative ion reactions and
destruction by photoabsorption and by reactions with positive ions.
The observed abundances of carbon monoxide can be reproduced by
a fully mixed model of the element distribution if He^+ ions undergo
rapid charge transfer or by a partly mixed model with slow charge
transfer. Silicon monoxide is next in abundance to carbon monoxide
Significant amounts of O_2, H_2, HeH^+ and OH are predicted.

On February 23, 1987 Supernova 1987A appeared in the sky. The supernova resulted from the collapse of the core of a blue supergiant star in the Large
Magellanic Cloud. The rebound shock heated the surrounding stellar material,
causing explosive nucleosynthesis, and ejected the layers of the progenitor star
into the interstellar medium of the galaxy. Initially the envelope was opaque
and the optical and infrared spectrum was featureless. As the ejecta expanded,
the envelope became increasingly transparent, the continuum emission gave way to
emission lines and the material from the core of the supernova could be detected
and. identified. In the spectrum appeared the vibrational bands of CO and SiO in
their ground electronic states (1-5).

At the time of the first appearance of the molecular bands, the temperature
was about 5000K and the fractional ionization perhaps about 0.01. Observations show
that the material was either neutral or singly-ionized with some indication of
the presence of Fe^{2+} (6).

The ionization is driven by the γ-rays released in the decay of ^{56}CO and
later of ^{57}CO and ^{44}Ti. The γ-rays lose energy in Compton collisions with
free and bound electrons and are degraded into X-rays. The X-rays are preferentially absorbed in K-shell ionizations of the heavy elements. Through Auger
processes, highly-ionized atomic systems are created.

The highly-ionized systems recombine by radiative and dielectronic recombination

$$X^{m+} + e \rightarrow X^{(m-1)+} + h\nu \tag{1}$$

but once some neutral material has formed subsequent recombination is dominated by
charge transfer processes such as

$$X^{m+} + He \rightarrow X^{(m-1)+} + He^+ \tag{2}$$

Processes of this kind in which $m \geq 2$ are almost always fast. The behavior of the
singly-charged ions is less certain. In many cases involving complex ions charge
transfer will be rapid but there may be exceptions. If they are all rapid, ioni-

zation will flow from the elements of high ionization potential like oxygen and carbon to elements of low ionization potential like silicon, iron, cobalt and nickel.

The charge transfer processes in which He^+ ions participate are of considerable importance to the supernova chemistry and are highly uncertain, though we know from the presence of He^+ ions in the topside ionosphere that

$$He^+ + O \rightarrow He + O^+ \tag{3}$$

cannot be very rapid.

The first models of the expanding ejecta indicate an homologous expansion in which the velocity increases linearly with distance outward from the center to its value of 10,000 km s^{-1} at the leading boundary and the elements are stratified as in the progenitor star. The models failed to reproduce the early emergence of γ-rays and X-rays and have been replaced by a mixed model in which some interpenetration of the layers occurs (7). We shall argue that the chemistry supports an intermediate model with small isolated regions of unmixed gas (8).

At 70 days after the explosion when the CO and SiO emissions were first detected, the density was about 10^{11} cm^{-3}, too low for three-body collisions to be effective. The formation processes were necessarily radiative either as direct radiative association or initiated by radiative attachment to form negative ions.

The initial suggestion (9,10) was the radiative association

$$C^+ + O \rightarrow CO^+ + h\nu \tag{4}$$

followed by

$$CO^+ + O \rightarrow CO + O^+ \ . \tag{5}$$

Only a small fraction of the CO^+ ions formed by reaction (4) lead to CO because of competition from

$$CO^+ + e \rightarrow C + O \ . \tag{6}$$

Further it happens that the rate coefficient of

$$C + O \rightarrow CO + h\nu \tag{7}$$

is an order of magnitude faster than that of (5) so that (7) is a more effective source of CO (8,10).

Potentially still more productive are sequences initiated by negative ion formation

$$e + O \rightarrow O^- + h\nu \tag{8}$$

followed by associative detachment

$$O^- + C \rightarrow CO + e \tag{9}$$

or

$$O^- + O \rightarrow O_2 + e \tag{10}$$

$$C + O_2 \rightarrow CO + O \ . \tag{11}$$

The rate coefficient of (8) is 1×10^{-15} cm^3 s^{-1}, a factor of 30 or so faster than that of reaction (7).

However, the negative ions can also be destroyed by photodetachment

$$O^- + h\nu \rightarrow O + e \qquad (12)$$

and by mutual neutralizations such as

$$O^- + O^+ \rightarrow O + O \qquad (13)$$

$$O^- + He^+ \rightarrow O + He \quad . \qquad (14)$$

If hydrogen is mixed back into the core region where the carbon monoxide is formed, yet another sequence could be significant. The radiative association

$$He^+ + H \rightarrow HeH^+ + h\nu \qquad (15)$$

has a rate coefficient of 1.0×10^{-15} $cm^3 s^{-1}$ at 500K (12). Photodissociation is not a significant loss process and probably neither is dissociative recombination. Reaction (15) is followed by

$$HeH^+ + H \rightarrow H_2^+ + He \qquad (16)$$

and

$$H_2^+ + H \rightarrow H_2 + H \qquad (17)$$

to form molecular hydrogen. With the formation of H_2, pathways found in dissociative shocks (13) are opened. Thus

$$O + H_2 \rightarrow OH + H \qquad (18)$$

$$C + OH \rightarrow CO + H \quad . \qquad (19)$$

Because of the variety of mechanisms for forming CO, its chemistry is robust and substantial amounts can be formed in diverse physical surroundings.
If the hydrogen is mixed back, SiO is formed by

$$Si + OH \rightarrow SiO + H \qquad (20)$$

and by

$$Si + H \rightarrow SiH + h\nu \qquad (21)$$

$$SiH + O \rightarrow SiO + H \quad . \qquad (22)$$

In the absence of hydrogen, radiative association

$$Si + O \rightarrow SiO + h\nu \qquad (23)$$

occurs but we estimate its rate coefficient to be less than that for the analogous carbon reaction (7). More efficient may be

$$Si + O_2 \rightarrow SiO + O \quad . \qquad (24)$$

Similar chemistries lead to CS and SO but the driving processes are usually slower and additional destructive reactions such as

$$CS + O \rightarrow CO + S \qquad (25)$$

$$SO + C \rightarrow CO + S \qquad (26)$$

occur to limit their abundances.

The CO and SiO molecules can be destroyed by photodissociation and photo-ionization but if helium is mixed in with the heavy elements, the dissociative charge transfer reactions

$$He^+ + CO \rightarrow He + C^+ + O \tag{27}$$

$$He^+ + SiO \rightarrow He + Si^+ + O \tag{28}$$

may be the most efficient.

Emission from carbon monoxide decreased with time after peaking in intensity at about 190 days (5). The diminishing intensity is a result of the decrease in temperature and the abundances of carbon monoxide may have increased slightly between 192 and 349 days (4). Detailed model calculations (8) can reproduce the abundances and the temporal variation under two alternative scenarios. If the destruction by He^+ is diminished by postulating a rapid charge transfer of He^+ to C, chemistry with a fully-mixed element distribution suffices, but if He^+ is not lost by reaction with C, a partly-mixed model appears to be necessary in which mixing occurs on large scales but not on small so that there are regions of heavy elements containing little helium. In either case the increase in CO abundance at later times can be attributed to the decreasing rate of ionization (8).

With either scenario our models predict an abundance of SiO that is consistent with its observation. More detailed calculations on SiO and other molecular constituents are in progress.

Acknowledgment. This research has been supported by the National Science Foundation, Division of Natural Sciences under Grant AST-8921939.

References

1)   Olivie, E. Moorwood, A. F. M. and Danziger, I. J. 1987 The Messenger 50, 18.

2)   Rank, D. M., Pinto, P. A., Woosley, S. E., Bregman, J. D., Witteborn, F. C. Axelrod, T. S. and Cohen, M. 1988 Nature 331, 505.

3)   McGregor, P. 1988 Proc. Astr. Soc. Australia 7, 450.

4)   Spyromilio, J., Meikle, W. P. S., Learner, R. C. M. and Allen, D. A. 1988 Nature 334, 327.

5)   Danziger, I. J., Lucy, L. B., Bouchet, P. and Gouiffes, C. 1990 in Supernovae, ed. S. E. Woosley (Springer: New York).

6)   Moseley, S. H., Dwek, E., Glaccum, W., Graham, J. R., Loewenstein, R. F. and Silverberg, R. F. 1989 Ap. J. 347, 1119.

7)   Pinto, P. A. and Woosley, S. E. 1988 Nature 333, 534.

8)   Lepp, S., Dalgarno, A. and McCray, R. 1990, Ap. J. in press.

9)   Lepp, S. Dalgarno, A. and McCray, R. 1988 Bull. A.A.S. 20, 671.

10) Petuchowski, S. J., Dwek, E., Allen, J. E. and Nuth, J. A., 1989 Ap. J. 342, 406.

11) Dalgarno, A., Du, M.L. and You, J. H. 1990 Ap. J. 349, 675.

12) B. Zygelman and A. Dalgarno 1990 Ap. J. submitted.

13) D. Neufeld and A. Dalgarno 1989 Ap. J. 340, 869.

The Chemistry of Dense Interstellar Clouds

T. J. MILLAR

Department of Mathematics, UMIST, PO Box 88, Manchester M60 1QD, England

I review the chemistries of sulphur, silicon, deuterium and phosphorus in cold dense interstellar clouds such as TMC-1. The availability of recent laboratory data and the results of chemical kinetic modelling indicates that carbon-chain molecules terminated by these elements might be abundant in cold clouds as long as the elemental depletions of S, Si and P are not too large.

1. Introduction

In recent years a number of carbon-chain molecules terminated by various ligands have been detected in dense clouds, in particular, TMC-1. These molecules include $C_5H(1)$, $C_6H(2)$, $C_2S(3)$ and $C_3S(4)$. In addition deuterated forms, such as $C_2D(5)$ and $C_4D(6)$, have been detected while the linear organo-silicon species $SiC(7)$ and $SiC_4(8)$ have been detected in the circumstellar envelope of IRC+10216. Along with these new observational results, there has been significant progress in the laboratory study of reactions involving 'metal' ions such as S^+, Si^+ and P^+ with hydrocarbon molecules. For example, the reactions

$$S^+ + C_2H_2 \longrightarrow HC_2S^+ + H$$

$$Si^+ + C_2H_2 \longrightarrow SiC_2H^+ + H$$

$$\text{and} \qquad P^+ + C_2H_2 \longrightarrow PC_2H^+ + H$$

all proceed efficiently at room temperature (9,10,11) and, upon dissociative recombination with electrons, can lead to the formation of C_2S, SiC_2 and C_2P, respectively. Indeed, the SiC_2H^+ and PC_2H^+ ions react with C_2H_2 to form even larger chains (10,11):

$$SiC_2H^+ + C_2H_2 \longrightarrow SiC_4H^+ + H_2$$

$$PC_2H^+ + C_2H_2 \longrightarrow PC_4H_2^+ + H$$

with rate coefficients of 2×10^{-11} cm^3s^{-1} and 9.1×10^{-10} cm^3s^{-1} at room temperature, respectively.

In this article, I shall report the results of pseudo-time-dependent chemical models of TMC-1, concentrating on the formation of the carbon-chain molecules. Results on the formation of the largest molecules, such as HC_9N, the role of PAHs in the chemistry, and on the influence of dynamics on cloud chemistry are covered in articles by Herbst, Black and Prasad, respectively, elsewhere in this volume.

2. Chemical Models

2.1 Sulphur

Millar and Herbst (12) have up-dated the earlier models of Prasad and Huntress (13) and Watt and Charnley (14) to include the formation of the organo-sulphur species H_2CS, C_2S, C_3S and C_4S and, in addition, re-calculated many of the important radiative association rate coefficients, for example those involved in the formation of H_2S and OCS. Their model, which incorporated the use of ion-dipolar rate coefficients, essential for explaining the HCS^+ observations, was applied to both TMC-1 and the Orion ridge cloud. Comparison with observations of TMC-1 showed good agreement for SO, SO_2, H_2CS and HCS^+ but the calculated CS abundance was a factor of about 5 too large. CS and SO are linked by the reaction

$$C + SO \longrightarrow CS + O$$
$$\longrightarrow CO + S.$$

Neither the rate coefficient nor the branching ratios are well known at low temperatures. This reaction cannot produce the two possible sets of products in their ground states without proceeding, at least partially, via a potential surface of triplet multiplicity - the stable ground state of OCS is a singlet - and this surface might be repulsive. Leen and Graff (15) have suggested that the dominant channel is to CS with a rapid rate coefficient of $1.7\ 10^{-10}\ cm^3 s^{-1}$, which was used in the calculations of Millar and Herbst, whereas the phase space approach of Galloway and Herbst (16) favours the CO channel. In addition, the argument of Graff (17) can be extended to suggest that the reaction between C and SO is not more rapid than $10^{-11}\ cm^3 s^{-1}$ at low temperatures. In either case, the implication is that the formation rate of CS and the destruction rate of SO have been overestimated, and changes in the rate coefficients adopted could lead to better agreement with the observations of CS. The calculated abundances of C_2S and C_3S were slightly smaller than those observed by Saito et al. (3) and Yamamoto et al. (4), but are in excellent agreement with those determined by Fuente et al. (18). The calculated abundance of H_2S is, however, an order of magnitude less than that observed (19), perhaps indicating that the recently calculated rate coefficient for the HS^+ - H_2 radiative association (20) is in error. Note that both SO and SO_2 abundances possess a severe time-dependence, increasing rapidly for $t > 5\ 10^5$ yr, an effect related to the rise in the O_2 abundance, which is involved in the formation of these molecules, and the fall in the C atom abundance which is effective in the destruction of SO. The C_4S molecule is predicted to have a fractional abundance of between $4\ 10^{-11}$ and $3\ 10^{-10}$, depending on the exact mode by which the HC_4S^+ ion recombines with electrons and might well be detectable in TMC-1 once laboratory frequencies become available.

2.2 Silicon

The chemistry of silicon in dense clouds has been studied recently by Herbst et al. (21) who found that around 90% of all available silicon is processed into the SiO molecule by 10^5 yr. The many ways by which the Si—O bond forms, both in ion-neutral and neutral-neutral reactions, makes it difficult to understand the low upper limit on its fractional abundance, $2\ 10^{-12}$, toward TMC-1. Herbst et al. suggested that an extremely large depletion of silicon was responsible, although the means by which this occurs - a method which obviously does not apply to sulphur - is not clear. On the other hand, Langer and Glassgold (22) have suggested that if H_2O is frozen onto grain surfaces and if neutral reactions involving the ground state fine-structure level of Si atoms possess an activation energy, then one can essentially switch off SiO production in cold clouds. Against this argument, however, methanol is observed in TMC-1 and conventional wisdom insists that gas phase H_2O is involved in its production. The room temperature rate coefficient of the Si^+ - H_2O reaction is a factor of 12 less than the collisional rate coefficient, perhaps indicating the presence of a small energy barrier in the reaction. A quantum mechanical study of this reaction shows no barrier and indeed predicts that the rate coefficient should <u>increase</u> at low temperatures (23). Perhaps some combination of both Si and H_2O frozen onto the grain surface can be made consistent with the observations. In any case, it is unlikely that silicon-bearing molecules, other than SiO, can be detected in cold, dark clouds. In warm clouds in which SiO is detected, organo-silicon species such as SiC_2, SiC_3 and SiC_2H_2 might be detectable if they are unreactive with O atoms. Note that while the stable form of SiC_3 appears to be linear (8), SiC_2H_2 is triangular (24).

In summary, there are still unresolved problems with our understanding of silicon chemistry in interstellar clouds. The recent detection of widespread SiO absorption in the Galactic centre will further complicate matters.

2.3 Deuterium

Millar, Bennett and Herbst (25) have recently studied deuterium chemistry in interstellar clouds and predicted significant fractionation in large hydrocarbon molecules. Their results are somewhat sensitive to the mode of dissociative recombination adopted although the recent laboratory study on H_3O^+ + e (26) seems to indicate that the older phase space approach of Green and Herbst (27) is more appropriate than that advocated recently by Bates (28,29). Laboratory data on other systems are needed to resolve this point. In any event, the calculations show that fractionation in hydrocarbons can result from the processes

$$CH_3^+ + HD \longleftrightarrow CH_2D^+ + H_2 + \Delta H_1$$
$$\text{and} \quad C_2H_2^+ + HD \longleftrightarrow C_2HD^+ + H_2 + \Delta H_2$$

where the reaction enthalpies are ΔH_1 = -370 K and ΔH_2 = - 550 K. These large enthalpies make fractionation via CH_2D^+ and C_2HD^+ more efficient at high temperatures than that via H_2D^+. Although the binary reactions of CH_2D^+ and C_2HD^+ with H_2 are inhibited at low temperatures by these large enthalpies, radiative association of CH_2D^+ and C_2HD^+ with H_2 can hinder fractionation, particularly at low temperatures due to the inverse dependence of radiative association rate coefficients on temperature. This is particularly important in understanding why the CCD/CCH ratio in TMC-1 is smaller than that in warm clouds. Turner's recent detection (6) of C_4D seems to agree fairly well with the theoretical predictions, but the degree of fractionation observed in C_3HD (30) is, if correct, much larger than the value predicted by the models. Even so, it appears that studies of deuterium fractionation in hydrocarbon molecules offer a potentially powerful tool for unravelling the means by which these species form.

2.4 Phosphorus

To date, PN is the only phosphorus-bearing molecule detected in the interstellar medium (31,32,33) while sensitive searches have been made for PO (29), PS (34) and CP (35) amongst other species. The chemistry of PN in Orion has been explained using neutral-neutral reactions (36). In this model, PO is readily formed through the reaction of P^+ and PH^+ ions with molecules such as O_2 and H_2O, but is rapidly destroyed by N atoms. A recent and comprehensive laboratory study on the reactions of PH_n^+ (n = 0-4) with 17 neutral molecules has been published (11), showing that, like Si^+ and S^+, P^+ ions can insert into hydrocarbon neutrals thus leading to the possible formation of organo-phosphorus molecules such as CCP and C_4P. The detection of such species may be more favourable than the corresponding silicon species because in diffuse clouds silicon is much more heavily depleted than phosphorus. It is also worth noting that some of the organo-phosphorus species might be cyclic - the ion $PC_2H_2^+$ is triangular (11).

Acknowledgements

I am grateful to the Royal Society and to the conference organisers for financial support which enabled me to attend this meeting.

References

1. Cernicharo, J., Guélin, M. and Walmsley, C.M. 1987, *Astr. Ap.*, **172**, L5.
2. Cernicharo, J., Guélin, M., Menten, K. and Walmsley, C.M. 1987, *Astr. Ap.*, **181**, L1.
3. Saito, S., Kawaguchi, K., Yamamoto, S., Ohishi, M., Suzuki, H. and Kaifu, N. 1987, *Ap. J. (Letters)*, **317**, L115.
4. Yamamoto, S., Saito, S., Kawaguchi, K., Kaifu, N., Suzuki, H. and Ohishi, M. 1987, *Ap. J. (Letters)*, **317**, L119.
5. Vrtilek, J.M., Gottlieb, C.A., Langer, W.D., Thaddeus, P. and Wilson, R.W. 1985, *Ap. J. (Letters)*, **296**, L35.
6. Turner, B.E. 1989, *Ap. J. (Letters)*, **347**, L39.
7. Cernicharo, J., Gottlieb, C.A., Guélin, M., Thaddeus, P. and Vrtilek, J.M. 1989, *Ap. J. (Letters)*, **341**, L25.
8. Ohishi, M., Kaifu, N., Kawaguchi, K., Murakami, A., Saito, S., Yamamoto, S., Ishikawa, S-I., Fujita, Y., Shiratori, Y., and Irvine, W.M. 1989, *Ap. J. (Letters)*, **345**, L83.
9. Smith, D., Adams, N.G., Giles, K. and Herbst, E. 1988, *Astr. Ap.*, **200**, 191.
10. Bohme, D.K., Wlodek, S. and Fox, A. 1988, in 'Rate Coefficients in Astrochemistry' eds. T.J. Millar and D.A. Williams (Kluwer), p.193.
11. Smith, D., McIntosh, B.J. and Adams, N.G. 1989, *J. Chem. Phys.*, **90**, 6213.
12. Millar, T.J. and Herbst, E. 1990, *Astr. Ap.*, in press.
13. Prasad, S.S. and Huntress, W.T., Jr. 1982, *Ap. J.*, **260**, 590.
14. Watt, G.D. and Charnley, S.B. 1985, *MNRAS*, **213**, 157.
15. Leen, T.M. and Graff, M.M. 1988, *Ap. J.*, **325**, 411.
16. Galloway, E. and Herbst, E. 1989, *Astr. Ap.*, **211**, 413.
17. Graff, M.M. 1989, *Ap. J.*, **339**, 239.
18. Fuente, A., Cernicharo, J., Barcia, A. and Gómez-González, J. 1990, *Astr. Ap.* in press.
19. Minh, Y., Irvin, W.M. and Zinrys, L.M. 1989, *Ap. J. (Letters)*, **345**, L63.
20. Herbst, E., DeFrees, D.J. and Koch, W. 1989, *MNRAS*, **237**, 1057.
21. Herbst, E., Millar, T.J., Wlodek, S. and Bohme, D.K. 1989, *Astr. Ap.*, **222**, 205.
22. Langer, W.D. and Glassgold, A.E. 1990, *Ap. J.*, in press.
23. Wlodek, S., Bohme, D.K. and Herbst, E. 1990, *MNRAS*, in press.
24. Cooper, D.L. 1990, *Ap. J.*, in press.
25. Millar, T.J., Bennett, A. and Herbst, E. 1989. *Ap. J.*, **240**, 906.

26. Herd, C.R., Adams, N.G. and Smith, D. 1990, *Ap. J.*, in press.
27. Green, S. and Herbst, E. 1979, *Ap. J.*, **229**, 121.
28. Bates, D.R. 1986, *Ap. J. (Letters)*, 306, L45.
29. Bates, D.R. 1989, *Ap. J.*, **344**, 531.
30. Bell, M.B., Avery, L.W., Matthews, H.E., Feldman, P.A., Watson, J.K.G., Madden, S.C. and Irvine, W.M. 1988, *Ap. J.*, **326**, 924.
31. Turner, B.E. and Bally, J. 1987, *Ap. J. (Letters)*, **321**, L75.
32. Ziurys, L.M. 1987, *Ap. J. (Letters)*, **321**, L81.
33. Matthews, H.E., Feldman, P.A. and Bernath, P.F. 1987, *Ap. J.*, **312**, 358.
34. Ohishi, M., Yamamoto, S., Saito, S., Kawaguchi, K., Suzuki, H., Kaifu, N., Ishikawa, S-I., Takano, S., Tsuji, T. and Unno, W. 1988, *Ap. J. (Letters)*, **329**, 511.
35. Saito, S., Yamamoto, S., Kawaguchi, K., Ohishi, M., Suzuki, H., Ishikawa, S-I. and Kaifu. N. 1990, *Ap. J.*, in press.
36. Millar, T.J., Bennett, A. and Herbst, E. 1987, *MNRAS*, **229**, 41P.

The Production of Large Molecules in Dense Interstellar Clouds

Eric HERBST

Department of Physics, Duke University, Durham, North Carolina 27706, U. S. A.

The production of many of the organic molecules seen unambiguously in dense interstellar clouds can occur via gas phase reactions. These reactions must be rapid at the low temperatures (10 - 100 K) of the clouds and must therefore possess no activation energy barriers. Although neutral-neutral reactions involving radicals play a role in the chemistry, exothermic ion-molecule reactions dominate. Syntheses for a variety of organic molecules are discussed and models of the chemistry presented. The successes and limitations of the models are discussed. The unique chemistries of star forming regions are mentioned and a specific model for the so-called Compact Ridge in the Orion Nebula is presented.

I. Pathways to Complex Molecules

Once H_2 is produced on the surfaces of interstellar dust particles, the production of organic molecules in dense interstellar clouds can occur via gas phase pathways, which are dominated by exothermic ion-molecule reactions because these processes are most often without activation energy and occur rapidly ($k > 10^{-9}$ cm^3 s^{-1}) (1). The initiating process is the cosmic ray-induced ionization of H_2 to yield mainly H_2^+:

$$H_2 + \text{Cosmic Ray} \longrightarrow H_2^+ + e^- + \text{Cosmic Ray}. \tag{1}$$

Immediately ($\approx$ 1 day) after the H_2^+ is formed, it reacts with H_2, the dominant species in dense clouds, to produce H_3^+, the simplest polyatomic ion:

$$H_2 + H_2^+ \longrightarrow H_3^+ + H, \tag{2}$$

which can then undergo reaction with a variety of atoms assumed to be present in the initial stages of the cloud. One primary reaction for hydrocarbon formation is the proton transfer process

$$C + H_3^+ \longrightarrow CH^+ + H_2 \tag{3}$$

which, although not yet measured in the laboratory, has been studied by *ab initio* methods (2). Another primary reaction for hydrocarbon formation involves ionized atomic carbon. Since this species cannot react exothermically with H_2, it is thought that the two reactants can undergo a "radiative association" reaction in which the CH_2^+ product is stabilized by emission of a photon; viz.,

$$C^+ + H_2 \longrightarrow CH_2^+ + h\nu, \tag{4}$$

at a calculated collision efficiency of $\approx 10^{-6}$ ($k \approx 10^{-15}$ cm^3 s^{-1}) (3,4). Recently, experimental verification of this efficiency has been reported (5).

The production of CH^+ and CH_2^+ leads to the production of the methyl ion via standard H-atom transfer reactions:

$$CH^+ + H_2 \longrightarrow CH_2^+ + H \tag{5}$$

$$CH_2^+ + H_2 \longrightarrow CH_3^+ + H \tag{6}$$

and the methyl ion can be converted into protonated methane by a well-studied radiative association reaction

$$CH_3^+ + H_2 \longrightarrow CH_5^+ + h\nu \tag{7}$$

for which there are two experimental measurements, albeit in some conflict with one another (6,7). Once CH_5^+ is synthesized, neutral methane can be formed by dissociative recombination; viz.,

$$CH_5^+ + e^- \longrightarrow CH_4 + H, \text{etc.}, \tag{8}$$

although the branching ratios are not yet known definitively, and via the reaction between CH_5^+ and CO. If PAH$^-$ is the dominant carrier of negative charge (8), reaction (8) would be expected to have similar products since neutral CH_5 is not stable in its ground electronic state.

After the formation of single-carbon-atom hydrocarbons, more complex hydrocarbons are produced via several pathways: C^+ and C insertion reactions, condensation reactions, and possibly radiative association reactions. Insertion of C^+ into hydrocarbons is well studied in the laboratory (9); the process

$$C^+ + CH_4 \longrightarrow C_2H_2^+ + H_2; \; C_2H_3^+ + H \tag{9}$$

is a simple example. Analogous insertion reactions involving neutral C and hydrocarbon ions are also expected to occur, but laboratory measurements have not yet been undertaken. Condensation reactions; e. g.,

$$CH_3^+ + CH_4 \longrightarrow C_2H_5^+ + H_2 \tag{10}$$

play a less important role, and the role of radiative association reactions is not yet accurately understood due to a paucity of measurements and calculations. Some preliminary evidence for the very efficient ($\approx 100\%$) radiative association process

$$C_4H_2^+ + C_2H_2 \longrightarrow C_6H_4^+ + h\nu \tag{11}$$

has been presented (9). Once complex hydrocarbon ions are produced, they can react with hydrogen, other neutral species of lesser abundance, and electrons (or PAH$^-$). Since hydrocarbon ion - H_2 reactions are often endothermic, the neutral products of dissociative recombination reactions tend to be very unsaturated, despite the large abundance of molecular hydrogen. As the size of the reactants increases, the availability of laboratory data decreases. In the latest models of Herbst and Leung (10,11), ion-molecule syntheses of hydrocarbons through nine carbon atoms in complexity are presented, although the syntheses of the more complex species have few reactions studied in the laboratory.

Organo-nitrogen species are produced via a variety of reactions. The dominant ion-molecule pathways to the cyanopolyynes involve N atom insertion into hydrocarbon ions. An example studied in the laboratory (12) is

$$N + C_3H_3^+ \longrightarrow HC_3NH^+ + H; \tag{12}$$

this reaction occurs efficiently despite the high electronic spin state of the nitrogen atom. Cyanoacetylene, the simplest cyanopolyyne, is presumably formed via dissociative recombination of the product ion in reaction (12) although the structure of this ion has not been elucidated. Analogous ion-molecule processes have been suggested to produce the radical CH_2CN as well as the more complex cyanopolyynes (10,11). Neutral-neutral reactions involving the CN radical may also play a role (11).

More saturated organo-nitrogen molecules are thought to be produced by radiative association reactions (10). Two important examples are

$$CH_3^+ + NH_3 \longrightarrow CH_3NH_3^+ + h\nu \tag{13}$$

$$CH_3^+ + HCN \longrightarrow CH_3CNH^+ + h\nu \tag{14}$$

which are calculated to be rapid at low interstellar temperatures (13) and which lead to methyl amine and acetonitrile via dissociative recombination (or neutralization) reactions.

The gas phase chemistry of "complex" oxygen-containing organic molecules such as alcohols appears to be dominated by radiative association reactions (10). In the main the rate coefficients for these processes at low temperature are obtained by theory but high pressure experiments, in which the complex is stabilized by collision rather than emission of radiation, often help in "normalizing" theoretical treatments (14). As the size of the reactants increases, calculated radiative association rate coefficients tend to increase in size and, at low temperature, can approach the collisional limit. The synthesis of protonated methanol is typical and proceeds via the reaction

$$CH_3^+ + H_2O \longrightarrow CH_3OH_2^+ + h\nu \tag{15}$$

with a calculated rate coefficient greater than 10^{-11} cm^3 s^{-1} at T < 100 K. Other important radiative association processes include

$$H_3O^+ + C_2H_2 \longrightarrow CH_3CHOH^+ + h\nu \tag{16}$$

$$CH_3^+ + CH_3OH \longrightarrow (CH_3)_2OH^+ + h\nu \tag{17}$$

which have been invoked to explain the syntheses of acetaldehyde and dimethyl ether, respectively (15). Interestingly, reactions (13) and (17) are thought to occur despite competition from normal exothermic product channels, due to so-called "bottlenecks" in the potential energy surfaces in the exit channels, as can be seen from phase space calculations (13,15). The best studied association reaction of this class is

$$CH_3^+ + CH_3CN \longrightarrow CH_3CNCH_3^+ + h\nu \tag{18}$$

for which high pressure (SIFT) data and low pressure (ICR) data at room temperature are available; the low pressure data show a radiative association channel in competition with several exothermic channels (16).

II. Models Including Complex Molecules

Chemical models of dense interstellar clouds consist of solutions of coupled differential equations called "kinetic" equations in which the time derivative of the concentration of any species in the model is equated to the sum of the source and sink terms for the molecule. In gas phase models, the source and sink terms are all gas phase reactions except for the formation of molecular hydrogen. Consider as a simple example the formation and depletion of the molecule C^+ via the reactions

$$A^+ + B \longrightarrow C^+ + D \tag{19}$$

$$C^+ + E \longrightarrow Products \tag{20}$$

which leads to the kinetic equation

$$d[C^+]/dt = k_{19}[A^+][B] - k_{20}[C^+][E] \tag{21}$$

where [ ] refers to concentration and the k's are rate coefficients. Solution of the coupled kinetic equations requires initial concentrations, elemental abundances, and a model for the physical conditions in the cloud. If the physical conditions remain constant, the model is often referred to as pseudo time dependent.

When modelling of interstellar clouds began over fifteen years ago, the steady-state approximation was invoked. In this approximation, the time derivatives are all set to zero and the coupled differential equations become coupled algebraic equations. Early steady-state models contained predictions for molecular abundances that are in reasonable agreement with observation (17). These models tended to be small by current standards and to contain only very small molecules (up to five atoms or so). The steady-state assumption for dense clouds is on shaky grounds because the time taken to reach steady state, as first shown by Prasad and Huntress (18) is roughly 10^7yr, which is long compared with the grain adsorption time of 3×10^5 yr. In the absence of an efficient desorption mechanism for species heavier than hydrogen, it is difficult to see how a gas phase can be maintained at times long enough to reach steady state. Some non-thermal desorption mechanisms have been suggested (19).

The first substantial disagreement with predictions from steady-state models occurred for neutral atomic carbon. Predicted to be at very low abundance at steady state, this species has been detected at high abundance in a variety of interstellar clouds. In pseudo-time-dependent models (10,11,18) in which the initial form of carbon is C^+, the abundance of neutral atomic C peaks at intermediate times (10^3 - 10^5 yr) before declining precipitously. Steady-state models can be brought into agreement with the observation of large amounts of C in a variety of ways. Photodissociation of CO on cloud edges and, to some extent, in the center of clouds due to cosmic rays, increases the predicted C abundance (20,21). A similar effect occurs if PAH's are included in the models (21). Use of carbon-rich rather than oxygen-rich elemental abundances accomplishes the same effect (22). However, these modifications do not address the fundamental weakness of steady-state models - the problem of adsorption onto the grains. It is more reasonable to abandon the steady-state hypothesis completely and to assume that the gas phase of clouds cannot reach steady state. If one chooses a time of 10^5 - 10^6 yr, then the agreement between theory and observation for most small molecules does not worsen and, in addition, a large abundance of neutral atomic carbon is predicted (10,11,18). It is not clear what a time scale of this length signifies since interstellar clouds, considered as a whole, are much older than this. Perhaps the clock can be said to begin anew after cataclysmic events such as dissociative shock waves.

Recent models containing the large organic molecules discussed in the previous section show a dichotomy between "early-time" and steady-state abundances not normally present for smaller species (10,11,23). The calculated abundances increase until they peak at a time $\approx 10^5$ yr, after which they decline to steady-state values. The more complex the species, the greater the dichotomy, which arises because organic molecule production requires carbon in the form of C^+ or C. Once CO is the dominant carbon-containing species, as it is at steady state in oxygen-rich models, there is not enough carbon produced in the reactive atomic forms to produce complex molecules at high abundance. To reproduce the observed complex molecule abundances in the unusually rich source TMC-1 requires calculated abundances at or near their peak values. Perhaps the clump of dense gas known as TMC-1 did form around 10^5 yr previously. In any case, reasonable agreement between observed and peak calculated abundances is achieved for almost all species in this source (10,11,24).

III. Physical Heterogeneity

The dichotomy between early-time and steady-state abundances of complex molecules in oxygen-rich pseudo-time-dependent models is a matter of concern to some. One method of reducing the size of this dichotomy is to treat the physical conditions in the clouds more accurately. Dense interstellar clouds are not homogeneous entities but rather complex heterogeneous ones in which clumps of dense gas exist in regions of lower density. This clumping occurs on a variety of length scales - at the smallest such scale, star formation occurs. However, clumping exists even at much larger scales. In a recent paper, it has been suggested by Chieze and Pineau des Forets (25) that dynamical mixing between the surrounding diffuse gas and the material in the dense clumps can relieve some of the dichotomy in calculated abundances of complex molecules in the dense gas between early time and steady state. The idea is that the diffuse gas will contain high abundances of both C and C^+ since it is subjected to the strong ultra-violet radiation field present in the interstellar medium. If large amounts of these atomic species are mixed into the dense clouds, they can affect the chemistry at steady state because they are precisely the species needed to initiate the syntheses of organic molecules. Recent work in progress by Pineau des Forets, Chieze, and myself shows however that the steady-state results obtained depend on what precisely is mixed in to the dense clouds and what the mixing time is. In Table I, the steady-state abundances of several complex molecules in dense clumps are shown as a function of the C/C^+ abundance ratio of the diffuse gas mixed into the dense clumps with a characteristic mixing time of 0.2 Myr. It can be seen that neutral C is more efficient than is C^+ at enhancing the abundances of complex molecules, due no doubt to the fact that increased abundances of C^+ also increase the ion-molecule depletion rate of molecules. The model with the highest C/C^+

ratio of diffuse cloud gas effectively removes much of the early-time steady-state dichotomy, as can be seen from the table.

How important is this result? It would seem that substituting the dichotomy between early-time and steady-state abundances for a rapid mixing rate raises as many problems as it answers. What is the cause of the rapid mixing? Can it have anything to do with sources such as TMC-1 in which narrow line widths constrain mixing rates? What about the problem of material sticking to the surfaces of dust particles? Can photodesorption occur due to the enhanced size of the radiation field in clumpy objects? These are some questions for which answers are currently unavailable.

Table I. Steady-State Fractional Abundances of Selected Complex Molecules as a Function of C/C$^+$ Ratio of Diffuse Cloud Gas Mixed into a Dense Cloud

Species	C/C$^+$ Abundance Ratio					
	2.6×10^{-3}	0.49	1.5	5.1	300	early-time
C_3H_2	3.3(-08)	4.7(-08)	8.0(-08)	1.3(-07)	1.2(-07)	1.7(-07)
CH_3C_4H	1.1(-13)	8.5(-13)	1.5(-11)	6.9(-10)	3.3(-09)	4.0(-09)
HC_9N	1.0(-14)	1.6(-14)	1.1(-13)	9.5(-12)	5.3(-11)	1.7(-10)
CH_3OH	1.6(-12)	4.6(-12)	2.6(-11)	5.0(-10)	4.0(-09)	5.2(-09)

IV. The Compact Ridge in Orion

In the immediate vicinity of star-forming regions, the heterogeneity discussed earlier occurs on small distance scales. A well-studied complex of such regions occurs in the Orion Molecular Cloud, in the direction of the Kleinmann-Low Nebula. Distinctive chemistries in three small areas have been detected - these areas are labelled the Plateau, Hot Core, and Compact Ridge sources (26). The Plateau source is thought to be dominated by a shock wave, which strongly affects the chemistry (27), and the Hot Core is thought to have its normal ion-molecule chemistry modified by the desorption of chemically saturated molecules formed on the mantles of dust particles by H atom addition reactions (28). The desorption occurs because of the formation of a star or proto-stellar activity nearby. The Compact Ridge is perhaps the most interesting of the three; it contains enhanced abundances of the oxygen-containing organic molecules methanol, methyl formate, and dimethyl ether. Interestingly, the species acetic acid and ethanol, isomeric to methyl formate and dimethyl ether respectively, have not been detected. The enhanced abundances are not explicable in terms of normal ion-molecule chemistry and Blake *et al.* (26) have suggested that the ion-molecule chemistry is modified by the mixing in of large amounts of water, streaming outward from the proto-stellar source IRc 2. A large abundance of water can lead to an enhanced abundance of methanol via reaction (15) and dissociative recombination, and the enhanced abundance of methanol can lead to enhanced abundances of the more complex oxygen-containing molecules detected.

Tom Millar, Steve Charnley, and I are currently undertaking a series of time-dependent model calculations to test the hypothesis of Blake *et al.* (26). In our approach, water is mixed into the Compact Ridge source at a rate derived from the known mass loss of IRc 2, the distance of the source from IRc 2, its spatial extent, and a water abundance in the outflow equal to the maximum allowed by elemental abundance considerations. The mixing is assumed to start soon after the normal early-time peak abundances of complex molecules are achieved in the pseudo-time-dependent chemistry governing the Compact Ridge source. This model does not appear to account for the enhanced abundances observed; the normal decrease of methanol after its peak early-time abundance is lessened somewhat by mixing in water but its abundance is certainly not enhanced. Detailed reasons for the failure will be explained in a forthcoming publication. One basic problem is that the large abundance of water leads to a large abundance of the reactive OH radical which tends to deplete the methyl ion.

In addition to injecting water into the Compact Ridge source, we have run a model in which methanol is injected into the source. This model is successful in reproducing the observed enhanced abundances of methyl formate and dimethyl ether. However, it is far from clear what the source of the methanol is since it is not observed in the outflow from IRc 2. Perhaps it is formed on dust particles by successive H atom addition reactions to formaldehyde (29), and then desorbed into the gas phase upon an increase in temperature due to nearby proto-stellar activity. If so, the model of the Compact Ridge is similar to an earlier model for the Hot Core (28).

V. Conclusions

Although the abundances of many molecules in dense interstellar clouds can be accounted for by simple, homogeneous, pseudo-time-dependent gas phase models of the chemistry, these models do not take into account the heterogeneous physical conditions prevailing. The dichotomy between so-called early-time abundances and steady-state abundances in these models can be alleviated by rapid mixing in models that allow for two or more phases of differing gas density. The chemistry in the vicinity of star forming regions of the interstellar gas is more complex and varied. Current model results of one such region - the Compact Ridge in Orion - indicate that the enhanced abundances of oxygen-containing molecules detected there are not explained by a simple mechanism in which copious amounts of water are injected from the outflow of IRc 2.

Acknowledgements

I wish to acknowledge the generous support of this work by the National Science Foundation via grant AST-8715446.

References

(1) Smith, D. and Adams, N. G. 1989, *Faraday Trans. II*, **85**, 1613.
(2) DeFrees, D. J., McLean, A. D., and Herbst, E. 1990, in preparation.
(3) Smith, I. M. W. 1989, *Ap. J.*, in press.
(4) Herbst, E. 1982, *Ap. J.*, **252**, 810.
(5) Gerlich, D., private communication.
(6) Barlow, S. E., Dunn, G. H., and Schauer, K. 1984, *Phys. Rev. Lett.*, **52**, 902.
(7) Gerlich, D. and Kaefer, G. 1989, *Ap. J.*, **347**, XXX.
(8) Lepp, S. and Dalgarno, A. 1988, *Ap. J.*, **324**, 553.
(9) Anicich, V. G. and Huntress, W. T., Jr. 1986, *Ap. J. Suppl.*, **62**, 553.
(10) Herbst, E. and Leung, C. M. 1989, *Ap. J. Suppl.*, **69**, 271.
(11) Herbst, E. and Leung, C. M. 1990, *As. Ap.*, in press.
(12) Federer, W., Villinger, H., Lindinger, W., and Ferguson, E. E. 1986, *Chem. Phys. Lett.*, **123**, 12.
(13) Herbst, E. 1985, *Ap. J.*, **292**, 484.
(14) Bates, D. R. and Herbst, E. 1988, in *Rate Coefficients in Astrochemistry*, eds. T. J. Millar and D. A.Williams (Dordrecht: Kluwer), p. 17.
(15) Herbst, E. 1987, *Ap. J.*, **313**, 867.
(16) McEwan, M. J. *et al.* 1989, *J. Phys. Chem.*, **93**, 4064.
(17) Herbst, E. and Klemperer, W. 1973, *Ap. J.*, **185**, 505.
(18) Prasad, S. S. and Huntress, W. T., Jr. 1980, *Ap. J. Suppl.*, **43**, 1.
(19) Léger, A., Jura, M., and Omont, A. 1985 *As. Ap.*, **144**, 147.
(20) Tielens A. G. G. M. and Hollenbach, D. J. 1985, *Ap. J.*, **291**, 722.
(21) Gredel, R., Lepp, S., Dalgarno, A., and Herbst, E. 1989, *Ap. J.*, in press.
(22) Langer, W. D., Graedel, T. E., Frerking, M. A., and Armentrout, P. B. 1984, *Ap. J.*, **277**, 581.
(23) Millar, T. J., Leung, C. M., and Herbst, E. 1987, *As. Ap.*, **183**, 109.
(24) Millar, T. J. and Herbst, E. 1990, *As. Ap.*, in press.
(25) Chieze, J. P. and Pineau des Forets, G. 1989, *As. Ap.*, **221**, 89.
(26) Blake, G. A., Sutton, E. C., Masson, C. R., and Phillips, T. G. 1987, *Ap. J.*, **315**, 621.
(27) Leen, T. M. and Graff, M. M. 1988, *Ap. J.*, **325**, 411.
(28) Brown, P. D., Charnley, S. B., and Millar, T. J. 1988, *M. N. R. A. S.*, **231**, 409.
(29) Tielens, A. G. G. M. and Allamandola, L. J. 1987, in *Interstellar Processes*, eds. D. J. Hollenbach and H. A. Thronson, Jr. (Dordrecht: Reidel), p. 397.

Grain Surface Chemistry

A. G. G. M. TIELENS

MS 245-3, NASA Ames Research Center, Moffett Field, CA 94035, U. S. A.

ABSTRACT: This paper reviews the composition of interstellar grain mantles and its implications for grain surface chemistry. The bulk composition of interstellar grain mantles is observed to be dominated by H_2O, CH_3OH, and CO and typically their abundances relative to H are comparable to those of gaseous CO. Furthermore, observations have shown that there are (at least) two independent grain mantle components along most lines of sight, dominated by H-bonding and non-polar molecules, respectively. The former is H_2O- and CH_3OH-rich, while the latter is dominated by molecules such as CO, O_2 or CO_2.

Theoretical studies suggest that interstellar grain surface chemistry is dominated by H addition and abstraction reactions and reactions with O. Because of its astronomical relevance, reactions of CO and its daughter products with H are discussed in some detail and it is suggested that the observed high abundance of solid CH_3OH results from these reactions. The observed H_2O- and CH_3OH-rich grain mantle component would thus reflect reducing conditions during their formation, while the non-polar component pertains to more inert or oxidizing conditions. Finally, the formation of ethane and ethanol from accreted acetylene is discussed and the expected large variations in the solid ethanol/methanol ratio is emphasized.

I   Observed Abundances in Icy Grain Mantles

Infrared observations of objects embedded in or located behind dense molecular clouds have revealed the widespread presence of absorption features due to simple molecules in grain mantles. Since the spectroscopic identifications have recently been reviewed elsewhere (Tielens 1989; Tielens and Allamandola 1987a) a short summary will suffice. Presently, H_2O, CO and CH_3OH are the three most abundant molecules known in icy grain mantles. Interstellar solid H_2O has been identified through its OH stretching and bending vibrations at 3.08 and 6.0 µm. The peak position and width of these features show that this H_2O is located in a strongly H-bonding environment, probably H_2O-rich. In contrast, based upon extensive laboratory studies of solid CO and mixtures of CO and other molecules, it has only recently been realized that interstellar solid CO is predominantly located in a non-polar grain mantle component (Sandford et al. 1988; Tielens et al. 1990). This component might be CO-rich or contain appreciable amounts of other non-polar molecules such as O_2 or CO_2. The fractional abundance of H_2O in this component is generally less than $\approx 20\%$. A small number of lines of sight also show evidence for a second solid CO component located in a H-bonding environment. Likely, this is the H_2O-rich component evident in the H_2O features. Finally, solid CH_3OH has been identified from its CH stretching and bending modes at 3.5 and $6.85\mu m$ (Tielens and Allamandola 1987a; Grim et al. 1990). Although the relative strength of the two modes does not agree well with lab studies (as with the 3.1 and $6.0\mu m$ H_2O bands), this is likely to be due to scattering effects (Pendleton et al. 1990) and the identification of methanol in interstellar grain mantles is quite solid. This solid CH_3OH is also in a H-bonding environment and, while not proven, is likely part of the H_2O-rich component. It certainly is not part of the non-polar CO component.

Table 1 lists derived abundances for solid H_2O, CO and CH_3OH relative to H in all forms (ie., $H+2xH_2$) derived for fieldstars behind the Taurus cloud (Whittet et al. 1985; 1989) and for the two protostars, BN and W33A (Tielens et al. 1984; Tielens and Allamandola 1987a). The two separate entries for solid CO refer to abundances derived for the two independent interstellar grain mantle components discussed above. The H column densities have been derived from extinction measurements towards these objects, assuming standard dust-to-gas conversion factors. The H column density towards the protostars may be somewhat uncertain due to the presence of local hot dust adversely affecting the derived dust column densities. For Taurus, the measured abundances are quite constant from one

Table 1: Solid and gas phase abundances@ in molecular clouds

Species	Taurus$		BN		W33A	
			solid state			
H_2O	8.	(-5)	6.	(-5)	1.5	(-4)
CH_3OH£	?		4.	(-5)	8.	(-5)
CO (H-bonding)#	<4.	(-6)	<2.	(-7)	2.	(-6)
CO (non-polar)¢	1.8	(-5)	<2.	(-7)	1.	(-6)
			gas phase			
CO	3.	(-5)	1.2	(-4)	1.	(-4)
CH_3OH	4.	(-9)	4.	(-7)	?	

notes: @ All abundances relative to H in all forms (ie., $H+2H_2$). See text for details. $Refers to fieldstars behind the Taurus dark cloud. £ Abundance derived from the 6.85 μm band. Analysis of the $3.5\mu m$ band gives an ≈ 5 times lower result for W33A. See text for details. # Derived from the broad ($10\ cm^{-1}$) CO feature centered $\approx 2136\ cm^{-1}$ characteristic for a H-bonding environment. ¢ Derived from the narrow ($\approx 5\ cm^{-1}$) CO feature at $2140\ cm^{-1}$ characteristic for a non-polar environment.

line of sight to another and, since these are background stars, presumably measure the average solid state abundances in this dark cloud.

For comparison, we also show derived gas phase abundances for CO and CH_3OH. The gas phase CO abundances towards the protostars have been derived from IR rovibrational studies and thus refer to the same pencil beam as the solid state species (Hall et al. 1978; Mitchell et al. 1988). Since the same H column density has been used for normalization, the gas phase and solid state abundances are directly comparable. For Taurus, the gas phase CO abundance was derived from a comparison of mm CO studies and star counts (Frerking et al. 1983) and presumably, like the solid state measurements for this region, reflect average abundances in the dark cloud. The gas phase methanol abundances refer to the TMC1 cloud in Taurus and the Orion hot core (Irvine et al. 1987).

Perusal of table 1 shows that H_2O dominates the bulk composition of interstellar grain mantles, containing about 10-25% of the elemental O. Although not evident from this particular set of objects, the abundance of solid H_2O towards embedded sources is generally somewhat less than towards the fieldstars in Taurus, possibly because the icy mantles have evaporated from the warmest (>100K) grains close to the protostar. Solid methanol may constitute about 30% of the grain mantle, corresponding to about 15% of the elemental C (however, see Grim et al (1990) for a different view). The solid CO abundance in the H_2O-rich component is generally less than 5% relative to H_2O. The total solid CO abundance can, however, be much larger, due to the presence of CO in the non-polar component. As table 1 shows the solid CO/H_2O ratio is quite variable, ranging from about 20% towards the Taurus fieldstars to less than 0.3% towards BN. Since these molecules are predominantly located in separate grain mantle components, variations due to differences in the chemistry involved in their formation are expected. Moreover, CO is more volatile than H_2O and this ratio may be further reduced towards embedded objects, such as BN where local heating is important. However, the solid CO/H_2O ratio is quite constant towards the Taurus fieldstars. Therefore, apart from sublimation, whatever chemical processes affect it operate on a global cloud scale.

Typically, the solid state species have comparable abundances to CO, the most abundant (heavy) gas phase species known (table 1) and thus grain mantles are an important reservoir of species. This also implies that grain surface reactions can potentially be as important as gas phase reactions. This is particularly well illustrated by the much higher CH_3OH abundance in the solid state than in the gas phase in Orion. Unfortunately, direct comparison of the gas phase and solid state H_2O abundance is hampered by telluric absorption. Nevertheless, theoretical studies suggest that H_2O is not the dominant gaseous oxygen reservoir in dark clouds, particularly when A_v is small. Yet, for Taurus, solid H_2O is quite abundant for A_v in excess of ≈ 4magn. Given the high reactivity of H with O and O_2, all expected to be very abundant in molecular clouds, this solid H_2O probably results from grain surface reactions (Tielens and Hagen 1982). The presence of solid CO in a non-polar environment attests that it can freeze out under interstellar conditions. Its general absence in H_2O-rich mantles, where it should be trapped much better than in non-polar matrices, coupled with its high gas phase abundance, suggests then that the accreted CO has reacted with other species. This is in line with theoretical studies which show that it can be readily reduced or oxidized on interstellar grain surfaces leading to (ultimately) CH_3OH and CO_2 (§2). Probably, these grain surface reactions are the origin of the much higher CH_3OH abundance in the solid state than in the gas phase in Orion. A search for solid CH_3OH in the Taurus cloud, where selective evaporation of very volatile species is probably of lesser importance than around protostars, may further illuminate this question.

Finally, various other molecules have been tentatively identified along the line of sight towards the protostar W33A, including H_2S, OCS and an otherwise unidentified nitrile, "XCN". These are however only trace constituents with abundances relative to solid H_2O at the percent level or less (Tielens 1989). Solid NH_3 was originally identified on the basis of an absorption feature around $2.9\mu m$, but this feature is now shown to result from scattering by H_2O ice grains around protostars (Knacke and McCorkle 1987; Pendleton et al. 1990) and the upper limit on the solid NH_3/H_2O ratio is now $\approx 10\%$. Likewise, upper limits to the abundance of solid H_2CO and CH_4 relative to H are presently of the order of 3×10^{-7} (Tielens 1989).

II Grain Surface Chemistry

Theoretical studies of interstellar grain surface chemistry show that their bulk composition is dominated by reactions of atomic H and O, which can readily diffuse at low temperatures, with other atoms and radicals, as well as with a selected set of closed shell species (see Tielens and Allamandola 1987b). In particular, since atomic H can readily tunnel even through a rather appreciable energy barrier ($\approx 3500K$), hydrogen addition and abstraction reactions with molecules such as CO, O_2, and H_2CO can play an important role. Reactions of atomic O are effectively inhibited for activation barriers larger than about 500K. However, this may allow the astrophysically important reaction with CO to occur on grain surfaces. Now, since the gas phase can contain apreciable amounts of atomic H and O at the same time, the resulting grain mantle can simultaneously show reduced and oxidized characteristics.

In view of its high gas phase abundance and the discussion in section 1, the fate of CO is particularly interesting. Laboratory studies have shown that H reacts rapidly with CO in matrices at 10K (van IJzendoorn et al. 1983). The resulting radical, HCO, can react on with H to form H_2CO. Formaldehyde itself is also unstable to H attack. At high temperature, H abstraction with reformation of HCO is known to dominate, but theoretical and some experimental studies suggest that at low temperatures H addition is favored (Harding and Schatz 1982; Okabe 1978). On a grain surface, stabilization of the excited intermediate complex CH_3O is then expected. Further reaction with H will then lead to methanol. Of course, intermediate products might also react with other abundant, mobile, accreted species leading to the formation of, for example, HCOOH and various deuterated molecules. Thus, it has been suggested that the same reducing conditions that convert O and O_2 into H_2O on grain surfaces might also lead to CH_3OH (Tielens and Hagen 1982). Consequently, the observed solid H_2O/CH_3OH ratio is then expected to reflect directly the gas phase abundance of their parent molecules. Further laboratory studies of this H addition reaction in low temperature matrices will be very important to test this grain surface reaction scheme.

Theoretical studies of grain surface chemistry have generally concentrated on the bulk composition of interstellar grain mantles and less attention has been devoted to trace constituents. However, based upon the high deuteration fractionation observed in the warm (>100K) gas surrounding various protostars, notably the Orion hot core, it has recently been suggested that these luminous objects may evaporate a large fraction of the grain mantles in the surroundings cloud. Studies of the gas phase composition of these regions may therefore present another tool, besides IR spectroscopy, to probe the composition of icy grain mantles, in particular its trace components (Walmsley 1989). One characteristic that seems to be emerging from this interpretation is a high chemical specificity of grain surface reactions. As an example of this specificity, we will here examine the solid ethanol/methanol ratio. H addition and abstraction reactions can also be of importance for trace species. Consider, for example, acetylene, which according to ion-molecule chemistry schemes may have a gas phase abundance up to 10^{-6} relative to H under some conditions (Leung et al. 1984). The energy barrier for H addition to C_2H_2 is only about 1250K, comparable to that of CO (Harding et al. 1982; Sugawara et al. 1981). Further hydrogenation of this radical to C_2H_4 should occur readily, followed again by H addition ($E_a \approx 1100K$; Lightfoot and Pilling 1987; Sugawara et al. 1981) to form C_2H_5 and ultimately C_2H_6. Of course, the intermediate radicals in this scheme could also be deuterated or oxidized. This is probably the dominant grain surface reaction route towards ethanol. As this example illustrates, grain surface chemistry is expected to be higly chemically specific. In particular, while the grain mantle end products, CH_3OH and C_2H_5OH, are chemically quite similar, their gas phase parent molecules, CO and C_2H_2, are not. Moreover, while methanol formation only entails reducing reactions, ethanol requires oxidation as well. Given that the abundances of the gas phase parents as well as the the gaseous H/O ratio can vary widely (Leung et al. 1984), large variations in the CH_3OH/C_2H_5OH ratio are expected to occur either in the solid or, after release, in the gas phase. Finally, in line with this acetylenic scheme, it may be speculated that other related species such as the cyanoacetylenes (ie., HC_3N) may be at least partially hydrogenated (ie., CH_3CH_2CN), deuterated, or oxidized. Further matrix isolation studies of acetylene and related compounds with atomic H and O may be very important to test these ideas and predict the effects of grain surface chemistry on these interesting interstellar compounds. Ultimately, such studies may form one test on the importance of ejection of molecules from grain surfaces.

References:

Frerking, M.A., Langer, W.D., and Wilson, R.W., 1982, *Ap. J.*, **262**, 590.
Grim, R., Baas, F., Geballe, T.R., Greenberg, J.M., and Schutte, W., 1990, *Astr. Ap.*, in press.
Hall, D.N.B., Kleinmann, S.G., Ridgway, S.T., and Gillett, F.C., 1978, *Ap. J. Letters*, **223**, L47.
Harding, L.B., and Schatz, G.C., 1982, *J. Chem. Phys.*, **76**, 4296.
Harding, L.B., Wagner, A.F., Bowman, J.M., Schatz, G.C., and Christoffel, K., 1982, *J Phys. Chem.*, **86**, 4312.
Irvine, W.M., Goldsmith, P.F., and Hjalmarson, Å., 1987, in *Interstellar Processes*, eds. D. Hollenbach and H. Thronson, (Reidel, Dordrecht), p.561.
Knacke, R.F., and McCorkle, S.M., 1987, *A. J.*, **94**, 972.
Leung, C.M., Herbst, E., and Huebner, W.F., 1984, *Ap. J. Suppl.*, **56**, 231.
Lightfoot, P.D., and Pilling, M.J., 1987, *J. Phys. Chem.*, **91**, 3373.
Mitchell, G.F., Allen, M., and Maillard, J.-P., 1988, *Ap. J. Letters*, **333**, L55.
Okabe, H., 1978, *Photochemistry of small molecules* (Wiley&sons, NY).
Pendleton, Y., Tielens, A.G.G.M., and Werner, M.W., 1990, *Ap. J.*, **349**, 107.
Sandford, S.A., Allamandola, L.J., Tielens, A.G.G.M., and Valero, G.J., 1988, *Ap. J.*, **329**, 498.
Sugawara, K., Okazaki, K., and Sato, S., 1981, *Bull. Chem. Soc. J.*, **54**, 2872.
Tielens, A.G.G.M., 1989, in *Interstellar Dust*, eds. L.J. Allamandola and A.G.G.M. Tielens, (Kluwer, Dordrecht), p.239.
Tielens, A.G.G.M. , and Allamandola, L.J., 1987a, in *Physical Processes in Interstellar Clouds*, eds. G. Morfill and M. Scholer, (Reidel, Dordrecht), p.333.
Tielens, A.G.G.M. , and Allamandola, L.J., 1987b, in *Interstellar Processes*, eds. D. Hollenbach and H. Thronson, (Reidel, Dordrecht), p.561.
Tielens, A.G.G.M., Allamandola, L.J., Bregman, J., Goebel, J., d'Hendecourt, L.B., and Witteborn, F.C., 1984, *Ap. J.*, **287**, 697.
Tielens, A.G.G.M., and Hagen, W., 1982, *Astron. Ap.*, **114**, 245.
Tielens, A.G.G.M., Tokunaga, A., Geballe, T.R., and Baas, F., 1990, in preparation.
van IJzendoorn, L.J., Allamandola, L.J., Baas, F., and Greenberg, J.M., 1983, *J. Chem. Phys.*, **78**, 7019.
Whittet, D.C.B., Longmore, A.J., and McFadzean, A.D., 1985, *M.N.R.A.S.*, **216**, 45p.
Whittet, D.C.B., Bode, M.F., Longmore, A.J., Adamson, A.J., McFadzean, A.D., Aitken, D.K., and Roche, P.F., 1988, *M.N.R.A.S.*, **233**,321.
Walmsley, C.M., 1989, in *Interstellar Dust*, eds. L.J. Allamandola and A.G.G.M. Tielens, (Kluwer, Dordrecht), p263.

A Model of the Distribution of Ionised Carbon in M17SW

Simon PINNOCK[1] and T. S. MONTEIRO[2]

[1] *Department of Physics, Queen Mary and Westfield College, Mile End Road, London E1 4NS, England*
[2] *Department of Mathematics, Royal Holloway and Bedford New College, Egham, Surrey TW20 0EX, England*

The observed distribution of ionised carbon with depth into the edge-on photodissociation region (PDR) in M17SW is not consistent with results from chemical models assuming a constant or smoothly varying density gradient. These observations have been interpreted to indicate that the PDR actually consists of many clumps embedded in a low density interclump medium (1). A chemical model is described which is designed to simulate such a clumpy PDR. Results of this model give a volume filling factor of 0.2, and a characteristic clump size of 0.4pc. The maximum density of the interclump medium is found to be $n_{H_2} \approx 500 cm^{-3}$.

Introduction

When a high mass star forms inside a molecular cloud it ionises the material around it to form an HII region. Between this and the cool molecular cloud exists a warm layer consisting of material that is partly ionised and photodissociated by far ultraviolet radiation (FUV) from the star, longward of the Lyman edge at 912Å. These photodissociation regions (PDR) are of great interest since they may be places where star formation is induced by the interaction of previously formed young stars with their parent molecular cloud. To interpret observations of these systems in terms of their physical characteristics, chemical models are used to predict the abundance, and to varying degrees the excitation, of the observed species. Applied to a PDR like the one behind θ^1 Ori C (2), which is essentially face-on to the observer, these models predict the observed column densities for simple species with reasonable accuracy. However, when applied to observations of an edge-on system like the HII region/molecular cloud interface in M17SW, it is found that the models cannot reproduce the observed distribution of species. Scans of [CII]158μm emission from M17SW (1,3) show the ionised carbon abundance decreases much less rapidly with distance from the cloud surface than is predicted by models of molecular clouds having a constant or slowly varying density. This implies that the FUV is penetrating to a greater depth than expected for the cloud density determined from molecular line observations (4). To account for this it has been suggested that, at least to a depth of a few parsecs, the molecular cloud surface has a clumpy or fragmented structure, with the interclump material having a low density and low far ultraviolet optical depth (1). Thus some of the FUV passes through gaps between the clumps to ionise the surfaces of fragments deep within the cloud, so increasing the depth to which [CII] is detected.

The Model

To simulate observations of an edge-on, fragmented PDR, a model is constructed from a spatially random distribution of cloud fragments. Each spherical clump consists of an optically thick, dense, cool core and a relatively less dense warm envelope. For each clump, its probability of being illuminated by radiation from the ionising source, being in the shadow of foreground cores, or being prescreened by some amount of envelope material, is calculated from the survival probability of a ray of light from the FUV source to the centre of the clump. In this way, calculating the distribution of conditions under which clumps exist, a one dimensional chemical model (5) is run for a representative population. The chemical model, which solves a network of 385 reactions, includes an accurate numerical treatment of the line transfer of the self and mutual shielding of CO and H_2 dissociating photons using data from the recent work of Yoshino et al.(6) for the CO lines. It also includes a wavelength dependent dust extinction law (7). This gives the average densities of say, C^+, for each of the envelopes and cores. Using these values the column density along the line of sight to the observer is calculated and convolved with a gaussian beam pattern. Using the transfer equation assuming a constant source function along each line of sight, and the relations of Crawford et al. (8) for the optical depth, the column density profile is converted to emergent intensity of [CII]158μm emission.

The Results Of The Model For CII In M17SW

Observations of the PDR in M17SW (1,9) have indicated the presence of warm dense gas T=250K, and $n_{H_2} \approx 10^{4.5} cm^{-3}$, coexistent with cooler denser material, T=50K, and $n_{H_2} \approx 10^{5.5} cm^{-3}$.

For the FUV radiation to penetrate two or three parsecs into the cloud, and still cause ionisation of carbon on the surfaces of clumps there, then the density of the interclump medium must be less than $n_{H_2} \approx 500 cm^{-3}$, assuming it is not dust depleted. Many high resolution single dish, and interferometer observations have revealed density enhancements of a few tenths of a parsec or less. These values are used for the standard clump density and temperature structure. With a clump diameter of 0.4 pc, an extent along the line of sight of 2pc for the PDR, and a volume filling factor of 0.2, the model gives the intensity distribution shown in Fig. 1.

The simulated $I_{[CII]}$ scan matches the observed one to about the accuracy of the observations. The [CII] intensity curve convolved with a 55″ beam, predicted for a PDR of uniform density $n_{H_2} = 10^{4.5} cm^{-3}$, is also plotted. The extended low intensity emission seen by Stutzki et al. (1) is supposedly due to embedded B-type stars, so one does not expect the models to reproduce this characteristic. Thus a constant [CII] column density corresponding to this emission has been added on to both model results before calculating the intensity. It is evident that the results of the clumpy model are a much better fit to the observations than the uniform density model.

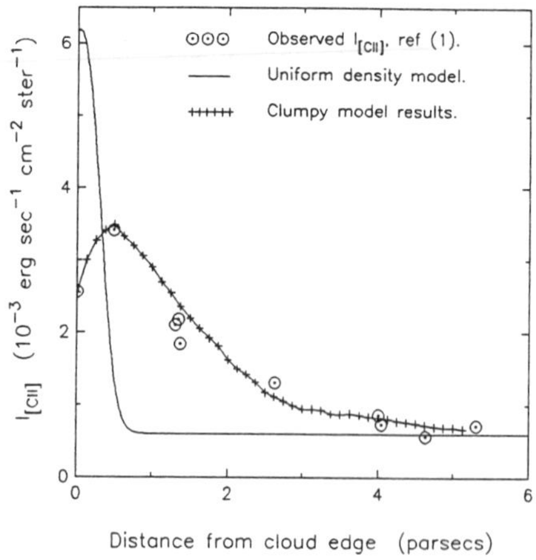

Fig. 1.
$I_{[CII]} 158\mu m$ in M17SW

Conclusion

For the clump density derived from observations, and the attenuation rate of [CII] emission with distance into the PDR, the model gives a clump size of $\approx$ 0.4pc and volume filling factor of 0.2. This gives an average of two clumps along the line of sight of a 50″ beam. This is not inconsistent with the results of the very different analysis of Martin, Sanders, and Hills (10), based on CO J=1–0 and 3–2, observations which require a number less than ten.

By using a model that attempts to mimic both the chemical and physical characteristics of a PDR, it has been possible to add weight to the interpretation of the observations in terms of a clumpy or fragmented PDR. If the M17SW PDR is not as closely edge-on as has been thought, then this conclusion is more tentative. A more complicated interface geometry may produce a similar CII distribution without clumping.

References

(1) Stutzki, J., Stacey, G.J., Genzel, R., Harris, A.I., Jaffe, D.T., Lugten, J.B., 1988, *Ap.J.*, **332**, 379.
(2) Tielens, A.G.G.M., and Hollenbach, D., 1984, *Ap.J.*, **291**, 747.
(3) Matsuhara, H., Takao, N., Shibai, H., Okuda, H., Mizutani, K., Maihara, T., Kobayashi, Y., Hiromoto, N., Nishimura, T., and Low, F.J., 1989, *Ap.J.Lett.*, **339**, L67.
(4) Snell, R.L., Mundy, L.G., Goldsmith, P.F., Evans, N.J., Erickson, N.R., 1984, *Ap.J.*, **276**, 625.
(5) Monteiro, T.S., and Pinnock, S., 1990, *in prep.*
(6) Yoshino, K., Stark, G., Smith, P.L., and Parkinson, W.H., 1988, *J. de Physique.*
(7) Seaton, M.J., 1979, *M.N.R.A.S.*, **187**, 73P.
(8) Crawford, M.J., Genzel, R., Townes, C.H., and Watson, D.M., 1985, *Ap.J.*, **291**, 755.
(9) Harris, A.I., Stutzki, J., Genzel, R., Lugten, J.B., Stacey, G.J., Jaffe, D.T., 1987, *Ap.J.Lett.*, **322**, L49
(10) Martin, H.M., Sanders, D.B., Hills, R.E., 1984, *M.N.R.A.S.*, **208**, 35.

The Abundance of SiO in Dense Interstellar Clouds

W. D. LANGER[1,2] and A. E. GLASSGOLD[2]

[1] *Princeton University, PO Box 451, Princeton, NJ 08543, U. S. A.*
[2] *Dept. of Physics, New York University, 2 Washington Pl., New York, NY 10003, U. S. A.*

<u>Abstract</u>

We discuss a new model of interstellar silicon chemistry that explains the lack of SiO in cold clouds and contains an exponential temperature for the SiO abundance. A key aspect of the model is the sensitivity of SiO production by neutral reactions to temperature and density, which arises from the dependence of the rate coefficients on the population of the fine-structure states of the silicon atom. This dependence was originally noted for carbon and oxygen by Graff, who pointed out that the leading term in neutral atom-molecule reactions involves the quadrupole moment of the atom, which depends on the fine-structure level.

It has probably been appreciated for some time that SiO is observed in the interstellar medium only under special conditions. At first sight this seems surprising because interstellar SiO was discovered (1) very soon after CO (2) and CO is very widely distributed. The large dipole moment of SiO cannot be the sole reason for the paucity of detections because many other molecules with large dipole moments are readily detected, e.g., HCN and CS. The general absence of SiO in quiescent clouds must have a chemical explanation. An obvious one is that essentially all of the silicon is depleted onto grains and that silicon compounds are only observable after the element has been removed from the dust, e.g., by heating. This idea fits in with a correlation of the SiO abundance with temperature reported by Ziurys, Friberg, and Irvine (3). However, in cold clouds, where Ziurys *et al.* obtained very small upper limits (fraction of SiO, $x(SiO) < 2\text{-}3 \times 10^{-12}$), depletion factors larger than 10^7 are implied, and this large factor seems hard to understand.

Expressing the temperature correlation in terms of $1/T$, Ziurys *et al.* obtained an $\exp(-90/T)$ fit for the abundance of SiO, which suggests a chemical process with a threshold of the order of 100K. However, this empirical correlation is not yet well established because the observational sample includes a diverse collection of clouds. In fact, all but one of the detections are (or are near) outflow sources or the supernova remnant IC 443.

Our proposal for explaining the SiO observations is based on the discussion by Graff (4) on the role of the atomic fine-structure in atom-neutral reactions for carbon and oxygen. At very low temperatures, the dominant interaction comes from the long-range force, e.g., the quadrupole-dipole force for the reaction channnels, C + OH and O + OH, that Graff analyzed in detail. In order for the atom to have a quadrupole moment in its ground electronic state, levels other than $J = 0$ must be populated. This requirement is no problem for O, where the lowest level in the ^3P_J fine-structure multiplet is $J = 2$, but it is a problem for C where the ground state is $J = 0$, as shown in Figure 1.

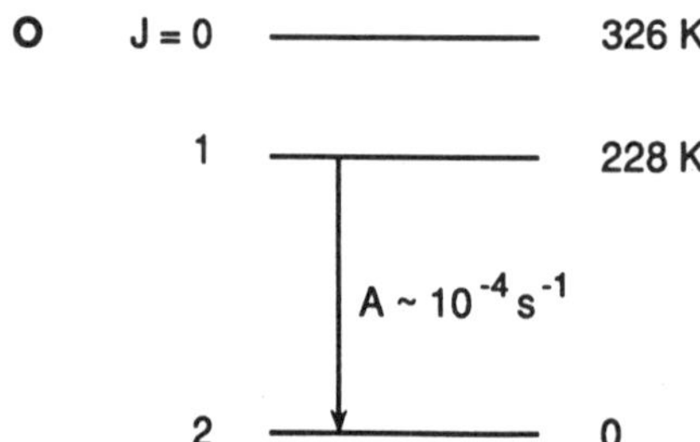

Figure 1. Schematic of the ground state fine-structure lines of C, O, and Si. Shown are the ordering of the J levels, the energy above the ground state in units of K, and the transition probabilities in s^{-1}.

Graff concluded that the rate coefficient for C + OH -> CO + H is much less than that for O + OH -> O_2 + H as T approaches zero. Following Graff's suggestion that similar concepts should apply to other neutral atomic reactions, we have considered their possible effect on the silicon chemistry of shielded interstellar clouds. Refering to Figure 1, we note that Si has the same ordering for its fine-structure levels as C and that the first level is 111 K above the ground state, roughly the barrier obtained by Ziurys *et al*. The relevant neutral reactions are

$$Si + OH \rightarrow SiO + H \tag{1}$$

$$Si + O_2 \rightarrow SiO + O \tag{2}$$

We propose that the reaction coefficients for these reactions are suppressed by a factor that is essentially the probability for occupation of the first excited state, i.e,

$$p_1 = 3 \, n \exp(-111/T) \, / \, [n \, (1 + 3 \exp(-111/T)) + n_{cr}] \tag{3}$$

where

$$n_{cr} = A_{10} / k_{10} \tag{4}$$

and A_{10} and k_{10} are the Einstein A-coefficient and the collisional de-excitation rate coefficient for the J = 1 - 0 transition, respectively. Because it is likely that $k_{10} < 10^{-11}$ cm^3 s^{-1} below 100 K (5,6), the critical density for Si is probably large, i.e., $n_{cr} > 10^6$ cm^{-3}. In any case, we see from equation (4) that there is a density as well as a temperature effect implied by Graff's considerations: Long range atomic quadrupole interactions will be reduced at low densities ($n < n_{cr}$) as well as at low temperatures ($T < T_{10}$). It is important to justify this extension of Graff's ideas to silicon atoms and to quadrupole-quadrupole interactions by fundamental theory.

In order to place this proposal in context, we have developed a simplified (analytic) form of interstellar silicon chemistry which includes ion-molecule pathways to SiO,

$$Si^+ + H_2O \rightarrow SiO^+ + H_2 \tag{5}$$

$$Si^+ + OH \rightarrow SiO^+ + H \tag{6}$$

(followed by reactions with H_2 to form $HSiO^+$ and then recombination to SiO) and relevant destruction mechanisms, as well as the neutral reactions already mentioned. The main production routes are shown in Figure 2.

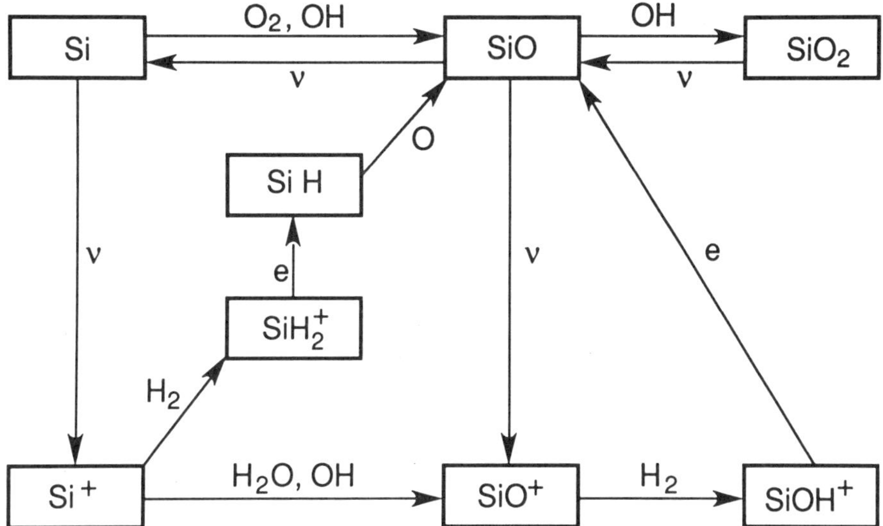

Figure 2.   Main production routes for SiO, omitting weak channels based on H_3^+ + Si and ionic destruction reactions, with v indicating photodestruction by cosmic ray produced UV .

Omitted from the diagram are relatively weak channels following the reaction

$$H_3^+ + Si \rightarrow SiH^+ + H_2 \, , \tag{7}$$

for example, the pathway suggested by Herbst *et al.* (7)

$$SiH^+ + O \rightarrow SiO^+ + H, \tag{8}$$

which then leads to SiO with reactions already given in Figure 2. Both reactions (7) and (8) are exothermic but unmeasured at present.

Further details on our model chemistry can be found in our recently published paper (8). The main results can be discussed in terms of the following steady state expression for the ratio of SiO to Si (both in the gas phase):

$$x(SiO) / x_{Si} = R \, f \, \exp(-111 / T) + Z \qquad (9)$$

where R gives the contribution of the neutral reactions (1) and (2) and Z the contribution of the ionic reactions (5) and (6); $f = p_1 / 3$ (see equation (3)); and R and Z have the approximate values

$$R = 16 \, n_5 \, [x(O_2)_{-6} + 0.1 \, x(H_2O)_{-8}] / \zeta_{-17} \qquad (10)$$

$$Z = 0.0025 \, x(H_2O)_{-6} / [x(PM_{-7}) \, T^{-0.5} + 0.005 \, x(H_2O)_{-6}] \qquad (11)$$

We use the notation $y_n = y/10^n$; PM stands for the large molecules or small grains that can neutralize Si^+ and ζ is the cosmic ray ionization rate. The occurrence of an exponential temperature dependence for x(SiO) depends on the relative sizes of the two terms in equation (9); approximate conditions are: $T > 30$ K and $n > 10^5$ cm^{-3}.

Equations (10) and (11) show that the SiO chemistry is intimately related to O_2 and H_2O, whose abundances have not yet been definitively measured. Associated with the very small predicted abundances of silicon molecules in shielded clouds are small chemical time scales; a more detailed analysis shows that this actually favors an exponential temperature dependence (8).

We have made the tacit assumption that the contribution of short range forces to the rate coefficients for reactions (1) and (2) are negligible. If not, they can contribute another term to equation (9) (without an exponential factor). The exponential temperature dependence can also be suppressed by other SiO formation pathways, as suggested by Herbst, Millar, Wlodek, and Bohme (7). Many if not all of the routes to SiO proposed by these authors involve hypothetical reactions, i.e., with unmeasured rate coefficients.

In conclusion, the non-detection of SiO in dark clouds can arise from several causes, possibly operating in concert:

(1) suppression of neutral reactions because of low temperature and/or hydrogen density,

(2) weakening of ionic synthesis by the freeze-out of H_2O,

(3) depletion of gaseous silicon onto dust grains.

To clarify these issues , it would be helpful to have the results of the following, challenging research projects: remeasurement of the rate coefficient of the basic reaction (1), especially for $T < 300$ K; further efforts to detect SiO in shielded clouds with different values of T and n; and definitive measurements of the O_2 and H_2O abundances, already the objects of space experiments.

References

(1) Wilson, R. W., Penzias, A. A., Jefferts, K. B., Kutner, M. L., and Thaddeus, P., 1971, Ap. J. Letters, 1970, **167**, L97.

(2) Wilson, R. W., Jefferts, K. B., and Penzias, A. A. 1970, Ap. J. Letters, **161**, L43.

(3) Ziurys, L.M., Friberg, P., and Irvine, W.M. 1989, Ap.J., **343**, 201.

(4) Graff, M.M. 1989, Ap. J., **339**, 239.

(5) Monteiro, T. S. and Flower, D.R. 1987, M.N.R.A.S., **228**, 101.

(6) Flower, D.R. 1989, preprint.

(7) Herbst, E., Millar, T.J., Wlodek, S., and Bohme, D. 1989, Astr. Ap.,

(8) Langer, W.D. and Glassgold, A.E., 1990 Ap. J. **352**, in press .

Isotope-Exchange Reactions as a Function of Temperature: Isotopic Fractionation in Interstellar Molecules

Michael HENCHMAN[1] and John F. PAULSON[2]

[1] Department of Chemistry, Brandeis University, Waltham, MA 02254-9110, U. S. A.
[2] Air Force Geophysics Lab (LID), Hanscom Air Force Base, MA 01731-5000, U. S. A.

A model is used to predict rate constants for primary deuteration reactions which fractionate deuterium in interstellar molecules.

For the molecular species found in dense, interstellar clouds, such as TMC-1 and Orion, the D/H relative abundances range from 0.002 to 0.05, compared to the cosmic D/H ratio of ~2 x 10^{-5} (1). Chemical models for the synthesis of these molecular species must also be able to predict isotopic fractionation. This latter requires rate constants for the isotope-exchange reactions which effect the isotopic fractionation. These rate constants are needed for the temperature range 10 - 100 K but few data are currently available. Modellers are forced to extrapolate from data measured in the range 80 - 300 K or otherwise to guess.

We are interested in predicting rate constants for isotope-exchange reactions as a function of temperature. For the representative isotope-exchange reaction

$$AH + BD = AD + BH \tag{1}$$

the rate constant k_f for the forward or exergonic direction [$\Delta G° < 0$], is related by thermodynamics to that for the back reaction k_b, by $k_b = k_f \exp(\Delta G°/RT)$. It has been standard to estimate k_f and k_b as follows (1,2),

$$k_f = k_c \tag{2}$$

$$k_b = k_c \exp(\Delta E/RT) \tag{3}$$

where k_c is the collisional rate constant (3). Our own approach is to model the temperature dependence of k_f and to relate k_b to k_f by thermodynamics. Our analysis replaces (2) by (4) and (3) by (5) as follows

$$k_f = k_c K/(K + 1) \tag{4}$$

$$k_b = k_c/(K + 1) \tag{5}$$

where K is the equilibrium constant for the reaction (1) written in the exergonic direction. Rate constants k may therefore be estimated as a function of temperature from thermodynamic estimates of equilibrium constants K as a function of temperature. We have used this approach for isotope-exchange reactions (4,5): Mautner has independently used the same approach for particle-transfer reactions (6). And the same result

$$k_f + k_b = k_c \tag{6}$$

had been identified prior to this — as an empirical relation from the analysis of experimental data (7-10).

Our previous study (5) examined several representative isotope-exchange reactions for which experimental rate data were available throughout the temperature range 80-300 K; and we showed how these data were well simulated by the model's theoretical predictions (4) and (5). We now use the model to analyze some isotope-exchange reactions which may fractionate deuterium in interstellar molecules. For some reactions, e.g.(7) and (8), the model seems to work but in others, (9) and (10), it does not. In the latter cases, it is important to be able to understand why.

For the comparison, we have selected primary deuteration reactions listed in Table 1 of Millar *et al.*'s model of deuterium fractionation in dense interstellar clouds (1). The scheme is as follows. The initial cosmic deuterium reservoir, consisting of deuterium atoms, is ionized by cosmic rays and reacts to form HD by reaction (7). Experimental measurements are restricted to the temperature range 200-300 K (8) and in this range the model predicts the data well (Table 1). At interstellar temperatures, the forward rate constant is effectively at the collision limit and the backward effectively zero. The model provides a reliable way of estimating the magnitude of these, in the absence of experimental data.

TABLE 1. Experimental rate constants (10^{-9} cm^3/molec s) predicted by the model

#	REACTION	$-\Delta H°/R$ (K)	TEMP.	k_f EXPT.	k_f MODEL	k_r EXPT.	k_r MODEL
(7) $D^+ + H_2 = HD + H^+$		462	205	2.2	2.0	0.11	0.14
			295	1.7	1.9	0.17	0.26
(8) $H_3^+ + HD = H_2D^+ + H_2$		227	80	1.3	1.5	0.29	0.26
			200	1.2	1.2	0.51	0.50
			295	1.1	1.2	0.56	0.58

The primary deuterating molecular species is HD, as formed by reaction (7). It reacts principally by reaction (8) to form H_2D^+ which, in turn, is a facile proton/deuteron donor — able to deuterate a range of interstellar molecules. Table 1 compares, for reaction (8), data down to 80 K (9) with model predictions, which rely on equilibrium constants determined by Herbst (11). These agree, as well for reaction (8) as for (7); and the same conclusions apply to both reactions.

After H_2D^+, the next most important deuterating species is DCO^+ — especially at low temperatures and early in the development of the galaxy (1). How is DCO^+ itself synthesized (12,13)? One route involves exchange of HCO^+ with the major deuterium reservoirs D and HD: a second is deuteron transfer from H_2D^+ to CO.

TABLE 2. Experimental rate constants (10^{-9} cm^3/molec s) not predicted by the model

#	REACTION	$-\Delta H°/R$ (K)	TEMP.	k_f EXPT.	k_f MODEL	k_r EXPT.	k_r MODEL
(9) $HCO^+ + D = DCO^+ + H$		796	120	0.2	1.4	<0.005	0.0055
			300	0.045	1.2	~0.01	0.26
			460	-	1.0	0.015	0.54
(10) $HCO^+ + D_2 = DCO^+ + HD$		301	80	<0.0005	1.1	<0.0005	0.02

Both routes to DCO^+ are summarized in Table 2, which again compares experimental data with predictions of the model. For the first route, involving the isotope-exchange reactions (9) and (10), the data are taken from refs.(14,15) and (9,16) respectively. The model uses equilibrium constants estimated as $K_9 = 0.46$ exp (796/T) and $K_{10} = 1.13$ exp (301/T) (17). Both reactions (9) and (10) show no agreement between experiment and theory. Reaction (10) (and others similar) are prevented by energy barriers which we have identified in the attractive reaction hypersurface (13). Reaction (9) is not "well behaved", in that it shows an anomalous negative temperature dependence; and thus the failure of our model is not surprising. Why is reaction (9) anomalous? We guess that it may derive from the need to conserve angular momentum, the moment of inertia of the products being half that of the reactants.

For the second route, our model predicts that H_2D^+ + CO yields a product ratio $HCO^+/DCO^+ = 3.8$ _under all conditions_. The reaction is equivalent to running the isotope-exchange reaction $DCO^+ + H_2 = HCO^+ + HD$ at a temperature of ~20,000 K, because the reaction $H_3^+ + CO = HCO^+ + H_2$ is exothermic by 1.7 eV (13)! [Thus for $DCO^+ + H_2 = HCO^+ + HD$, K = 3.78 exp (-380/T) (17).] The same argument applied to the reaction $CH_2D^+ + O = DCO^+ + H_2$ ($HCO^+ + HD$), gives the same product ratio $HCO^+/DCO^+ = 3.8$ _under all conditions_.

In conclusion, we have discussed isotope-exchange reactions where the rate constants can be predicted accurately and others where the agreement is poor. Where the model fails, we can often identify this in advance and understand why. Is the model useful for modelling isotopic fractionation in interstellar molecules? For the reactions listed in Table 1, the answer is no. The difference between the rate constants used by the modellers from (2) and (3), and our values from (4) and (5) is negligible compared with the many uncertainties in the astrophysical models (1). On the other hand, the differences can be much larger for the reactions in Table 2. Thus if our guess about reaction (9) is correct, the same behavior may well apply to $H_3^+ + D = H_2D^+ + H$, a reaction which is probably important (12). In that case the estimated rate constant may be much too large (1). Finally we suggest that our model will be useful for estimating rate constants for fractionating heavier isotopes such as ^{13}C, ^{15}N, ^{18}O etc. It will be hard to measure these for polyatomic reactants at interstellar temperatures.

REFERENCES

1.   Millar,T.J.,Bennett,A.,and Herbst,E.1989,*Ap.J.*,**340**,906.
2.   Watson,W.D.1976,*Rev.Mod.Phys.*,**48**,513.
3.   Su,T.,and Chesnavich,W.J.1982,*J.Chem.Phys.*,**76**,5183.
4.   Villinger,H.,Henchman,M.J.,and Lindinger,W.1982,*J.Chem.Phys.*,**76**,1590.
5.   Henchman,M.,and Paulson,J.1989,*J.Chem.Soc.,Faraday Trans.2*,**85**,1673.
6.   Sieck,L.W.,and Meot-Ner,M.1982,*J.Phys.Chem.*,**86**,3646. M. Meot-Ner in *Structure/Reactivity and Thermochemistry of Ions*, ed.P.Ausloos and S.G.Lias (Reidel,Dordrecht,1987),p.383.
7.   Smith,D.,and Adams,N.G.1980,*Ap.J.*,**242**,424.
8.   Henchman,M.J.,Adams,N.G.,and Smith,D.1981,*J.Chem.Phys.*,**75**,1201.
9.   Adams,N.G.,and Smith,D.1981,*Ap.J.*,**248**,373.
10. Smith,D.,and Adams,N.G.1984 in *Ionic Processes in the Gas Phase*, ed.M.A.Almoster Ferreira (Dordrecht:Reidel),p.41.
11. Herbst,E.1982,*Astron. Astrophys.*,**111**,76.
12. Dalgarno,A.,and Lepp,S.1984,*Ap.J.*,**287**,L47.
13. Henchman M.,Smith,D.,Adams,N.G.,Paulson,J.F.,and Lindinger,W.1988 in *Rate Coefficients in Astrochemistry*, ed. T.J.Millar and D.A.Williams (Dordrecht:Kluwer), p.201.
14. Adams,N.G.,and Smith,D.1985,*Ap.J.*,**294**,L63.
15. Federer,W.,Villinger,H.,Tosi,P.,Bassi,D.,Ferguson,E.,and Lindinger,W.1985, in *Molecular Astrophysics*, ed. G.H.F.Diercksen,W.F.Huebner and P.W.Langhoff (Dordrecht:Reidel),p.649.
16. Huntress,W.T.1977,*Ap.J.Suppl.*,**33**,495.
17. Henchman,M.1984,*unpublished results*.

Shock Chemistry in Interstellar Clouds

George F. Mitchell

*The Canada-France-Hawaii Telescope Corporation, Waimea, Hawaii and Department of
Astronomy, Saint Mary's University, Halifax, Canada*

Some recent theoretical and observational work in shock chemistry is summarized.
While observational upper limits to species such as SH^+ and C_3 are beginning to
constrain diffuse cloud shock parameters, the sole raison d'être for diffuse cloud
shock models continues to be the molecular ion CH^+. The dense C-shock models, which
were successful in explaining the intensity of emission from hot H_2, CO and OH in
OMC-1, are being challenged by extensive new infrared observations of H_2 emission
lines. New calculations of molecule formation behind dissociative J-shocks have
pointed to potential diagnostics for such shocks.

1. Introduction

Because the speed of sound in the cold component of the interstellar medium is
low (≤ 1 km s^{-1}), interstellar clouds are readily subjected to supersonic impacts.
Events which will drive shocks into gas clouds include supernovae explosions,
stellar winds and outflows, cloud-cloud collisions, and the expansion of ionization
fronts. A hydrodynamic (i.e. single fluid) shock produces a discontinuity of gas
properties at a shock front. The well known jump conditions require compression,
heating, and acceleration of gas by the shock. An interstellar hydrodynamic shock
consists, in general, of three regions: (1) a radiative precursor, in which
preshock gas may be heated, dissociated, and/or ionized; (2) the shock front, in
which ordered motion is converted into heat; and (3) a cooling zone in which the
gas cools to a new equilibrium value. At interstellar cloud densities, the shock
front is negligibly small compared to the cooling region.

Because interstellar gas clouds contain a magnetic field and are weakly ionized,
it is necessary, in general, to consider neutral and charged species separately
(Draine 1980). Such shocks are known as multifluid or magnetohydrodynamic shocks.
In diffuse clouds, it is sufficient to consider three fluids, namely neutrals,
ions, and electrons. In dense clouds, drag due to charged grains becomes
significant. In most dense cloud shock calculations to date, grains have been
included in an approximate manner. Recently, Pillip, Hartquist, and Havnes (1990)
included grains in a four fluid model, showing that, for some regions of the
relevant parameter space, the exact treatment of grain drag is important. They
suggest that it is necessary to go further and include an appropriate spectrum of
grain sizes and charges in dense cloud shock models.

The structure of MHD shocks is very different from that of hydrodynamic shocks.
The difference can be understood by considering two different characteristic
speeds. Compressive disturbances propagate in the charged component at the ion
magnetosonic speed, $v_{i,mag} = B/(4\pi\rho_i)^{\frac{1}{2}}$, whereas compressive disturbances travel in
the neutral gas at the neutral sound speed, c_n. For shocks with speeds, v_s, in the
range $c_n < v_s < v_{i,mag}$, the charged fluid is accelerated through the neutrals. The
neutrals, which are the dominant component in mass, are accelerated and heated by
collisions with the ionized component. This extended region of ion-neutral

streaming is called the "magnetic precursor". If the magnetic field is too low or
the fractional ionization is too large (or zero), the magnetic precursor is small
and the shock is J-type, like a single fluid shock. For ranges of magnetic field
strength and fractional ionization which can exist in the interstellar gas, all
variables remain continuous throughout the shock, and the shock is called C-type.
Fast shocks, with $v_s > 50$ km s^{-1}, are always J-type.

2. Shocks in Diffuse Clouds

The development of diffuse cloud shock models was motivated by the failure of
quiescent cloud models to account for the abundance of the ion CH^+. CH^+ is
destroyed by rapid reactions with H_2, H, and electrons while the direct, potentially
fast formation reaction, $C^+ + H_2 \rightarrow CH^+ + H$, is endothermic by 0.4eV. The original
shock model of Elitzur and Watson (1978) appeared to produce, within the
uncertainties, sufficient CH^+. Shock calculations with more complete chemistry
(Michell and Watt 1985, Pineault des Forêts et al. 1986) failed, however, to
reproduce observed CH^+ abundances. C-type shock models (Draine 1980) were,
therefore, invoked. Flower, Pineault des Forêts, and Hartquist (1985) showed that
the same reaction which forms CH^+ controls the ionization and, therefore, the shock
structure. From MHD shock models with reasonably comprehensive chemistry, Pineault
des Forêts et al. (1986) found that the CH^+ observations could be reproduced.
Another constraint on shock models is, however, the observed populations of
rotational levels of H_2. It proved difficult to fit CH^+ and excited H_2
simultaneously.

In an independent MHD shock modeling effort, Draine and Katz (1986a,b) and
Draine (1986) were able to reproduce the observed abundances of CH^+ and excited H_2
in the line of sight to ζ Oph. They used, however, incorrect products for the
photodissociation of CH^+. With the correct reaction, their model for ζ Oph yields
less CH^+ by a factor of 3, so that the exact agreement is lost. Draine and Katz
also found that large quantities of OH, CH, CO, and HCO^+ were produced in the
shocked gas, as already pointed out for single fluid shocks by Mitchell and Watt
(1985). High abundance of these latter species does not necessarily point to a
shock, however, because they may be abundant in unshocked gas. Clearer shock
diagnostics are required.

In a diffuse shock calculation with sulphur chemistry, Pineault des Forêts,
Roueff, and Flower (1986) confirmed a suggestion of Millar et al. (1986) that SH^+
will be more abundant in C-shocks than in J-shocks, being formed via the reaction
$S^+ + H_2 \rightarrow SH^+ + H$. An endothermicity of 10^4 K means that this reaction is more
effectively driven by ion-neutral streaming than by thermal collisions unless shock
speeds are quite high. In an unsuccessful search for $SH^+(\lambda 3363)$ toward ζ Oph,
Millar and Hobbs (1988) obtained an upper limit of $N \approx 10^{13}$ cm^{-2}, very close to the
predicted 1.1×10^{13} cm^{-2}. Magnani and Salzer (1989) failed to find SH^+ toward 9
other sources, to a similar upper limit. An improvement in the upper limit to SH^+
by a factor of a few should be attempted.

The molecule C_3 is a potentially very powerful shock diagnostic because, unlike
C_2, it is expected to have the very low column density of $\approx 10^7$ cm^{-2} in quiescent
diffuse clouds but to be abundant in shocks. Snow, Seab, and Joseph (1988) failed
to detect the molecule C_3 in absorption toward 8 sources. They placed a low upper
limit, 5×10^{10} cm^{-2}, on its abundance. Shock calculations show that the abundance
of C_3 is very sensitive to gas density. In a hydrodynamic shock with speed 20 km
s^{-1}, initial density 10 cm^{-3}, and a ratio of molecular to atomic hydrogen of 0.05,
Mitchell and Watt (1985) found $N(C_3) = 1.3 \times 10^{10}$ cm^{-2}. Pineault des Forêts et al.
(1987) found, in an MHD shock with $v_s = 15$ km s^{-1}, $n_0 = 20$ cm^{-3}, and $H_2/H = 2$, $N(C_3)$
$= 7 \times 10^9$ cm^{-2}. In shocks with higher preshock gas densities, the C_3 abundance
becomes very high: Mitchell (1983), in a J-shock calculation with $v_s = 15$ km s^{-1}
and $n_0 = 100$ cm^{-3}, predicted a column density of C_3 hydrocarbons of 2×10^{13} cm^{-2},
Pineault des Forêts et al. (1987), in a C-shock with essentially the same

parameters, found $N(C_3) = 4 \times 10^{13}$ cm^{-2}. The C_3 limit of Snow, Seab, and Joseph (1988) already rules out preshock densities much greater than 10 cm^{-3} for diffuse cloud shocks.

All of the MHD calculations mentioned above have the defect that the initial fractional abundance of molecular hydrogen is too high, $H_2/H \approx 1$. In gas with an initial density of 10 cm^{-3}, we expect only $H_2/H \leq 0.1$. The assumption of a high molecular fraction was made for computational reasons. For low values of H_2/H, a J-shock occurs in the neutral gas and the numerical treatment of a J-shock with magnetic precursor is difficult. The treatment of diffuse cloud shocks is further complicated by the existence of a third class of shock, called C*-shocks. Details of the neutral cooling determine which type will occur (Chernoff 1987). If the gas remains everywhere supersonic, the result is a C-shock. If the gas heats up and becomes subsonic but remains everywhere continuous, the result is a C*-shock. If the neutral gas becomes subsonic producing a viscous subshock, the result is a J-shock with precursor. Roberge and Draine (1990) present numerical methods for computing the structure of J- and C*-shocks for diffuse cloud parameters. They compute a series of models with $H_2/H = 0.035$, $n_0 = 25$ cm^{-3}, and B = 5 μG, for shock speeds between 5 and 30 km s^{-1}. They find that the shock is C-type for shock speeds less than 5 km s^{-1}, the shock is J-type for speeds from 5 to 18 km s^{-1}, and the shock is C*-type for speeds from 18 to 30 km s^{-1}. The maximum temperature of the neutral gas increases with shock speed to $\approx 10^4$ K for a 30 km s^{-1} shock. The neutral gas temperatures are much higher than found in previous diffuse cloud C-shock models. Chemical predictions using these new models have not yet been made. They will be most interesting.

It is disquieting that CH$^+$ remains the only chemical species for which shock synthesis is required. A second type of shock signature is kinematical. The velocity differences between cold preshock gas, hot shocked gas, and cold postshock gas may be detectable. About half the diffuse cloud lines of sight do show a difference in velocity of 1 to 2 km s^{-1} between CH$^+$ and CH. The usual interpretation is that CH$^+$ resides in the hot shocked gas while CH is in the cold, postshock gas. The existence of small scale clumping (e.g. Falgarone and Pérault) 1988) complicates the interpretation, opening up the possibility that the two species occur in different clumps along the same line of sight. Emission lines of CO at millimeter wavelengths can be used to probe the velocity structure because these transitions are optically thin in diffuse clouds. High velocity resolution observations of CO(J=1-0) towards ζ Oph (Langer, Glassgold, and Wilson 1987) showed four components, but did not suggest an obvious identification with a shock. Bourlot, Gérin, and Pérault (1989) used CO(1-0) and CO(2-1) observations of the ζ Oph cloud to infer a strongly inhomogeneous component. The CO emission observations of ζ Oph do not confirm the presence of a simple planar shock of the type assumed in models of shock chemistry.

3. Dense Cloud Shocks and the Orion Outflow

Dense C-shock models (Draine and Roberge 1982; Chernoff, Hollenbach, and McKee 1982) were invoked to account for the hot H_2, CO, and OH found in OMC-1. A two shock model for the Orion outflow can explain many of the observations. In this picture, a shell is driven supersonically into the ambient medium by a fast wind (see Figure 1). The shell is bounded by an inner "wind shock" and an outer "ambient shock". The wind shock decelerates the wind as it encounters the shell. It is modeled as a dissociative, J-type, shock with a speed of $\approx$ 70 km s^{-1} and a preshock (i.e. wind) density of $\approx 3 \times 10^4$ cm^{-3}. The ambient shock, which accelerates the ambient cloud gas to the shell velocity, is C-type, with a velocity of $\approx$ 36 km s^{-1} and a preshock density of 2×10^5 cm^{-3}.

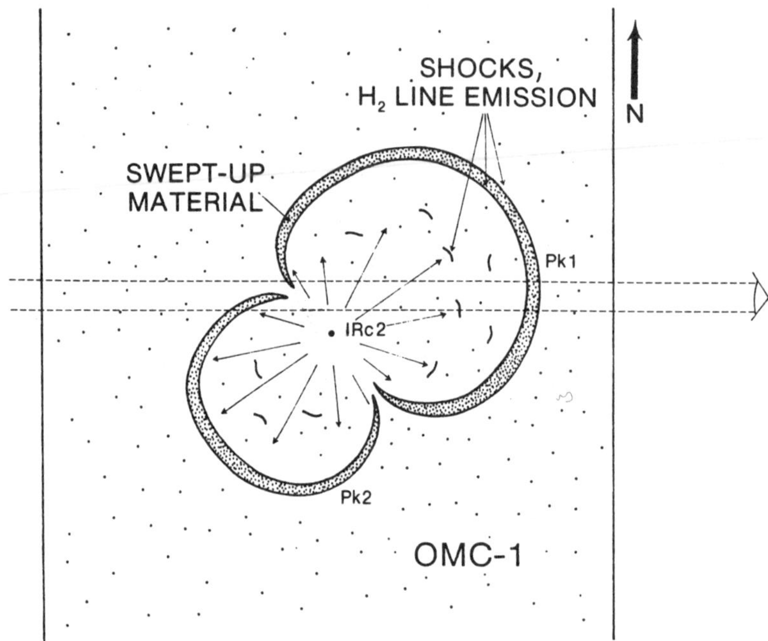

Figure 1:   A sketch of the central region of OMC-1 showing shells driven by a bipolar outflow from the embedded young stellar object IRc2.   Pk1 and Pk2 refer to maxima in H_2 line emission.   (From Geballe et al. 1986)

Two fine structure transitions, OI at 63 μm (Werner et al. 1984) and SiII at 35 μm (Haas, Hollenbach, and Erickson 1986), have been attributed to the dissociative Orion shock.   The OI line is $\approx$ 50 km s^{-1} wide at the peak of hot H_2 emission, and has a blueshifted wing, like the H_2 line.   The SiII peak is unresolved at an observed resolution of 114 km s^{-1}.   The H51α recombination line of hydrogen (Hasegawa and Akabane 1984) may also occur in the J-shock.   An observed upper limit for the intensity of the Brα emission line (Geballe and Garden 1987) is just equal to the predictions of J-shock models.   More sensitive searches for Brα, and other recombination lines, are clearly desirable.   The above evidence for a dissociative shock is not unambiguous, however, because the Orion core lies behind an HII region and its associated photodissociation region (ion-neutral interface).   Atomic fine structure emission is produced in photodissociation regions and hydrogen emission lines are produced in HII regions.

Molecule formation behind fast dissociative shocks has recently been considered in detail by Neufeld and Dalgarno (1989a,b) and Hollenbach and McKee (1989).   They considered shocks with speeds from 50 to 120 km s^{-1}, the range in which dissociation and ionization by a radiative precursor increases from slight to complete.   Behind fast shocks, the chemistry is fundamentally different from cold cloud chemistry because of the high fractional ionization and the intense uv radiation field (much of which is in Lyα photons).   All chemistry follows the formation of H_2 which is first formed from H$^-$, after the temperature has dropped to $\approx 10^4$K, by H + e $\rightarrow$ H$^-$ + hν, followed by H$^-$ + H $\rightarrow$ H_2 + e and H$^-$ + H$^+$ $\rightarrow$ H + H.   When the fractional ionization falls, formation of H_2 on grains dominates the H-route.   OH and H_2O, for example, then form via the endothermic H atom abstractions O + H_2 $\rightarrow$ OH + H, and OH + H_2 $\rightarrow$ H_2O + H.   They are destroyed by the reverse reactions and by photodissociation.   OH is important in the formation of CO and SiO, by C$^+$ + OH $\rightarrow$ H$^+$ + CO and Si + OH $\rightarrow$ SiO

+ H. Cooling is by atomic fine structure transitions and molecular rovibration
ransitions. When the temperature decreases to a few hundred kelvins, molecular
synthesis becomes slow but photodestruction does not, so that molecular abundances
drop to low values.

Neufeld and Dalgarno (1989b) suggest several new diagnostics for fast J-shocks.
Molecular ions become abundant in hot (6000-10000 K) gas, so that vibrationally
excited molecular ions (e.g. OH^+, SO^+, HeH^+) may be a unique signature of fast
shocks. Intensities of v=1-0 transitions of OH^+ and HeH^+ in the range 3.3 to 3.7
μm were calculated and may be detectable. SiO, becomes very abundant ($N \approx 10^{16}$ cm$^-$
^{2}). The J=19-18 transition of SiO, which lies in the atmospheric window at 364 μm,
is a prime candidate for observation. The HCO+ ion also becomes abundant ($N \approx 10^{13}$
cm^{-2}), making a search for high J submillimeter transitions attractive.

4. Molecular Hydrogren Emission from the Orion Shock

The intensities of excited H_2 emission lines from Orion can be reproduced rather
well by a non-dissociative C-shock. The shapes and widths of the H_2 lines remain,
however, difficult to understand. Toward Peak 1 the 1-0 S(1) line in Orion has a
FWZI > 140 km s^{-1}, with no structure seen in the line at a resolution of 12 km
s^{-1}. Brand et al. (1989a) show that a bow shock due to a fast wind impacting a
dense clump yields concave profiles, unlike the observed profiles. They also
considered multiple shocks from an ensemble of emitters with a range of velocities
up to 140 km s^{-1}. The observed smoothness of the line requires 30 to 40 cloudlets
in a line of sight 2500 AU in diameter. The symmetry of the H_2 lines about zero
velocity is a severe problem because the H_2 position is displaced from the centre
of the outflow. Given the geometry, most clumps along this line of sight should
be blueshifted. A single fast J-shock (to produce broader lines) will not work
because the predicted 1-0 S(1) intensity is too low by a factor of $\approx$ 1000.

Several arguments favour J-shocks in OMC-1: (1) A Boltzmann plot for 19
infrared H_2 lines with E/k from 6500 K to 25000 K is well fit by a simple
approximation to a hydrodynamic (single fluid) shock with H_2 cooling. A C-shock
is a poor fit, underpopulating the higher levels (Brand et al. 1988); (2) The
temperature sensitive line ratio, 0-0 S(13)/1-0 O(7), is constant over the outflow
region (Brand et al. 1989b). This is expected in gas cooling slowly after prompt
heating (J-shock), but not in the extended parameter-sensitive heating zones of C-
shocks; (3) The three lines 1-0 S(1), 2-1 S(1), and 3-2 S(3), have very different
upper energy levels (7000 K, 12500 K, and 19000 K) but identical profiles. Behind
J-shocks the temperature distribution of H_2 is almost independent of shock speed,
whereas behind C-shocks the H_2 temperature is sensitive to shock speed and other
parameters. Thus, line profiles from multiple J-shocks will be similar for all
upper state energies. This is not so for multiple C-shocks (Moorehouse et al.
1990).

Smith and Brand (1990a,b) have introduced analytical simplifications into C-
shock modeling and have used these methods to compute H_2 line profiles, line
intensities, and line ratios. They find that (1) H_2O cooling, as expected in C-
shocks, implies very narrow lines ($\approx v_s/2$), (2) H_2 cooling yields broader lines ($\approx$
v_s), (3) lines from lower energy levels (< 7000 K) are double-peaked, while lines
from higher levels (> 12000 K) have a single peak at low velocity and a high
velocity wing, (4) the H_2 excitation temperature should vary greatly across an
extended source, in contradiction to the constancy of the 0-0 S(13)/1-0 O(7) ratio
in Orion, and (5) computed column densities agree with observations. They
conclude that line ratios, widths, and shapes cannot be explained by a single
planar C-shock.

References

Brand, P.W.J.L., Moorhouse, A., Burton, M.G., Geballe, T.R., Bird, M., and
     Wade, R. 1988, Ap.J. (Letters), **334**, L103.
Brand, P.W.J.L., Toner, M.P., Geballe, T.R., and Webster, A.S., 1989a,
     M.N.R.A.S., **237**, 1009.
Brand, P.W.J.L., Toner, M.P., Geballe, T.R., Webster, A.S., Williams, P.M.,
     and Burton, M.G. 1989b, M.N.R.A.S., **236**, 929.
Chernoff, D.F. 1987, Ap.J., **312**, 143.
Chernoff, D.F., Hollenbach, D.J., and McKee, C.F. 1982, Ap.J. (Letters), **259**,
     L97.
Draine, B.T. 1980, Ap.J., **241**, 1021.
_______ 1986, Ap.J., **310**, 408.
Draine, B.T. and Katz, N.S. 1986a, Ap.J., **306**, 655.
_______ 1986b, Ap.J., **310**, 392.
Draine, B.T. and Roberge, W.G. 1982, Ap.J. (Letters), **259**, L91.
Elitzur, M. and Watson, W.D. 1978, Ap.J. (Letters), **222**, L141.
Falgarone, E. and Pérault, M. 1988, in *Physical Processes in Interstellar
     Clouds*, eds. G. Morfill and M. Scholer (Dordrecht: Reidel).
Flower, D.R., Pineault des Forêts, and Hartquist, T.W. 1985, M.N.R.A.S., **216**, 75.
Geballe, T.R. and Garden, R. 1987, Ap.J. (Letters), **317**, L107.
Geballe, T.R., Persson, S.E., Simon, T., Lonsdale, C.J., and McGregor, P.J. 1986,
     Ap.J., **302**, 693.
Haas, M, Hollenbach, D.J., and Erickson, E.F. 1986, Ap.J. (Letters), **301**, L57.
Hasegawa, T. and Akabane, K. 1984, Ap.J. (Letters), **287**, L91.
Hollenbach, D.J. and McKee, C.F. 1989, Ap.J., **342**, 1989.
Langer, W.D., Glassgold, A.E., and Wilson, R.W. 1987, Ap.J., **322**, 450.
Le Bourlot, J., Gérin, M., and Pérault, M. 1989, Astr. Ap., **219**, 279.
Magnani, L. and Salzer, J.J. 1989, A.J., **98**, 926.
Millar, T.J. and Hobbs, L.M. 1988, M.N.R.A.S., **231**, 953.
Millar, T.J., Adams, N.G., Smith, D., Lindiner, W., and Villinger, H. 1986,
     M.N.R.A.S., **221**, 673.
Mitchell, G.F. 1983, M.N.R.A.S., **205**, 765.
Mitchell, G.F. and Watt, G.D. 1985, Astr. Ap., **151**, 121.
Moorhouse, A., Brand, P.W.J.L., Geballe, T.R., and Burton, M.G. 1990,
     M.N.R.A.S., in press.
Neufeld, D.A. and Dalgarno, A. 1989a, Ap.J., **340**, 869.
_______ 1989b, Ap.J., **344**, 251.
Pilipp, W., Hartquist, T.W., and Havnes, O. 1990, preprint.
Pineault des Forêts, G., Flower, D.R., Harquist, T.W., and Dalgarno, A. 1986,
     M.N.R.A.S., **220**, 801.
Pineault des Forêts, G., Roueff, E., and Flower, D.R. 1986, M.N.R.A.S., **223**, 743.
Pineault des Forêts, G., Flower, D.R., Hartquist, T.W., and Millar, T.J. 1987,
     M.N.R.A.S., **227**, 993.
Roberge, W.G. and Draine, B.T. 1990, Ap.J., in press.
Smith, M.D. and Brand, P.W.J.L. 1990a, M.N.R.A.S., in press.
_______ 1990b, M.N.R.A.S., in press.
Snow, T.P., Seab, C.G., and Joseph, C.L. 1988, Ap.J., **335**, 185.
Werner, M.W., Crawford, M.K., Genzel, R., Hollenbach, D.J., Townes, C.H., and
     Watson, D.M. 1984, Ap.J. (Letters), **282**, L81.

Circumstellar Chemistry

A. E. Glassgold[1] and G. A. Mamon[2]

[1] *New York University, New York, NY 10003, U. S. A.*
[1] *Observatoire de Paris, 92195 Meudon, France*

Recent theoretical studies of circumstellar chemistry are discussed for both red-giant and protostellar winds. The generalized photochemical model is able to account for the recently discovered silicon-bearing molecules in the prototypical, C-rich, AGB star IRC +10216. The surprising occurrance of CO in protostellar winds that are largely atomic is interpreted to be the result of the high density and the rapid decrease of the temperature with distance that is expected for such winds.

I Introduction

The objects of main concern to circumstellar chemistry are stellar winds that occur at critical stages of stellar evolution, stellar birth and death. An important property of these winds is the small characteristic time scale, e.g., for red giants

$$t = r/V = 300 \text{ yrs } r_{16}/V_6$$

measuring distance r in 10^{16} cm and expansion velocity V in 10^6 cm s^{-1}. The small time scales imply that we can extrapolate back to the initial conditions fairly directly, in contrast to the situation for interstellar clouds. Another aspect of circumstellar chemistry is that the flows are well-defined, which simplifies the modeling. In the following sections, we report briefly on silicon chemistry in C-rich winds from evolved stars and on the outflow chemistry of deeply embedded, low-mass protostars.

II IRC +10216

This nearby object is the best studied, evolved star with a massive stellar wind. At the Kona symposium, the list of Lucas and Guelin (1) contained 14 more molecules than the 22 given in the review by Glassgold and Huggins (2), not counting numerous isotopic variants and vibrationally excited molecules. If we add the most recent discoveries, C_5 (3), SiC (4), SiC_4 (5), and CP (6), 40 molecules have been detected to date in IRC +10216. The detection of C_3 (7) and C_5 (3) demonstrate the power of IR absorption techniques, especially the capability to provide temperature information (which is how we know that C_3 and C_5 are located in the outer envelope). The detection of HCO$^+$ at the 20 mK level (1) agrees with theoretical predictions (8,9) and confirms both the penetration of cosmic rays and the operation of ion-molecule reactions in circumstellar envelopes. The detection of metal halides (10), organic sulfur compunds (11), and numerous silicon-bearing molecules all attest to the continuing vitality of circumstellar chemistry research. Intimately connected with the question of <u>how</u>

these molecules are formed, is <u>where</u>. Of invaluable help in this regard are interferometer maps, e.g., those recently obtained by Bieging and Rieu at Hat Creek (12,13).

We discuss the chemistry of IRC +10216 from the point of view of the photochemical model; a recent review by Millar (14) provides a good point of departure. In the generalized photochemical model, the following ideas are invoked to explain molecular abundances outside of the region of dust formation (8,9,15):

1. Equilibrium, established interior to or in the dust shell, implies the existence of several well-bound molecules with high abundances, i.e., H_2, CO, C_2H_2, N_2, HCN, and SiS (progenitor molecules).
2. Interstellar radiation penetrates from the outside, destroys the progenitor molecules, and produces radicals and ions in shells at intermediate distances.
3. The photo-fragments react to form more complex molecules, which are photodestructed in turn. Except for dust grains, all circumstellar molecules are eventually reduced to the atoms and ions characteristic of the diffuse interstellar medium.

A characteristic prediction of the model is that a temporal production sequence translates into a spatial sequence, e.g., the species in the primary acetylene "photochain",

$$C_2H_2 \rightarrow C_2H \rightarrow C_2 \rightarrow C \rightarrow C^+ \ ,$$

are located at progressively larger distances. The first (C_2H_2) and last (C^+) are, of course, special, i.e., have constant abundances at small and large r, respectively. This picture is somewhat oversimpliied because the products of <u>all</u> of the photochains react with one another and produce more complex species in overlapping regions. In addition, the acetylenic ion is formed in the C2H2 photo-chain, and together with the other abundant ion C^+, generates a rich ion-molecule chemistry. If, as is likely, $C_2H_2^+$ reacts with H_2, then the active molecular ion is $C_2H_3^+$ and some minor changes occur in the chemistry as presently formulated. The pattern of abundances observed by Bieging and Rieu (12) provides strong support for the picture derived from the photochemical model.

One of the most interesting aspects of the chemistry of IRC +10216 is the detection of six silicon-bearing molecules, two in the last year: SiO (16), SiS (16), SiH_4 (17), SiC_2 (18), SiC (4), and SiC_4 (5). A qualitative understanding of the synthesis of the carbides in IRC +10216 can be obtained from the photochemical model starting with the SiS photochain (19). It is well known that SiS is the most abundant silicon-bearing molecule in the inner envelope and that its abundance decreases with distance (13), presumably due to dust depletion and, at larger distances, photodissociation. Because $C_2H_2^+$ (or $C_2H_3^+$) is the most abundant molecular ion (with peak abundances reaching 10^{-7}), we have proposed the following reactions to produce the silicon carbides:

$$C_2H_2^+ + SiS \rightarrow SiC_2H^+ + SH \qquad (1a)$$

or

$$C_2H_3^+ + SiS \rightarrow SiC_2H^+ + SH_2 \qquad (1b)$$

$$C^+ + SiS \rightarrow SiC^+ + S \qquad (2)$$

$$SiC_2H^+ + C_2H_2 \rightarrow SiC_4H^+ + H_2. \qquad (3)$$

Only the last reaction has been measured (19). We start with an initial abundance of SiS obtained from the measurements of Bieging and Rieu (13) and allow SiS to

be depleted onto grains at a rate determined by the geometric cross section of
the dust grains. Together with other familiar processes, these reactions produce
the observed molecules according to the following <u>schematic</u> diagram:

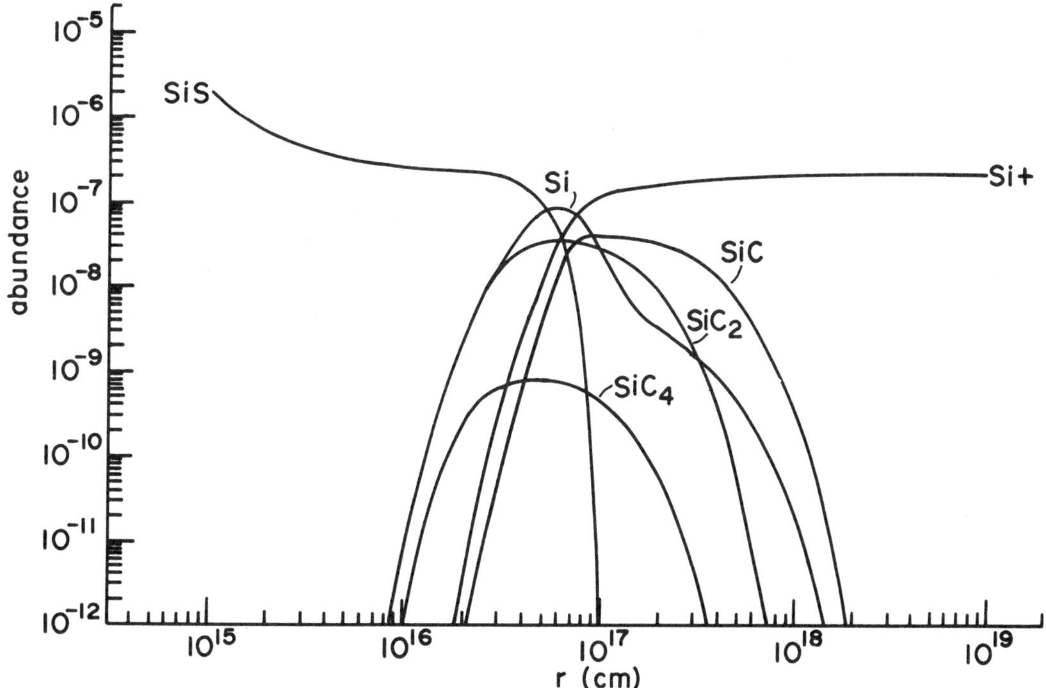

The variation of the abundances of the silicon species with distance is shown
below in Figure 1. The results agree roughly with the observations, in particular
with the fact that SiS is an "interior" molecule and that the other species are
probably located in "shells". Reactions of Si+ and acetylene (20) and of Si with
the acetylenic ion (21) may also contribute to the chemistry. Considerations
similar to the above have also been advanced recently by Howe and Millar (22),
who assume that SiC_2 is a progenitor molecule. The theory of the SiS photochain
chemistry is preliminary because of the lack of basic laboratory data and
theoretcal calculations, as well as of observations. Further confirmation of the

Fig. 1. Abundance variation of silicon molecules with distance, based on the
ion-molecule chemistry of the SiS photochain (19). Some SiC_3 will be produced by
photodissociation of SiC_4.

theory requires measurement of the spatial variation of the brightness disributions of SiC_2, SiC, and SiC_4. Laboratory measurements of reactions (1) and (2) and other ion-molecule reactions of SiS are also needed, as are calculations of photorates for SiS and the silicon carbide species.

Similar qualifications apply to the photochemical model in general. Although its predictions of cutoffs and peaks in the abundances of progenitor and product species have been confirmed in many cases, a uniform analysis of the observations has never been been attempted. High quality data obtained with large telescopes and interferometers are required for this purpose. Additional theoretical improvements are also needed, e.g., incorporation of a more complete neutral chemistry (22) and improved treatment of the far UV dust opacity, which is not known a priori. However, we do not favor adding additional progenitor molecules simply to improve the agreement with observations. Our list (H_2, CO, C_2H_2, N_2, HCN, and SiS) is based on thermodynamic equilibrium calculations. For the pressure and temperature range expected for the upper atmosphere of IRC +10216 (23, 24), these are the only molecules that have large and approximately constant abundances. For example, the assumption (15) that CH_4 and NH_3 are injected into the outer envelope with abundances greater than 10^{-6} is not well founded, especially for NH_3 where equilibrium abundances are typically $< 10^{-12}$. This issue again underscores the importance of observations that provide information on the spatial variation of circumstellar molecules.

PROTOSTELLAR WINDS

The challenges of protostellar winds differ from those of red giants because the size scale is so much smaller, typically by a factor of a thousand. In a red-giant wind, the largest telescopes and interferometers are already able to the detect the spatial variation of the abundances of both progenitor molecules and the products of in-situ molecular synthesis. In contrast, radio observations of protostars refer to phenomena that occur far from the protostar and are averages over relatively large regions; they often sample interactions of the wind with ambient material, e.g, shock phenomena. Our concern here is with molecular processes that occur in the wind itself, particulary those associated with the generation of the wind. Observations at an angular resolution of 5" probe a distance of 1000 AU in objects that are 200 pc distant, so that direct imaging of the vital arts of the protostar is infeasible at present. Nonetheless, there is direct spectroscopic evidence of primary protostellar winds in the form of extremely broad atomic and molecular line profiles, indicative of wind speeds of several hundred km s^{-1}. Of particular interest is the detection of CO in winds from low-luminosity protostars that are primarily atomic in character (25).

The chemistry in the near regions of a protostellar wind are quite different from interstellar chemistry, as the following limits to the relevant parameters attest:

Physical Conditions at the Base of a Protostellar wind

Density	$n_H < 10^{14}$ cm^{-3}
Temperature	$T < 10^4$ K
Photo Rates	$G < 10^6$ s^{-1}
Atomic component	$x(H_2) \ll x(H)$
Ionization	$x_e < 10^{-2}$
Velocity Gradient	$d/dr\ v < V/R_* \sim 10^{-5}$ cm s^{-1}

These conditions have the following implications for the chemistry:

1. Neutral reactions of atoms and radicals provide the most important synthesis routes.
2. Collisional dissociation is important for the destruction of non-polar molecules, i.e., H_2, CO, etc.
3. Charge exchange controls the ionization of the wind and also destroys low ionization-potential molecules (but not H_2 and CO).
4. Protostellar radiation and its shielding are important.
5. Radiative association provides a minimum molecular content.

Because the winds are hydrogen rich, the formation of H_2 is the first step toward molecular synthesis. It can be produced three ways: from the progenitor ions H^- and H_2^+ and by three-body reactions, such as $3H \rightarrow H_2 + H$. Heavy molecules like CO can then be made most directly from neutral chemical reactions in two additional steps, e.g.,

$$O + H_2 \rightarrow OH + H \qquad C + OH \rightarrow CO + H.$$

The first reaction has a barrier of about 4000 K so that the wind must be warm - but not be too warm, otherwise collisional dissociation of H_2 becomes effective.

In a preliminary report (26), we concluded that this type of scheme could be efficient if the temperature decreased sufficiently rapidly. We used an an-hoc temperature distribution with free parameters and assumed that the wind was spherically symmetric and started out at terminal speed. The basis for our chemical modeling has been greatly improved by the first calculations of the heating and cooling of protostellar winds (25). We have recently finished a study of the chemistry of low-luminosity protostellar winds (28) that uses this theory and that also includes a more complete formulation of the chemistry, including radiation transfer. Because of space limitations, we refer the reader to reference (27) for the details of the hydrogen chemistry and to reference (28) for the heavy element chemistry. The earlier qualitative conclusions are confirmed by our new calculations, i.e., molecular synthesis can be quite efficient in these winds, especially for the higher densities expected in the acceleration region and in collimated winds. The results are rather remarkable even for a completely isotropic wind. For example, for a mass-loss rate of 3×10^{-6} M_0 yr^{-1}, one quarter of the available carbon is in CO but only 10^{-5} of the hydrogen is in molecular hydrogen.

ACKNOWLEDGEMENTS

The authors wish to acknowledge the partial support of a NASA grant, the early receipt of a preprint (22) from Tom Millar, and the help of their collaborators, particularly Patrick Huggins and Alain Omont.

REFERENCES

1. Guelin, M. and Lucas, R. 1988, Kona symposium, "Submillimeter and Millimeter Astronomy".
2. Glassgold, A.E. and Huggins, P.J., in "M, S, and C Stars", Eds. H. Johnson and F. Querci (NASA and CRNS, 1987), p. 114.
3. Bernath, P.F., Hinkle, K.H., and Keady, J.J. 1989, Science 244, 562.
4. Cernicharo J. et al. 1989, Ap. J. (Letters), 341, L25.
5. Ohishi. M. et al. 1989, Ap. J. (Letters) 345, L86.
6. Guelin M. et al. 1990, Astr. Ap. in press.
7. Hinkle, K.W., Keady, J.J., and Bernath. P.F. 1988, Science 241,1319.
8. Glassgold, A.E., Lucas, R., and Omont, A. 1986, Astr. Ap. 157, 35.

9. Glassgold, A.E. Mamon, G.A., Omont, A., and Lucas, R. 1987, Astr. Ap. <u>180</u>, 183.
10. Cernicharo J. and Guelin, M. 1987, Astr. Ap. <u>183</u>, L10.
11. Cernicharo, J., Guelin, M., Hein, M., and Kahane, C. 1987, Astr. Ap. <u>181</u>, L9.
12. Bieging, J.H. and Rieu, N.-Q. 1988, Ap. J. (Letters), <u>329</u>, L107.
13. Bieging, J.H. and Rieu, N.-Q. 1989, Ap. J. (Letters), <u>343</u>, L25.
14. T. Millar 1989, in "Rate Coefficients in Astrophysics", Eds. T.J. Millar and D.A. Williams, (Kluwer), p. 287.
15. Nejad, L.A.M. and Miller., T.J. 1984 Astr. Ap. <u>183</u>, 279.
16. Morris, M., Gilmore, W., Palmer, P., Turner, B.E., and Zuckerman, B. 1975, Ap. J. (Letters), <u>199</u>, L47.
17. Goldhaber, D. and Betz, A. 1984, Ap. J. (Letters) <u>279</u>, L55.
18. Thaddeus, P., Cummins. S.E., and Linke, R.A. 1984, Ap. J. (Letters), <u>283</u>, L45.
19. Glassgold, A.E., Mamon, G.A., and Omont, A. 1989, in preparation.
20. Bohme, D., Wlodek, S., and Fox, A. 1988, in "Rate Coefficients in Astrophysics", Eds. T.J. Millar and D.A. Williams, (Kluwer), p.193.
21. Herbst, E., Millar, T.J., Wlodek, S., and Bohme, D.K. 1989, Astr. Ap. in press.
22. Howe, D.A. and Millar, T.J. 1989, preprint.
23. Lafont, S., Lucas, R., and Omont, A. 1982, Astr. Ap. <u>106</u>, 201.
24. Liu, W. and Glassgold, A.E. 1990, in preparation.
25. Lizano, S. et al. 1988, Ap. J. <u>328</u>, 763.; Koo B.-C., 1989, Ap. J. <u>337</u>, 318; Murgulies, M. and Snell, R.L. 1989, Ap. J. <u>343</u>, 779.
26. Glassgold, A.E., Mamon, G.A., and Huggins, P.J. 1989, Ap. J. (Letters) <u>339</u>, L29.
27. Ruden, S.P., Glassgold, A.E., and Shu, F.H. 1990, Ap. J. in press.
28. Glassgold, A.E., Mamon, G.A., and Huggins, P.J. 1990, Ap. J., in press.

Exotic Chemistry in Star Forming Regions

C. M. WALMSLEY

Max-Planck-Institut für Radioastronomie, Auf dem Hügel 69, 5300 Bonn 1, F. R. G.

A brief review is given of the chemistry in "hot core" type molecular clumps situated in the immediate vicinity of young massive newly formed stars. In particular, I discuss the hypothesis that what is seen in hot cores is basically newly evaporated dust mantle material and that one can therefore deduce the constitution of dust mantles from the observation of the molecular abundance distribution in hot cores. Special attention is given to oxygen bearing species and the question of the water abundance.

1. Exotic or not exotic

Exotic or not exotic : that is the question. If one takes a running mean of Websters and the Concise Oxford dictionaries, one finds that the first meaning of exotic is "introduced from abroad"' but that it can also be used in reference to strip-tease dancers. In the following review of hot core chemistry, I will discuss the evidence for the proposition that many of the molecules observed in hot cores have recently been stripped from the surfaces of dust grains. If this is correct, the molecules concerned are presumably exotic in that they are introduced from another phase of the interstellar medium. However , as usual, other interpretations of the data are possible. As an aid towards deciding this question, I give in section 2 a short description of what we know about hot cores with references to other more detailed accounts. In section 3, I summarise recent work on oxygen bearing molecules and in section 4 I give a brief conclusion.

2. Introduction to Hot Cores

In a review of interstellar molecular abundance measurements carried out a few years ago, it was concluded (1) that "hot" molecular clouds (mainly near HII regions) and cold clouds had a similiar abundance distribution. This could be interpreted as meaning that gas phase chemistry in general and ion-molecule chemistry in particular was the dominant process determining molecular abundances. Subsequent reviews of this topic (e.g.2) confirm this general conclusion. However, it has become increasingly clear that like all good rules, this one has exceptions. One of these exceptions is a hot dense pocket of gas immersed in the Orion Kleinmann-Low nebula which has become known as the Orion Hot Core.

The existence of this entity first became apparent in measurements of ammonia towards Orion (see 3,4) where it was found that in addition to the velocity component at around 8 kms^{-1} characteristic of the ambient Orion molecular cloud gas, there was a spatially compact region with it's peak emission at around 5 kms. This was one of the first regions to be mapped in spectral lines by the VLA. The results showed (5,6) that the region had an angular extent of roughly 10 arc sec. corresponding to a linear diameter of roughly 0.02 pc. From the brightness temperature of the lines observed

with the VLA as well as from an excitation analysis, one could conclude that the
temperature was of order 200 K. By molecular cloud standards, this is extremely high
and suggests that the region lies close to the young stars embedded in and
responsible for the strong observed infrared emission. More recently, the Orion hot
core has been detected in a variety of other molecular line transitions as well as by
it's dust emission at 3 mm (see eg 7,8,9,10,12,13,14). From the dust emission
observations (9,10), one can conclude that the column density of the hot core is of
order $10^{24} cm^{-2}$ and the mean hydrogen density $10^7 cm^{-3}$. Thus it is indeed a hot
dense clump of gas whose origin is presumably related to the star formation process in
the region. A useful general account of the earlier ammonia observations of the Orion
hot core is given in the review by Ho and Townes (11). The more recent data is
reviewed by references 12 and 13 while ref.13 also discusses observations of sources
more distant than Orion. Whether the Orion hot core is typical is unclear.

The chemical constitution of the Orion hot core also marks it out (see 2). In contrast
to the situation in cold dark clouds such as TMC1, one finds predominantly saturated
species (C_2H_5CN, HC_3N, CH_3CN) and no radicals or ions. CH_3CN for example is found in
the hot core but not in the adjoining ridge gas and indeed, the difference between the
observed hot core abundance and the upper limit for the ambient gas is two orders of
magnitude. Figure 1, taken from reference 14 shows a graphical comparison of hot core
and Orion ridge abundances.

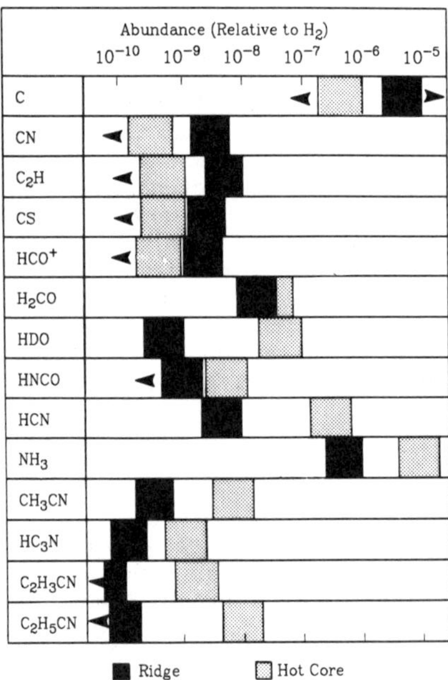

*Fig. 1   A comparison of observed abundances
in the Orion hot core region with those
determined in the more extended ridge gas.
(from ref.14).*

How can these "anomalies" be interpreted ? Although, the details are unclear, it seems
likely that part of the explanation is that while the timescale for achieving gas phase
chemical equilibrium is long (more than 10^6 years), the age of the Orion hot core is
short. If one defines a rough "dynamical age" for the region as $l/\Delta v$ where l is the hot
core diameter and Δv is the observed line width, one finds a timescale of a few

thousand years. Another approach is to note that the hot core has certainly only existed in it's present form in the time since the star responsible for heating the gas and dust to the observed high temperatures became luminous. The star in question (12) seems likely to be the luminous infrared source IRc2. Associated with this source is outflowing gas seen in water masers, CO, and SiO emission. The dynamical age of these flows is also a few thousand years and so it is reasonable to conclude that IRc2 has not existed for much more than 10^4 years. Certainly, its age is likely to be much smaller than that necessary for the evolution of much gas phase chemistry. It follows that the hot core abundances probably reflect the conditions imposed at the time of the switch-on of IRc2. This could happen either due to evaporation of grain mantles or shocks or a combination of these effects. Grain mantle evaporation seems in fact inevitable at the temperatures (above 150 K) estimated for hot core dust. Shocks are also highly likely due to the interaction of the flows from IRc2 with its surroundings. Various authors (14,15) have compared what is known about grain mantle composition with observed molecular abundances in the hot core. Detailed models of the chemical evolution have also been attempted (16,17). These follow the evolution of gas phase and dust mantle abundances in a collapsing cloud. There is competition between depletion onto grain surfaces and the normal evolution of gas phase chemistry. On the grain surfaces, light atomic species (H and D in particular) are mobile and can react producing water and ammonia. In the gas phase, ion-molecule reactions can give rise to deuterium fractionation and produce hydrocarbons. At a certain density ($10^7 \mathrm{cm}^{-3}$), the evolution is supposed halted by the switch-on of the central star which causes the evaporation of the ice mantles thus throwing heavy elements back into the gas phase. The gas phase evolution is then followed for a time (10^4yrs) comparable to the supposed age of hot core regions. In general, one finds that one can explain the over-abundance of simple hydrides (HDO, NH_3) relative to cold clouds but the more complex species represent a much more difficult problem. In the next section, we discuss recent results relative to oxygenated species and in particular to water.

3. Oxygen compounds in the hot core

If one accepts the hypothesis that in hot core type regions grain mantles have been largely evaporated, it is plausible that the gas phase elemental distribution in hot cores should resemble that observed in diffuse clouds. In the latter situation, it is thought that the action of the background ultra-violet radiation field prevents the accretion of mantles. In hot cores, it is reasonable to suppose that the elevated dust temperature achieves the same result. Hence, one might expect broadly similiar elemental abundances in the two cases. In diffuse low density clouds (see ref.18 and references therein), the observed depletions vary between 0 and 0.5 dex for carbon, nitrogen, and oxygen but are much larger for heavier elements such as silicon and iron. However, depletion correlates with density and these numbers are lower limits to the depletions in clouds of densities of order 100 cm^{-3} . The regions where one has direct evidence for the presence of mantles have probably still higher densities (19,20) but unfortunately, we have no good determinations of the gas phase abundances in those lines of sight. A priori however, it seems reasonable to suppose that most of the available C, N, O will be in the gas phase under hot core conditions.

The particular case of oxygen is interesting. If oxygen is in the gas phase in hot cores, what form does it take? In the Orion hot core, measurements of ^{13}CO are available which seem consistent with the canonical result $[CO]/[H_2]=10^{-4}$ (see Refs. 21, 22). However, there is no reason to expect oxygen to be more depleted than carbon (although see 23) and hence, even if all gas phase hot core carbon is in the form CO, there still may be other abundant oxygen-bearing molecules. The various possibilities are well discussed in ref.14. Briefly, one might imagine that oxygen is either atomic,or in the form of O_2, H_2O, CO_2 or perhaps in hydrogenated forms of some of the above. Models for dust mantle chemistry give us some guide as to which of these possibilities are likely (see 15, 23 and refs. therein). From these models (eg 16, 24, 25, 26), one can conclude that at low densities ($10^3-10^4 \mathrm{cm}^{-3}$) where atomic hydrogen is abundant in the gas phase, most free oxygen is channelled into water and indeed in general, the simple hydrides (methane and ammonia) appear to dominate. At higher densities, the situation is less clear but O_2 and CO_2 are likely to be important.

What can one learn from the observations about hot core abundances of oxygen bearing species? For molecular oxygen, upper limits are available (27, 28) which indicate that O_2 is less abundant than CO. For the case of atomic oxygen, measurements of the fine structure lines at 63 and 144 microns allow in principle estimates of the abundance but the confusion due to foreground shocked and photo-heated gas renders a column density determination for the hot core component very difficult. Moreover, the models mentioned above suggest that atomic oxygen is not liable to be abundant. CO_2, like many species of interest in the hot core , has no dipole moment and hence cannot be observed directly. One possible approach in general to determining the gas phase CO_2 abundance is to use measurements of its protonated form (29) but this method appears to fail in the hot core due to the lack of molecular ions (in particular, the result of (29) that $[CO_2]/[CO]$ in Orion is less than 10^{-2} is not applicable to the hot core). Hence, one must currently reserve judgement concerning this species.

But the one molecule above all others which one might hope to see in hot cores is H_2O. The most direct approach to obtaining the water abundance is certainly (for ground based observers) to observe $H_2^{18}O$ but unfortunately only the $3_{13}-2_{20}$ transition at 203 GHz can be readily observed and measurement of this line is problematic due to blending with other species. For this reason, it is fortunate that HDO appears to be readily observable under hot core conditions with several transitions available for analysis. Recent measurements of both these species are discussed by references 30 and 31. With considerable uncertainty, one finds that $[H_2O]/[H_2]$ is approximately 10^{-5} and that $[HDO]/[H_2O]$ is typically $3\ 10^{-4}$. In a sense, the surprising thing about this result is that the water abundance is so low. It implies that only a small fraction of the hot core oxygen is in the form of water and hence other possible oxygen reservoirs such as CO_2 may be important. The detection of a solid-state feature of this species (32) with an abundance similiar to that of CO renders this possibility attractive. On the other hand the water-ice column densities in the sources studied by ref. 32 (see also 23) are an order of magnitude larger than found for either CO or CO_2 and hence from this point of view also, the low hot core H_2O abundances are a puzzle. While the uncertainties on these abundance estimates are large, it seems hard to believe that they can be sufficient to explain such discrepancies. One possible explanation (see 31) is that complete heavy element depletion it never occurs or in any case has not occurred prior to the switch-on of the stars responsible for heating the hot core gas. In this scenario, atomic oxygen (which is thought to be abundant in gas phase models prior to freeze-out) might still be an abundant hot core species for example.

4. Conclusions

The best test of the hypothesis that hot core molecular abundances essentially reflect grain-mantle composition might be expected to be simply a comparison of the observed abundance distributions in the two cases (say a comparison of table 1 from the review by Tielens (23) with the results in fig.1). By this standard, the grain-mantle evaporation hypothesis fails in that CO appears to be more abundant than H_2O in hot cores as well as some more minor problems. However, the evidence here is skimpy (and perhaps unreliable as far as the CO and H_2O hot core abundance measurements are concerned) and the grain-mantle hypothesis can explain the qualitative feature of high abundances for saturated molecules combined with the absence of radicals and ions. Deuterium chemistry, which I have not discussed, can also be reasonably well understood (17, 31). Hence, I feel that a final conclusion on "exotic or not exotic" is far from having been reached.

References

1. Watson W.D., Walmsley C.M. 1981 in Regions of recent Star
   Formation (eds. Roger R.S., Dewdney P.) (publ. D.Reidel, Dordrecht)
2. Irvine W.M., Goldsmith P.F., Hjalmarson A. 1987 in
   Interstellar Processes, eds Hollenbach D.J., Thronson H.A.
   (Dordrecht,Reidel)
3. Barrett A.H., Ho P.T.P., Myers P.C. 1977 Astrophys. J **211**, L139.
4. Wilson T.L., Downes D., Bieging J.H. 1979 Astron.Astrophys. **71**, 275.
5. Genzel R., Downes D., Ho P.T.P., Bieging J. 1982 Astrophys. J
   259, L103.
6. Pauls T.A.,Wilson T.L., Bieging J.H., Martin R.N. 1983
   124, 123.
7. Plambeck R.L., Wright M.C.H. 1987 Astrophys. J **317**, L101.
8. Masson C.R., Mundy L.G. 1988 Astrophys. J **234**, 538.
9. Masson C.R. et al 1985 Astrophys. J **295**, L47.
10. Wright M.C.H., Vogel S.N. 1985 Astrophys. J **297**, L11.
11. Ho P.T.P, Townes C.H. 1983 Ann.Rev.Astron.Astrophys. **21**, 239.
12. Genzel R., Stutzki J. 1989 Ann.Rev.Astron.Astrophys. **27**, 41.
13. Wilson T.L.,Walmsley C.M. 1989 The Astron. and Astrophys. Review **1**, 141.
14. Blake G.A., Sutton E.C., Masson C.R., Phillips T.G. 1987
   Astrophys. J **315**, 621.
15. Walmsley C.M. 1989 in "Interstellar Dust" 1989
   IAU Symposium 135 (L.J.Allamandola and A.G.G.M.Tielens eds.)
   (publ. Kluwer, Dordrecht).
16. Brown P.D., Charnley S.B., Millar T.J. 1988 Monthly Notices Roy.
   Astron. Soc. **231**, 409.
17. Brown P.D., Millar T.J 1989 Monthly Notices Roy. Astron. Soc. **237**, 661
18. Jenkins E.B. in "Interstellar Dust" (p 23) 1989
   IAU Symposium 135 (L.J.Allamandola and A.G.G.M.Tielens eds.)
   (publ. Kluwer, Dordrecht).
19. Whittet D.C.B. etal 1988 Monthly Notices Roy. Astron. Soc. **233**, 321.
20. Whittet D.C.B. etal 1989 Monthly Notices Roy. Astron. Soc. **241**, 707.
21. Masson C.R. etal 1984 Astrophys. J **283**, L37.
22. Masson C.R. etal 1985 Astrophys. J **295**, L47.
23. Tielens A.G.G.M in "Interstellar Dust" 1989
   IAU Symposium 135 (L.J.Allamandola and A.G.G.M.Tielens eds.)
   (publ. Kluwer,Dordrecht).
24. Tielens A.G.G.M., Hagen W. 1982 Astron.Astrophys. **114**, 245.
25. d'Hendecourt L.B., Allamandola L., Greenberg J.M. 1985
   Astron. Astrophys. **152**, 130.
26. Jones A.P., Williams D.A. 1984 Monthly Notices Roy. Astron. Soc.
   209, 955.
27. Goldsmith P.F., Snell R.L., Erickson N.R., Dickman R.L.,
   Schloerb F.P., Irvine W.M. 1985 Astrophys. J **289**, 613.
28. Liszt H.S., Vanden Bout P.A. 1985 Astrophys. J **291**, 178.
29. Minh Y.C., Irvine W.M., Ziurys L.M. 1988 Astrophys. J **334**, 175.
30. Jacq T., Jewell P.R., Henkel C., Walmsley C.M., Baudry A.
   1988 Astron.Astrophys. **199**, L5.
31. Jacq T., Walmsley C.M., Henkel C., Baudry A., Mauersberger R.,
   Jewell P.R. 1990 Astron.Astrophys. (in press).
32. d'Hendecourt L.B., Jourdain de Muizon M. 1989 Astron. Astrophys.
   223, L5.

Chemical Evolution from Dust and Gas Interactions: Photoproduction of Solid CO_2 and Production of HCO_2^+ in the Gas*

R. J. L. H. Breukers[1], L. B. d'Hendecourt[2], and J. M. Greenberg[1]

[1] *Laboratory Astrophysics, P. O. Box 9504, 2300 RA Leiden, The Netherlands*
[2] *Groupe de Physique des Solides, Université Paris, Tour 23, 2 place Jussieu, Paris, France*

More gas phase CO_2 is required to make, together with H_3^+, the recently observed large amounts of the HCO_2^+ ion towards the galactic centre (1) than can be accounted for by pure gas phase chemistry. Laboratory studies have shown that solid CO_2 is abundantly produced in photoprocessing of solid ice mixtures containing CO and H_2O, which are generally dominant in interstellar dust. Recently solid CO_2 has been detected in the IRAS Low Resolution Spectra (LRS) of embedded protostars (2,3) confirming this process. The introduction of grains and the photoproduction of solid CO_2 in the grain mantle and desorption into the gas phase appears to provide a ready basis for explaining the production of HCO_2^+ in the gas.

1. INTRODUCTION

Recently the CO_2 molecule has been detected in the IRAS LRS spectra of embedded protostars as part of the icy mantles of interstellar grains (2,3). Its abundance is shown to be roughly equal to that of solid CO and is at least two orders higher than the abundances predicted by pure gas phase chemistry models. Laboratory studies have shown that CO_2 is an effectively produced molecule in photoprocessing of solid ice mixtures.

The explosive mechanism (4-7) releases the solid CO_2 from the grains into the gas. The abundant gas phase CO_2 so produced may combine with H_3^+ to produce the HCO_2^+ which is observed in the radio spectrum. Recently the HCO_2^+ ion has been observed towards Sgr B2(2N) in large amounts (1). A fractional abundance of $F(HCO_2^+) > 1.5.10^{-10}$ with respect to all H-atoms has been calculated which is over 2 orders of magnitude higher than the predictions of the gas phase ion- molecule chemistry of Herbst and Leung (8). It is further stated in ref.1 that, using the gas phase model of Herbst et al. (9), a ratio $[CO_2]/[CO] \geq 0.06$ is needed to get the observed HCO_2^+ abundance, whereas Herbst&Leung (8) have a ratio of 10^{-4}-10^{-3} under the given conditions. Here we want to show the influence of the photoproduction of solid CO_2 in the grain mantle on the gas phase abundances of CO_2 and HCO_2^+ and on the composition of the grain mantle.

2. MODEL

The gas phase reaction network used has been adopted from de Jong et al (10) and Boland et al (11) and is limited to molecules containing the most abundant atoms H, C, O and N, and metals. The solid grain chemistry can be divided in four processes in a way similar to the model of d'Hendecourt et al (12): accretion of gas phase species onto the grains, surface reactions,

---

* The investigations were supported (in part) by the Netherlands Foundation for Astronomical Research (ASTRON) with financial aid from the Netherlands Organization for Scientific Research (N.W.O)

photoproduction of solid CO_2 in the grain mantle and desorption from the grains into the gas triggered by grain-grain collisions in order to prevent depletion of the gas phase species.

The time dependent solution of the set of differential equations describing the evolution of the chemical species is calculated using a variable order, variable step Gear method. As initial conditions depleted cosmic abundances, consistent with depletion of the elements in the interstellar dust (13) and approximately equal to those of ς Oph, have been used. These initial abundances are used in the standard model (STM): $n_o = 2.10^4$ cm^{-3}, $A_v = 4$, $T_{gas} = T_{dust} = 10K$.

3. PRELIMINARY RESULTS

In this section we present preliminary results from both our pure gas phase model and models in which the interaction between dust and gas has been taken into account. We will compare our results at 10^7 and 10^8 years with the results of Herbst&Leung (8).

Fig. 1 shows the time evolution of CO, CO_2 and HCO_2^+ from our pure gas phase chemistry model (dashed line) and from the standard model (STM ; continuous line). At the right of the figure the predictions of Herbst&Leung (8) and the observed abundances can be found. The corresponding fractional abundances and ratios are outlined in table 1. The differences between the calculated abundances from our pure gas phase model and those of the Herbst&Leung model are an effect of the different gas phase chemistry scheme and not of initial conditions. Introducing the grains and solid CO_2 photoproduction in our gas phase chemistry scheme greatly enhances the calculated abundances of gas phase CO_2 and HCO_2^+. Fig. 2 shows the time evolution of the most important grain mantle constituents for the STM.

When reducing the photoproduction of solid CO_2 in the grain mantle by a factor of 10, the $[HCO_2^+]/[CO_2]$ ratio remains 10^{-6} (table 1 ; STM2), still a factor of 10 lower than the ratio from the model of Herbst&Leung (8). However the $[CO_2]/[CO]$ ratio is now 0.08, which is about equal to the value Minh et al (1) suggest is needed to get the observed amount of HCO_2^+

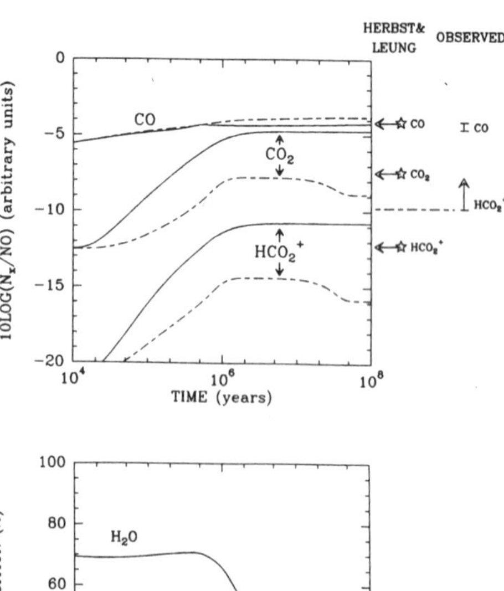

Fig.1

Chemical evolution of CO, CO_2 and HCO_2^+ from pure gase phase model (dashed line) and STM (continuous line) as fractional abundance vs. time. At the right of the figure the predictions of Herbst&Leung and the observed abundances are indicated.

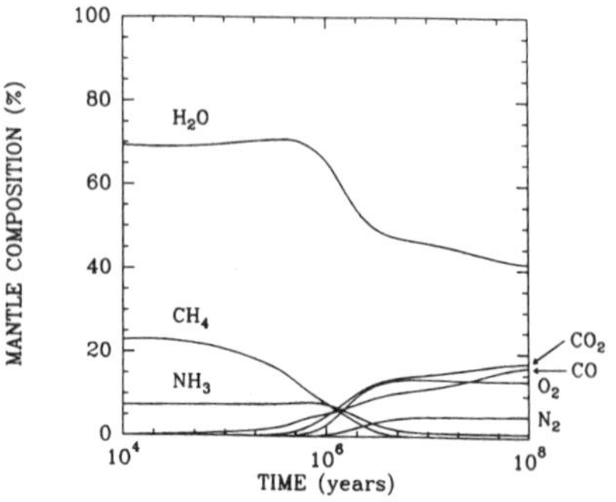

Fig.2

Evolution of the most important mantle species, given in percentages for the STM.

Table1

Fractional abundances and corresponding ratios

| | OUR gas phase model | | STM | STM2 | Herbst&Leung |
	10^7 yrs	10^8 yrs	$\geq 10^7$ yrs	$\geq 10^7$ yrs	$> 10^7$ yrs
$F(CO)$	-3.9	-3.8	-4.3	-4.2	-4.1
$F(CO_2)$	-7.8	-8.9	-4.7	-5.3	-7.4
$F(HCO_2^+)$	-14.5	-15.8	-10.8	-11.4	-12.2
R1	-3.9	-5.1	-0.4	-1.1	-3.3
R2	-6.7	-6.9	-6.1	-6.1	-4.8

$R1 = {}^{10}\log([CO_2]/[CO])$; $R2 = {}^{10}\log([HCO_2^+]/[CO_2])$

4. CONCLUSIONS

Pure gas phase chemistry can not produce the observed abundance of HCO_2^+. Our simple gas phase chemistry produces less CO_2 and consequently less HCO_2^+ than the more sophisticated gas phase model of Herbst&Leung (8).

Including grains and photoproduction of solid CO_2 in the icy grain mantles in the gas phase scheme produces: enough solid CO_2 to compare with the observed value (2,3), a factor of $1.6.10^4$ increase for the CO_2 abundance in the gas and a factor of 1.10^5 increase for the abundance of HCO_2^+ to a fractional abundance of 2.10^{-11}, which is already within a factor of 10 of the observed abundance. We note that our simplified gas phase chemistry without the solid state processes produces much less HCO_2^+ and therefore a $[HCO_2^+]/[CO_2]$ ratio lower by a factor of 10-100 than the Herbst&Leung model (8). Therefore it is strongly indicated that a combination of a more sophisticated gas phase chemistry with the solid state processes can simultaneously satisfy both the gas phase abundance of HCO_2^+ and the solid state abundance of CO_2.

REFERENCES

(1) Minh,Y.C, Irvine,W.M., Ziurys,L.M., 1988, Astrophys. J. <u>334</u>, 175

(2) d'Hendecourt,L.B., Jourdain de Muizon,M., 1989, Astron. Astrophys. <u>223</u>, L5

(3) d'Hendecourt,L.B., these proceedings

(4) Greenberg,J.M., Yencha,A.J., 1973, in <u>Interstellar dust and related topics</u>, J.M. Greenberg and H.C. van de Hulst, eds, I.A.U. Symp. no. 52, Reidel, Dordrecht, 309

(5) Greenberg,J.M., 1979, in <u>Stars and Star Systems</u>, B.E. Westerlund, ed., Reidel, Dordrecht, 173

(6) d'Hendecourt,L.B., Allamandola,L.J., Baas,F., Greenberg,J.M., 1982, Astron. Astrophys. <u>109</u>, L12-L14

(7) Leger,A., Jura,M., Omont,A., 1985, Astron. Astrophys. <u>144</u>, 147

(8) Herbst,E., Leung,C.M, 1986, Mon.Not.R.astr.Soc. <u>222</u>, 689

(9) Herbst,E., Green,S., Thaddeus,P., Klemperer,W., 1977, Astrophys. J. <u>215</u>, 503

(10) de Jong,T., Dalgarno,A., Boland,W., 1980, Astron. Astrophys. <u>91</u>, 68

(11) Boland,W., de Jong,T., 1982, Astrophys.J. <u>261</u>, 110

(12) d'Hendecourt,L.B., Allamandola,L.J., Greenberg,J.M., 1985, Astron. Astrophys. <u>152</u>, 130

(13) Greenberg,J.M., 1982, in <u>Comets</u>, ed. L. Wilkening, U. of Arizona Press, 131

Balloon Observations of Diffuse Interstellar C^+ Emission: M17 Region

Hideo MATSUHARA[1], Haruyuki OKUDA[2], Hiroshi SHIBAI[2], Takao NAKAGAWA[2],
Toshinori MAIHARA[3], Kohei MIZUTANI[3], Yukiyasu KOBAYASHI[4],
Norihisa HIROMOTO[5], Frank J. LOW[6], and Tetsuo NISHIMURA[6]

[1] *Department of Astrophysics, Nagoya University, Furōchō, Higashi-ku, Nagoya 464-01, Japan*

[2] *The Institute of Space and Astronautical Science, Nakanodai, Sagamihara 229, Japan*

[3] *Department of Physics, Kyoto University, Kyoto 606, Japan*

[4] *Institute of Astronomy, University of Tokyo, Mitaka, Tokyo 181, Japan*

[5] *Communications Research Laboratory, Koganei, Tokyo 181, Japan*

[6] *Steward Observatory, University of Arizona, U. S. A.*

Balloon observations of [CII] $158\mu m$ and [OI] $63\mu m$ have been made toward the M17 HII/molecular cloud complex. Around the M17 HII region, the CII region extends far beyond the hydrogen ionization front. The [OI] line emission is much stronger than the [CII] line emission, indicating that the CII region is rather dense ($> 10^4 cm^{-3}$) over several parsecs. A low-brightness, diffuse [CII] emission component toward M17 SW molecular cloud has been found. The [CII] emission feature roughly follows the molecular cloud distribution. Possible energy sources for the [CII] emission are discussed.

I. INTRODUCTION

Recently, the physics and the chemistry of dense, warm interface regions between HII regions and molecular clouds have been extensively studied(1,2). Such interface regions are called photodissociation regions (PDRs), since hydrogen and many other kind of molecules are dissociated by far-ultraviolet (FUV) radiation longward of the Lyman edge of hydrogen. In PDRs, carbon atoms are mostly ionized(CII regions), and their far-infrared fine structure line([CII] $^2P_{3/2} \to \ ^2P_{1/2}$ $157.7\mu m$) emission and the neutral oxygen line([OI] $^3P_1 \to \ ^3P_2$ $63.2\mu m$) emission are dominant coolants of PDRs(2,3) and serve as good diagnostic tools of the physical condition in PDRs(4). One of the important physical parameters is the FUV flux illuminating PDRs, which is conventionally expressed by G_0, the flux normalized to that at the solar neighborhood(2). G_0 is quite large in regions close to the massive stars($10^4 - 10^5$), while it is expected to be $1 - 10$ in general interstellar space. It should be noted that CII regions/PDRs can be formed without massive O stars; early B stars can also supply sufficient FUV radiation to form CII regions/PDRs. Therefore, the observations of the [CII] and the [OI] lines could provide us with information of numerous B star populations in the Galaxy.

In the following, we present a part of the results of our recent balloon observations, especially that of large-scale observations toward the M17 HII/molecular cloud complex.

II. OBSERVATIONS

The instrumentation used for the observations was a balloon-borne infrared telescope(BIRT) with a liquid helium cooled Fabry-Perot spectrometer(5,6). BIRT was a 50cm telescope mounted on an alt-azimuth pointing system. The spectrometer had two channels, one for the [CII] line and the other for the [OI] line with a resolving power of $\sim$ 2000. The beam was a $3'.7$ diameter disk for sources with uniform brightness. The system in-flight sensitivity was $I_{min} \simeq 6 \times 10^{-5} ergs\, s^{-1}\, cm^{-2}\, sr^{-1}\, Hz^{-1/2}$ for the [CII] line.

We used two observational modes: one was a conventional spatial chopping mode with a chopping width of $15'.6$, and the other was a frequency switching mode(6,7) to obtain the spatial distribution more effectively and reliably. The latter mode is suitable for balloon observations where good atmospheric conditions can be achieved, and was especially used for the [CII] line.

The balloon flight for the present work was made on June 5, 1988(UT) from the National Scientific Balloon Facility at Palestine, Texas. The ceiling altitude was 31km and stable.

III. RESULTS AND DISCUSSION

In Figure 1, distribution of the [CII] emission around the M17 HII region is shown(7). The [CII] emission region extends widely beyond the hydrogen ionization front, as is clearly seen especially to the northeast front. In Figure 2, the result of observations made by the spatial chopping mode is shown. The values in the circles are ratios of the [OI] line brightness to the [CII] line brightness. Clearly, the [OI] emission is much stronger than the [CII] emission($I([OI])/I([CII]) \sim 4$). Considering the required densities for the excitation of these lines, this brightness ratio indicates that the [CII] region is quite dense, probably $> 10^4 \, cm^{-3}$(4). On the other hand, the scale height of the [CII] emission region from the filled circle in Figure 1 is as large as $8'$ (5pc). If the CII region is uniform, it is quite unlikely that FUV radiation can reach through such large distances and form the CII regions without being attenuated. Therefore, the CII region must be clumpy on a large scale. The [CII] and [OI] line luminosities from the regions around the HII region are estimated to be $\simeq 5000 L_\odot$, $\simeq 8000 L_\odot$, respectively.

We also made [CII] mapping observations toward the nearest cloud of the M17 SW molecular clouds(10). The result is shown in Figure 3. Clearly, there exists a low-brightness, diffuse [CII] emission component which roughly follows the molecular cloud distribution. Filled circles in Figure 3 are locations of IRAS point sources(11). It is notable that there are 2 bright sources close to a southwest bright spot of the [CII] emission. These point sources are considered to be embedded B0V stars(12) and are candidates to be energy sources for the [CII] bright spot. It should be noted, however, that in this case the ratio of [CII] line luminosity to the total luminosity of the point sources becomes $\sim 0.5\%$, which is significantly larger than that of the regions around the HII region(0.1%).

The diffuse, low-brightness [CII] emission feature may originate from the diffuse CII regions forming on the surfaces of the molecular clouds illuminated by the general interstellar radiation field. A piece of evidence is that the C^+ column density estimated from the observed [CII] brightness($N(C^+) \simeq 1 - 2 \times 10^{17} \, cm^{-2}$) agrees well with the theoretical column density computed using the physical parameters: $A_V^{tot} \simeq 4 - 5$, $G_0 \geq 1$, $\delta(C) = 0.1 \sim 0.4$(3). However, considering the energy balance in the CII regions, the observed [CII] line brightness indicates that the FUV radiation illuminating the CII regions is quite large: the observed [CII] brightness should be roughly equal to (or smaller than) the heating energy integrated along the line of sight. Using the grain photoelectric heating rate, if there are m spherical clouds with CII envelopes whose thickness is $A_V(C^+)$ in visual extinction in the line of sight, we obtain $I([CII]) \simeq 6.4 \times 10^{-6} m G_0 (1 - exp(-1.8 A_V(C^+))) \, ergs \, s^{-1} cm^{-2} sr^{-1}$(13). From the observed brightness, $m G_0 \simeq 30 - 40$. Therefore, there should be several clouds in the line of sight, and/or the FUV field is much stronger than that at the solar neighborhood.

We thank M. Narita, H. Takami, and the balloon launching staff of ISAS who worked with us to develop and improve the balloon telescope system. We are grateful to the staff of the National Scientific Balloon Facility at Palestine, Texas(flight number 1465p).

REFERENCES

1)Genzel, R., Harris, A. I., and Stutzki, J. 1989, in *ESLAB Symposium 22, Infrared Spectroscopy in Astronomy,* ed. B. H. Kadeich (ESA SP-290), p.115.
2)Tielens, A. G. G. M., and Hollenbach, D. 1985, *Ap. J.,* **291**, 722.
3)Van Dishoeck, E. F., and Black, J. H. 1988, *Ap. J.,* **334**, 771.
4)Watson, D. M. 1984, in *Galactic and Extragalactic Infrared Spectroscopy,* eds. M. F. Kessler and J. P. Phillips (Dordrecht: Reidel), p.195.

5)Shibai, H., *et al.* 1990, in *Instrumentation in Astronomy VII*, ed. D. L. Crawford, *Proc. SPIE*, vol.**1235**, 108.

6)Nakagawa, T., Okuda, H., Shibai, H., Matsuhara, H., Kobayashi, Y., and Hiromoto, N. 1990, in *Instrumentation in Astronomy VII*, ed. D. L. Crawford, *Proc. SPIE*, vol.**1235**, 97.

7)Matsuhara, H., *et al.* 1989, *Ap. J. (Letters)* **339**, L67.

8)Felli, M., Churchwell, E., and Massi, M. 1984, *A. Ap.* **136**, 53.

9)Gatley, I., and Kaifu, N. 1987, in *IAU Symposium 120, Astrochemistry*, eds. M. S. Vardya and S. P. Tarafdar(Dordrecht: Reidel), p153.

10)Elmegreen, B. G., and Lada, C. J. 1976, *A. J.* **81**, 1089.

11)*IRAS Point Source Catalog.* 1985, Joint IRAS Science Working Group (Washington:GPO).

12)Elmegreen, D. M., Phillips, J., Beck, K., Thomas, H., and Howard, J. 1988, *Ap. J.* **335**, 803.

13)Shibai, H., *et al.* 1991, *Ap. J.* **374**, 522.

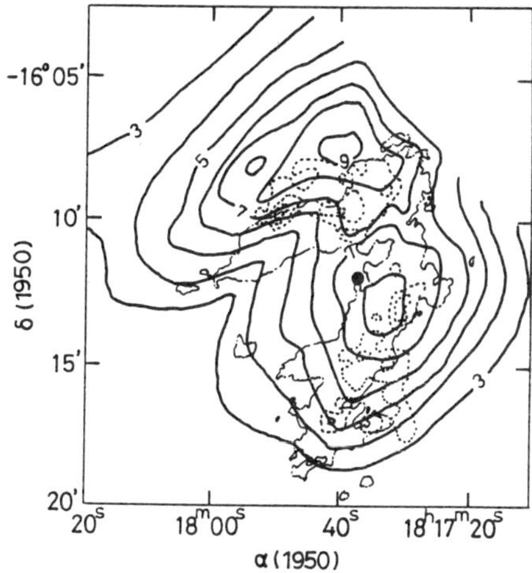

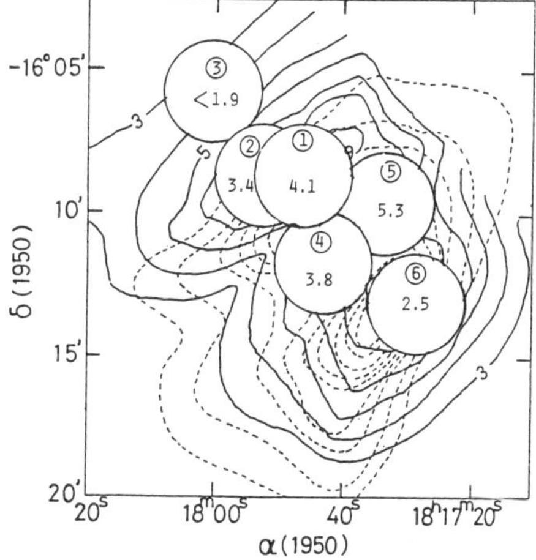

Fig. 1 –Comparison of distribution of [CII] 158μm line emission (7; solid contour lines) and locations of the hydrogen ionization fronts(8; dash-dotted lines). Unit of the [CII] contour intervals is $2.4 \times 10^{-4}\ ergs\ s^{-1}cm^{-2}sr^{-1}$. Dotted contour lines indicate the location of H$_2$ $v = 1 \rightarrow 0$ S(1) emission(9). Filled circle indicates the location of the ionizing stars of the HII region(8).

Fig. 2 –Positions where both [OI] and [CII] lines are observed are shown by circles on the [CII] contour map(same as in Fig. 1). Ratios of the [OI] line brightness to the [CII] line brightness are shown in the circles.

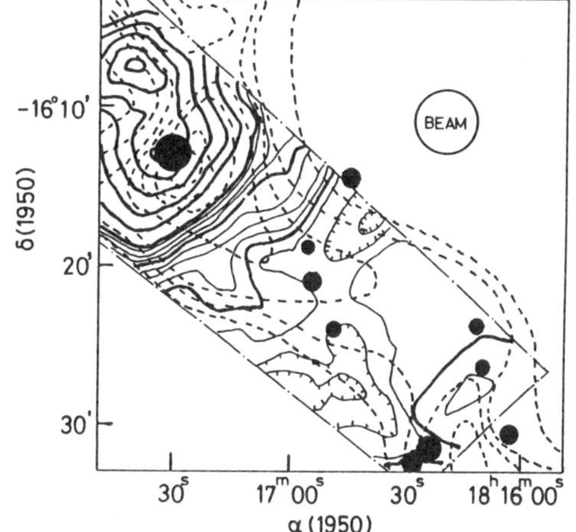

Fig. 3 –Distribution of low-brightness, diffuse [CII] emission toward the M17 SW molecular cloud. Thick solid lines are the same as those in Fig. 1,2, and for the two lowest contour intervals, each interval is divided by 4 and shown by thin contour lines. Dashed contour lines show distribution of ^{12}CO $J = 1 \rightarrow 0$(10). Filled circles indicate locations of the IRAS point sources(11), whose sizes are proportional to logarithm of 60μm fluxes.

INDEX OF AUTHORS